GORBACHEV'S REVOLUTION

Also by Anthony D'Agostino

MARXISM AND THE RUSSIAN ANARCHISTS

SOVIET SUCCESSION STRUGGLES

Gorbachev's Revolution

Anthony D'Agostino

NEW YORK UNIVERSITY PRESS
Washington Square, New York

First published in the U.S.A. in 1998 by
NEW YORK UNIVERSITY PRESS
Washington Square
New York, N.Y. 10003

This book is printed on paper suitable for recycling and
made from fully managed and sustained forest sources.

Library of Congress Cataloging-in-Publication Data
D'Agostino, Anthony, 1937–
Gorbachev's revolution / Anthony D'Agostino.
p.   cm.
Includes index.
ISBN 0–8147–1898–1 (alk. paper)
1. Soviet Union—Politics and government—1985–1991.   I. Title.
DK288.D34 1998
947.085'4—dc21   97–30236
                                                      CIP

Printed in Great Britain

To Alexander Leo

Contents

Acknowledgements

It is a pleasure to recognize intellectual and other debts incurred in the writing of this study, much of it done as the events themselves were unfolding. During those years it was sensed everywhere that world history was passing a divide. We had to change every predisposition and throw out all our old maps, especially the intellectual ones. Every thinking person in the world breathed the air of the Gorbachev revolution – as their counterparts must have in the days of the French Revolution. It made sense to say: 1789, 1848, 1917–18, 1989–91.

So in a larger sense than usual, this work bears the imprint of valuable encounters with students, friends, colleagues and, as well, people in the media and the world of affairs, Soviet *glasnost* intellectuals, and internet contacts. I can mention by name only a small number of them. Jack Boas, Marek Chodakiewicz, Jerald Combs, and Norair Taschian read the whole manuscript and gave me the benefit of copious comments. Walter LaFeber contributed thoughtful remarks about the chapter on the Soviet side of the Cold War, while Theodore Karasik, Boris Kagarlitsky, Ronald Bee, Ciro Zoppo, and Maziar Behrooz took me up on various other chapters and topics. My notions about Gorbachev and the ex-USSR were probed by Michael Krasny, Kevin Pursglove, Philip Maldari, Chris Welch, and Bill Shechner in searching radio and television interviews. Generous support for the research and writing was provided by the offices of Faculty Affairs and Professional Development and of Research and Sponsored Programs of San Francisco State University. Dariusz Salata, Attila Gabor, Virginia Wright, Kevin Clarke and Karen Graham helped me to locate and copy many important items. Many thanks also to my editors at Macmillan, T. M. Farmiloe, Aruna Vasudevan, and John Smith, and to the intrepid copy-editor, Penny Dole. In the transliteration of Russian names, the approach that I have adopted is to use a newspaper system for the text (thus Yeltsin rather than Elt'sin) and a modified Library of Congress system for the notes. Countries whose names were changed are given according to the name prevalent at the time in question (for example, Belorussia up to 24 August 1991 and Belarus thereafter). My thanks to the Russians, Belarusians, and Balts who showed me around the ex-Soviet Union and helped me to better understand the end of the Soviet era.

My wife Susan Fiering and our children helped me to keep things in perspective in their own way. My daughter Martine gladdened my heart by turning toward the study of American literature rather than some less promising alternative, such as a life of following the Grateful Dead. And most of all, my son Alexander lightened things by regarding my deepest thoughts with the high irony that only eight-year-olds seem to be able to muster, and by pressing his points home physically with his highly ineffectual wrestling.

1 A Gorbachev Epiphany?

Somebody asked the American President whether he still considered the Soviet Union to be an 'evil empire.' He said *no*, and he said that within the walls of the Kremlin, next to the Tsar cannon, right in the heart of the 'evil empire.' We take note of that. As the ancient Greeks say, 'everything flows, everything changes. Everything is in a state of flux.'

The confident words were those of Mikhail S. Gorbachev, who was speaking in Moscow in June 1988 to a gathering of journalists and assessing his just-concluded summit meeting with American President Ronald Reagan. Only five months before, the two leaders had signed a treaty on intermediate range nuclear forces, an important benchmark in the history of US–Soviet relations, one that appeared at the time to have closed off what some had called 'Cold War Two' – the period of tense confrontation between the superpowers in the early eighties – and opened up a new period of understanding, not the Detente of the Nixon–Kissinger period, but perhaps 'Detente Two'.

These days we have no Detente, no Cold War, no Soviet threat. Soon it will be necessary to teach young people the meaning of these musty historical terms. Nevertheless, I ask the reader to think back on that frightful time and its unique menaces, when half a million Soviet troops stood over eastern Europe, backed by weapons with more than ten thousand nuclear warheads. The problem of the western 'Euromissiles' had then been at the centre of the worries of the Soviet leaders since the NATO decision in 1979 to match the Soviet missile buildup in eastern Europe by new western missiles. Soviet defence specialists were concerned about the American Pershing Two rocket, part of the NATO deployment, with its reputed ten minute flight time from a European site to a Moscow target. They also feared the ground-launched cruise missiles that were to be put into western positions, weapons that were invisible to Soviet radar and seemingly perfect for a surprise attack against 'time-urgent targets'.

Worst of all was a dangerous logic that they thought was being set in motion. The Soviet side reserved the right to match all the NATO nuclear weapons in Western Europe, including British and French ones, while the Americans reserved the right to match the Soviet weapons, irrespective of the British and French. This seemed to be a formula for a perpetual ratcheting-up of the arms race in its crucial European theatre.

But Gorbachev had found the answer to that problem where others had failed. He had cut the Gordian knot simply by accepting Ronald Reagan's earlier proposal for a 'Zero Option': to eliminate all the American and Soviet missiles in question, and to forget about the British and French weapons. The Zero Option had up to then been dismissed by arms control advocates on both sides as a propaganda ploy. The astute Soviet General Secretary Yuri Andropov, who had carried on painstaking discussions on the subject in 1982–83, had been unable to accept it. Yet Gorbachev had found his way to a solution in short order, only two years after coming into office, and a solution moreover that would stand as a milestone in the history of arms control – the first treaty to scrap an entire class of nuclear weapons – actually not just arms control, but disarmament.

Now, in June 1988, in announcing the demise of Reagan's 'evil empire' rhetoric, the wonderful Gorbachev was permitting himself a small boast: that he understood how history can often transform things into their opposites; and moreover, because of his superior understanding and his bold initiatives, he had been able to tame the hard-bitten anti-Communist President of the United States, in the process removing a major nuclear threat against Soviet soil.

For Gorbachev's Soviet supporters the achievement seemed still the more wondrous for having been the result of diplomatic moves that one could hardly have expected from his predecessors. The overtures that had led to success in removing the American missiles from Europe were made against a background of far-reaching social and political changes in Soviet life, changes by then in motion for three years. The message Gorbachev was issuing to his global audience, as he had explained it in his book, *Perestroika: New Thinking for Our Country and the World*, was that Soviet foreign policy was being changed in order to advance more fully the cause of Soviet domestic reform, and that the real proof that he was not merely engaged in another of the 'peace offensives' of the Brezhnev era was the depth and scale of the USSR's internal transformation. Gorbachev admitted that many in the west would continue to doubt his sincerity. That was why he insisted that the only way to make the reforms credible was to make them genuine.

The 'taming' of Reagan was interpreted by Soviet observers as powerful and solid evidence of Gorbachev's success, at once a vindication of *perestroika* and an advertisement of the indispensability of his hand at the controls. Throughout the struggles that engulfed the leadership in 1987–88, it was impossible to deny that, whatever the failures and difficulties of Gorbachev's specific agricultural or industrial policies, his colleagues and even his critics simply lacked the capacity to think in terms of doing without him. It seemed that he could be limited 'only' by the belated resistance of a few hidebound Stalinists, defending the old ways of Kremlin political life, but at the same time no one doubted his ability to keep them at bay by his brilliant innovations.

Gorbachev's most important convert was Ronald Reagan. The American President seemed convinced that the new Soviet foreign policy was propelled by far-reaching Soviet internal changes, and he was sure that these included changes in Communist ideology itself. He had told a *Washington Post* reporter in February 1988 that

> [Gorbachev] is the first leader that has come along who has gone back before Stalin and he is trying to do what Lenin was teaching ... with Lenin's death, Stalin actually reversed many of the things. Lenin had programs that he called the new economics and things of that kind. And I've known a little bit about Lenin and what he was advocating, and I think that this, in *glasnost* and *perestroika* and all that, this is much more smacking of Lenin than of Stalin.[1]

A subsequent editorial in the same paper criticized this statement for failing to recognize the organic links between Leninism and Stalinism. Had the President forgotten that the Communist system of Lenin's time was already a thoroughgoing dictatorship, one that had only been perfected by Stalin? But the White House stood by its analysis, for understandable reasons. Reagan and his advisors wanted to show that the dense ideological debates in Moscow were not merely a charade, and that they understood their ultimate meaning for US policy.

They also felt that they deserved credit for what was happening in the Soviet Union. Their foreign policy of supporting armed opposition to Communist regimes, under the rubric of the Reagan Doctrine, and their defence policy providing for research on ballistic missile defence, as described in the President's Strategic Defense Initiative, had applied such pressure on the Soviets, so the Reagan people thought, as to prompt Gorbachev's initiatives as a response. They had been tough and the Soviets had now begun to say 'uncle'. They had every reason to think that they themselves had, however indirectly, fathered the Gorbachev revolution.

Western suspicion of Gorbachev's motives practically disappeared. The old debate about whether one could deal with the Soviets had already changed into a debate on the question whether Gorbachev was perpetrating a grand deception. Now that debate also seemed to have been transcended. Even the most sceptical were forced into a position of watchful waiting. No one predicted the collapse of Communism. A new prestige surrounded those western Sovietologists who had maintained for years that pluralistic forces had always seethed beneath the surface of Soviet totalitarianism. They could now argue that Gorbachev's was only the latest of two powerful 'waves of reform', the first being Khrushchev's in the 1950s.

This seemed to suggest that Soviet history had not been one unbroken line of totalitarian continuity up to Gorbachev's accession in April, 1985.

There had always been a reform current in Soviet politics, it was widely argued, and now that current was carrying things along, perhaps for no other reason than that the regime had to seek ways to address the problem of the economy's declining growth rates. Now the reformers led by Gorbachev would have to tackle their domestic problems by shifting investment away from heavy industry and defence and toward the satisfaction of the consumer. De-emphasizing 'ironmongery', they would soon find that central planning was not the best way to distribute consumer goods, and they would see anew the virtues of the market. Once the process had begun it would only be a few short steps to the full integration of the Soviet system into the world economy.

One reason all this seemed so attractive was that the new Detente, as Gorbachev was describing it, promised a real relaxation of tensions between the nuclear superpowers without disturbing the basic balance of power in Europe that had followed on the results of the Second World War. Germany would remain divided. The security of Europe's various states, who in the final analysis had not managed to live in freedom and at peace with a powerful united Germany since it had first appeared in 1871, would continue to be guaranteed by the imperfect but proven device of an east–west separation of the continent. 'If not an inevitable outcome of history', one respected analyst had called the division of Europe, 'it was at least a natural one'.[2] Few wanted to say it openly and crudely, but the thought was taking hold that Europe had not really fared so badly under the international regime of the Cold War.

Retaining the broad framework of a divided Europe around a divided Germany, a framework that, if it was not particularly just, was at least stable, the West might assist Europe in adjusting to a graceful decline of Soviet power and a gradual release of the Soviet hold over Eastern Europe. That the Soviets showed willingness to end the era of military confrontation already revealed that they had in general lost the desire to pose a threat to the West. This was what was meant when one heard it said, only half-jokingly, that the Soviets had finally admitted that the West had won the Cold War. The European observers and pundits were even more hopeful about these prospects than American ones.[3]

It was a kind of zenith of Gorbymania. Polls showed his popularity in the United States to be higher than any Soviet chief had ever enjoyed. *Time* had declared Gorbachev its 'Man of the Year', judging that he had 'reinvented the idea of a Soviet leader'.[4] A *New York Times*/CBS poll taken in May 1989 would show that two out of three Americans no longer considered the Soviet Union to be an immediate military threat.[5] In West Germany, Gorbachev's prestige was greater than that of Reagan, and it was he, rather than Reagan, who was given the most credit for progress on arms control. One heard it argued that Gorbachev's policies were providing, not only a solvent to east–west tensions, but a kind of new world synthesis, according to his idea of 'Universal Human Values', of the main features of the two social systems that had been locked in political combat since the beginning of the Cold War.

There seemed to be only one small problem: would Gorbachev succeed in defeating the recalcitrant conservatives at home who were trying to slow the process of reform? It was treated as a footnote. Western opinion, having witnessed the rise and demise of a series of leaders from Brezhnev through Andropov and Chernenko to Gorbachev, had somehow come to consider Soviet succession to be a relatively straightforward matter of establishing the authority of a new leader, something that usually took four or five years and consisted mainly in getting the right team into the right positions. The institutional structure of Communist party and Soviet state would be essentially the same one the West had learned, often painfully, to do business with over the last 40 years. No one thought Gorbachev's opponents to be as formidable as he. One respected Sovietologist discounted their chances of stopping him, because, he said, next to Gorbachev, they were 'pygmies'.

That was in June 1988, after the first three years of Gorbachev's leadership. The next four years were to show all the suppositions of that time to be illusory. The world witnessed, instead of Detente, the collapse of the Warsaw Pact; instead of managed reform in the Soviet Union, a vast democratic revolution that banished the Communists to oblivion; instead of gradual and measured change in Europe, the sudden appearance of more than two score new states, in a new Europe being reshaped from top to bottom; instead of peace and stability, a power vacuum east of Berlin and a landscape of potential and real conflict in east central Europe and the Balkans, areas where the struggles that led to the two world wars were born.

How quaint the hopes and ambitions of Gorbachev's *perestroika* seemed when viewed from the later perspective of a continually unravelling former Soviet world. Could all this have been a natural and logical extension of the course of reform that Gorbachev began to plot in 1985? It was tempting and comforting to think that the whole thing was inevitable because of the long-run weaknesses of the Soviet economy, or perhaps because the whole Soviet experiment dating back to 1917 was never viable. On this reading, one must describe the whole history of Soviet Russia as a process of steady decline, with Gorbachev's *perestroika* a mere detail in the last stages of the story.

And how did Gorbachev himself explain the career of his reforms? At a meeting I attended in San Francisco in 1992, he blithely took credit for the end of Communism, explaining that, as he had got further into the process of Soviet reform, he had come to realize that the problems in the Soviet system were greater than he had first supposed! Yet, two years later, at the Moscow trial of General Varennikov in 1994, he said something quite different. During the coup of August 1991, Varennikov had been in charge of the detail assigned to arrest Gorbachev. Varennikov had refused the amnesty later granted to the conspirators by parliament, in order to present his case against Gorbachev before the public. Varennikov argued that Gorbachev had been intent on breaking up the Soviet Union, as a logical consequence of the reform campaign, when the coup plotters attempted to intervene and save their country. It was Gorbachev, not he, who was guilty of treason.

Gorbachev testified, however, that the breakup of the Union was not at all the result of historical necessity, but the consequence of a power struggle and intrigues in the corridors of power. 'I reject all attempts', he said indignantly, 'to accuse me of bringing down the Soviet Union.'[6] Well, which way was it? Did he knowingly preside over the changes of 1985–91 or were they done against his best intentions? Both of Gorbachev's lines of argument cannot be correct, and in fact this study leans toward the latter interpretation, that the end of Communism was never the goal of Gorbachev but, like so much else in the history of Soviet Russia, was the unplanned and unintended result of an intense and many-faceted struggle for power.

By contrast, the pervasive image of stable in-system reform, which seemed so compelling when Gorbachev began his reforms, in fact ignored the evidence from Soviet history. In the past, inauguration of a new Soviet leader has been no tea party, but has usually involved a political struggle so vast as to revolutionize the entire country and cripple the Communist party itself. Succession struggles have been wars of ideas in which the victors got their way, not only by bureaucratic wire-pulling, but by challenging their opponents' interpretations of Soviet history, in turgid, poisonous disputes that usually resulted in the loss of careers and lives. The rise of Stalin involved the reduction of the party to insignificance and the eventual shooting of most of its leaders. On a far different scale, Khrushchev was also turning the party and state structures upside down before his ouster in 1964. The rise of Gorbachev saw many of the same patterns of political struggle that had already shown their features in the 1920s and the 1950s. So it was an illusion for the West to suppose in the heyday of *perestroika* that Gorbachev was firmly in control of affairs at home. In fact, at the height of his international popularity, he was stating openly that he was still a long way from the position of personal hegemony that he thought necessary.

Gorbachev was actually gearing up for a series of daring blows against the collective leadership that had elected him in 1985. Between the spring summit of 1988 and the second Russian revolution that swept the country in August 1991, he would drive his opponents out of the leadership, then crack the power of both the Politburo and the Central Committee. Removing the party from its 'leading role' in January 1990, he would, by August 1991 be enforcing a ban against it, all the while proclaiming his fidelity to the 'socialist choice' that he said the Russian people had taken in 1917. In an interview given amid the wreckage of his regime after the failure of the coup against him in 19–21 August, he declared his refusal to resign as President of what was already being called by many the 'former Soviet Union', claiming that his work was not yet done! This from a man who had assumed the post of party General Secretary six years before with a pledge to 'further the perfection' of the Soviet system.

There is also good reason, then, to doubt the version of Gorbachev's revolution eventually favoured by those of us who were once so much in the grip of Gorbymania, what might be called the thesis of the Gorbachev Epiphany, according to which Gorbachev at some point in the first years of his power went through a conversion experience that revealed to him that Communism had been a bad idea all along, with the consequence that he set about dismantling the Soviet *bloc* and the legacy of the Russian Revolution of 1917.

Searching the record of the revolution he set in motion, we find no evidence of a Gorbachev Epiphany. What we do see is a bitter power struggle, of necessity a war of ideas, in which Gorbachev sought to build his personal power by advancing a political course of zig-zags, often in the direction of reform comparable to what he knew of the work of former right-wing Communist moderates, people such as Bukharin in the 1920s and Malenkov in the 1950s, but sometimes in the opposite direction, opposing those who advocated such reforms. He made his revolution under the pressure of this complex struggle and against the background of international competition with the United States of Reagan and Bush. Success in US–Soviet relations was always his best trump card against opponents.

A revisionist account, assuming the power struggle itself to be the engine of his reform projects, is at least better than trying to prove that Gorbachev intended the liquidation of the state that he was chosen to rule. It is probably also better than supposing that Gorbachev at some point turned against the Communist idea, which he was in fact still defending the day before he resigned his position as head of the party, on 24 August 1991. It permits us to tell the story of Gorbachev's extraordinary career by charting the course of his rise to power and of his fall, observing the mechanics of his struggle, taking note of the way Gorbachev made his coups against opponents by timely arms control initiatives, or internal reforms, or by unleashing the debate on Stalin and Stalinism in the literature of *glasnost*. We can give proper attention to the interplay between the ideas that came forward in the debates on party history and Gorbachev's changing conceptions of his own role in the Communist movement and Soviet state. We can note the changes in the meaning of the various discourses on socialism and democracy. The fascination of the process for the historian lies in the seeming paradox that the great changes for which he will be remembered were actually the result of the failure of his most treasured projects.

Putting the struggle for power at centre stage also offers a chance to demonstrate what I consider to be the fatal weakness of Soviet Communism, the fact that, while it attempted to exercise a one-party dictatorship, it could never permanently resist the conquest of the party itself by personal despotism. The ideal of collective leadership in the end proved to be a utopia.

Lenin alone had been able to live with the idea of being only first among equals because of his intellectual authority as the architect of the whole system. He knew that leading the party to power had invested him with a halo of infallibility and a reputation, as his associate Lunacharsky once put it, of 'a man who could see ten feet into the earth'. One Comintern official who had occasion to observe Politburo meetings of the first years of Soviet power related that a typical meeting would begin with Trotsky making a characteristically bold suggestion, against which the other members would make their various ingenious criticisms. After a time they would sense that they lacked the knock-out punch, and one of them would say, 'Let's ask Ilyich (Lenin).' Lenin would reply, 'I agree with Trotsky.' The measure would pass unanimously. The anecdote illustrates not only how Trotsky stood with the long-time colleagues of Lenin (that is, very poorly), but also how Lenin ruled.

Stalin, who was in the inner circle of the leadership from the earliest days of the Revolution, stood in awe of Lenin's ability to attract what must seem to us as the bizarre and almost pathological deference of intelligent people, a deference which he, Stalin, passionately wanted for himself. In the end he would get it only from those who yielded it out of prudence. As George Kennan has remarked of Lenin's relations with the top party leaders, 'He could rule them through the love they bore him, whereas Stalin was obliged to rule them through their fears.'

Stalin's initial programme was a moderate one permitting him to appear the least threatening of the collective leadership that succeeded Lenin. He gained his most important early successes by opposing the more radical plans of Trotsky and others. But the completion of his campaign for unlimited hegemony over his opponents required a revolutionary transformation of the country and a campaign of terror with few equals even in our hard century. After Stalin's death, Khrushchev too was on the verge of tearing party authority to shreds as Stalin had, with his plan for its division into industrial and agricultural branches, when he was removed from power in 1964. But Khrushchev thought himself a liberator freeing the country from the worst of Stalinism. His downfall was the result of the fact that he had spent the first part of his rise building up his power through the party, only to have the party, in defence of its ideal of collective leadership, turn against him.

Gorbachev alone has played to perfection the role that proved to be beyond Khrushchev, that of a 'Good Stalin'. Like Stalin and Khrushchev, a centrist who made timely turns to outwit those who sought to limit him, he subordinated and finally crushed the party and its collective controls. However, where Stalin replaced the dictatorship of the party by that of the police, aided by a kind of lynch-law populism, Gorbachev unhinged the party by confronting it with genuinely open elections and *glasnost*, with the intended aim, to be sure, of its democratic and socialist renewal and ultimate salvation in another form.

His promise was to cleanse and perfect the Russian Revolution and to bring it into the rest of the world by liquidating the world's fear of Soviet threat. He did not promise to bring, in effect, a new edition of the French Revolution to all of Eastern Europe, nor to partition his country, disperse its nuclear weapons, and unhinge its economy. Neither did he offer to liberate Russia according to the Western model of a multiparty democracy and a market economy, nor to turn a depressed, humiliated, and devolving rump Russia, despite its rapturous liberation from the grinding tyranny of the old system, into a site of deepened poverty and poisonous nationalist, even anti-semitic, moods. He did all these things despite himself. Because of him, the track on which the Twenty-First Century Limited was scheduled for departure has been changed, and now it will embark on another. Revolutions are usually made by people who cannot know exactly what they are doing. How they change the world – according to their own designs and according to the world's designs on them – is the theme of this study of the great revolutions of the end of the twentieth century.

The story of the last Soviet power struggle is not, I believe, one that is best understood in terms of an irrestible unfolding of large historical forces and trends. On the contrary, it is in many respects the most curious story in modern history. No determinism can properly appreciate the role of contingency and individual frailty, of chance and irony, of the heroic and the tawdry, that comprise the stuff of this drama. The historians of the twenty-first century will no doubt pour over this tale, that is, if they still want to know how they got where they are. And I think they will pursue the approach employed here. Memoirs of participants, despite the inevitable apologetic distortions, will no doubt help illuminate their path, as they have mine. They will also profit from consulting valuable journalistic works. Yet these sources will probably have to be challenged by those who will cite and further explicate the texts of the period, the documents of the Gorbachev revolution that are found in the footnotes of this volume. Future historians will want a close account of exactly how Gorbachev gathered up the power and how and why he lost it. And they will not fail to notice that Gorbachev's struggle for the succession was at bottom a war of ideas. No doubt a longer view will give them a better picture of the larger meaning of the whole thing. Even so, they may lose sight of facts and perspectives that we have in focus presently. At any rate, this is a discussion that will certainly not end in the twentieth century, or even the twenty-first, and it can not be amiss for contemporaries to open it now.

2 The Old Regime of the Soviet Communists: Foreign Policy in the Cold War

The cold war launched by militaristic circles in the west was nothing less than an attempt to revise the results of the second world war, to deprive the Soviet people, the world forces of progress and democracy, of the fruits of their victory.

Mikhail S. Gorbachev, 1985

Revolutions may end by tearing up the deepest roots of the former institutions, but they start by action from above. The French monarchy of Louis XVI was first set upon, not by the most miserable of its subjects but by those who most benefited from it, whose fortunes were on the rise and whose prospects seemed limitless, whose eyes were opened after a long slumber to the possibility of change, and whose hearts were hardened to the old regime by its very attempts at reform. 'A grievance comes to appear intolerable', says Tocqueville, 'once the possibility of removing it crosses men's minds.'

In Gorbachev's revolution, it was the Communists, particularly their salaried intellectuals, 'the Soviet Intelligentsia', who turned first and most sharply against the Stalinist order, who found its meagre improvements inadequate, who under conditions of reform and amelioration came to find the system itself intolerable. So we must reject the idea that the Soviet leaders who brought Gorbachev to power in 1985 resembled the hard-bitten conservatives that Gorbachev would later make them out to be. And with it, the notion that there was a significant difference between his thinking and theirs. For all their commitment to continuity with the past, they had come to see the need for a change of line in foreign policy. The most influential figure among them, longtime foreign minister Andrei Gromyko, spoke frankly about the disadvantages of the heightened Cold War tension of the early eighties and the need to return to an international arms control regime like the Nixon–Kissinger Detente. For many years the polished executor of a Soviet foreign policy devised by others, Gromyko finally got his chance at the helm during the Chernenko interregnum in 1984, while Gorbachev was second secretary. He put an end to the freeze in US–Soviet relations, moving to resume the Geneva talks on Intermediate Nuclear Forces, which had been broken off by the Soviets in November 1983 when NATO began its European deployment of Pershing 2s and cruise missiles. His meeting with Reagan in September 1984 dramatized the new approach, while the simultaneous demotion of Chief of Staff Nikolai Ogarkov, at that time the most consistently hawkish Soviet military voice, helped clear away opposition at home.

Gromyko thought it was time for a turn toward a more moderate foreign policy line, time to begin another of the periods of consolidation and retrenchment that have alternated in the decades of the Cold War with more active periods of forward policy. Usually a turn to a new policy had been brought about by the victory of one or another faction in the Soviet leadership. Zhdanov in the 1940s, Malenkov, Molotov, and Khrushchev in the immediate post-Stalin years, Kozlov, Suslov and others thereafter, had fought for power offering conflicting and successive policy choices. When Gorbachev took office in 1985, he and his colleagues could look back on six phases of Soviet foreign policy in the Cold War: three of forward policy, which they would not have described as necessarily aggressive or confrontational in the military sense, but simply as active political reconnaissance with all the available instruments – usually including a confident analysis of the economic 'crisis' in the west; and three periods of *recueillement*, a more cautious line prescribing withdrawal from the theatres of sharpest conflict, a withdrawal usually accompanied by a quiet build-up of military strength and a reappraisal of the international balance. The phases may be listed as follows:

I. Forward Policy: Before the Cold War, 1944–47
II. *Recueillement*: Closing off Eastern Europe with covering action in Asia, 1948–54
III. Forward Policy: Khrushchev courting the non-aligned, 1955–63
IV. *Recueillement*: Withdrawing from excess commitments, 1964–72
V. Forward Policy: Seeking revolutionary advances, 1973–80
VI. *Recueillement*: Gains of the seventies under attack, 1980–85.

The term *recueillement* goes back to the nineteenth century and to the foreign policy of Prince Aleksandr Mikhailovich Gorchakov, Foreign Minister to Tsar Aleksandr II, who took office after Imperial Russia's disastrous defeat at the hands of a broad coalition of European powers in the Crimean War. Gorchakov's policy was based on his famous statement: 'La Russie ne boude pas, mais se recueille' (Russia is not sulking, she is gathering her strength). Gorchakov judged that his country had gone too far in the 1840s in trying to hurry along the partition of the empire of Ottoman Turkey, and had frightened all Europe, prompting in response an unlikely anti-Russian front of great powers. His idea was to turn Russia toward a more or less prolonged phase of *recueillement*, in effect a semi-retirement from direct involvement in European conflicts, thus to avoid actions likely to provoke an alliance of Britain and France, who, he judged, had their own grave differences. A quieter policy would permit Russia to make much-needed internal reforms centring around the abolition of serfdom, reforms that Gorchakov considered necessary to the empire's inner harmony.

Gorchakov judged that as long as Russia did not provoke and threaten, the anti-Russian cause would lose its urgency and the European powers would pursue their other conflicts of interest. Russia could then force the revision of the onerous impositions dictated by the Crimean War victors, in particular the clauses of the Paris peace agreement of 1855 which kept Russian ships from the Black Sea. At the same time, Russia could expand into areas of lower political pressure, Central Asia and the Far East, as long as she was careful to avoid direct confrontation with the Western powers. Gorchakov's Russia would no longer help to defend the status quo in Europe in the spirit of the conservative Holy Alliance, but instead encourage the Bonapartist France of Napoleon III in its promotion of revolutionary movements in Serbia, Greece, and Italy. She even showed benevolent neutrality in the French war against Austria for Italian unification. The new unified Italy in turn provided a front against the Habsburg armies in the Austro-Prussian war of 1866, the key stage in the unification of Imperial Germany. After the German defeat of France in 1870–71, Russia unilaterally abrogated the Black Sea Clauses. Gorchakov had spoken openly of looking for someone to help him to do this, and the man had turned out to be German Chancellor Bismarck. So Gorchakov's policy of *recueillement* was a major influence in changing the map of Europe and introducing a new balance of power. Russia got what she wanted from the sweeping political changes of the 1860s and she enhanced her standing among the great powers. Nor was Gorchakov awed by the newly united Imperial Germany. He showed that he could still restrain Bismarck's plans simply by counselling moderation, as he did in the Iron Chancellor's attempt to intimidate France in the 'war scare' of 1875. 'Peace is assured', said Gorchakov at that time, 'thanks to me.'[1]

Soviet diplomats who studied tsarist foreign policy considered Gorchakov's policy a striking precedent of successful action in relation to the West, action predicated on a turn from aggressive posturing to quiet watchful diplomacy. Georgii Chicherin, People's Commissar for Foreign Affairs from 1918 to 1930, once wrote a detailed account of the foreign policy of Gorchakov. Gromyko was known to be an admirer of Gorchakov's, who spent many hours reading the latter's notes and dispatches in the diplomatic archives, where he developed an appreciation for the policy of *recueillement*.[2] He wanted to profit by the lessons of the past while at the same advancing the ideological goals of the Brezhnev foreign policy. In his mind, the global advantage of missile parity with the United States depended above all on a return to detente. Because of his passion for continuity and stability in the leadership, he threw his weight behind the election of Gorbachev in 1985, and later endorsed the policies of *glasnost*, *perestroika*, and democratization as a necessary response to what he called 'the serious inadequacies which had led to the failure of our develop-ment plans in the seventies and early eighties'.[3] Gromyko assumed that Gorbachev would continue the turn to a new *recueillement* that had already been under way since 1984. He and the other Soviet leaders endorsed Gorbachev precisely because they saw him as the best man to create a new detente with the West.

Gorbachev, however, burned to be master of his own foreign policy. No sooner had he been elected than he removed Gromyko from the Foreign Ministry and took up what turned out to be a protracted struggle to lessen his influence. Gromyko tried feebly to broker the Politburo con-flicts that developed in 1987 between Gorbachev and his opponents, but Gorbachev only intensified the anti-Gromyko campaign. The process of eliminating Gromyko's influence, only complete by 1989, at a time when Gorbachev lost control of the Soviet bloc, had by then turned into a fight to transcend the established policy alternance. Instead of a turn to detente within the context of the established East–West relations, Gorbachev plunged into a series of initiatives that would end the Cold War – and everything else.

While Gromyko had looked back fondly to the detente period, Gorbachev and his bloc allies, especially the Hungarian Communists, had erected their own myth about the more ambitious policy of the pre-Cold War 1944–47 period. As they saw it, this was a time when, in a setting of 'Peoples' Democracy', with coalition governments and 'party pluralism', Communists had enjoyed a dominant influence in East Central Europe without a hostile western alliance facing them across an Iron Curtain, while at the same time advancing the socialist idea in western Europe. The clash of these two visions was a key feature of Gorbachev's struggle for power. Before we describe that struggle, it would be well to take a closer look at the stages of Soviet foreign policy in the Cold War in order to show what Gorbachev was ranged against and what he could summon up as historical precedent for his own major departures.

I. FORWARD POLICY: BEFORE THE COLD WAR, 1944–47

Gorbachev and his generation of Soviet politicians were trained to see the world as locked in a fierce class struggle between competing social systems, but they were also nurtured on the memory of broad wartime solidarity with the West against fascism. They liked to see Communism as it appeared in romantic cinematic and literary images from those days, as part of a world-wide 'democratic' front, a kind of fruition of Soviet efforts in the 1930s to create Popular Fronts in western countries against the extreme right. Even while they accepted the hard regimen of the fight against 'western Imperialism,' they remembered the war effort as a solvent of the tensions of the international class struggle and as a positive, even civilizing, experience that transformed their own society for the better. This memory, reinforced by the oft-repeated suggestion in their propaganda that the Cold War had not been inevitable, led to a rosy mental picture of the moment of victory over Hitler, a time when everything had seemed possible but for the interruption of the Cold War. If we appreciate the sentimental attraction of the post-war period, it may help us to understand some of the deeper drives, and the naiveté, of the Gorbachev revolution.

It was a time of Communist revival. In western histories it is a commonplace that Stalin had not waged war against Hitler's Germany solely as a Bolshevik. When German tanks were in the Moscow suburbs, and from atop their turrets their commanders could see the spires of the Kremlin through their field glasses, Stalin did not confine his exortations to reminders of the Great October Revolution and the heritage of Lenin. Instead he urged that 'the manly images of our great ancestors – Aleksandr Nevsky, Dmitry Donskoi, Kuzma Minin, Dmitry Pozharsky, Aleksandr Suvorov, Mikhail Kutuzov – inspire you in this struggle.' After the Germans had been turned back at Moscow, and a year later at Stalingrad, Stalin reactivated the Orthodox church, perhaps thinking of the liberation of Balkan territories where eastern Christianity was the predominant faith. The Communist International, the symbol of the revolutionism of 1917, was duly dissolved in May 1943. After VE day, Stalin toasted with effusive praise the heroic war efforts of the 'Russian people,' who had fought, he seemed to imply, with more dedication than the other Soviet peoples.

Russian themes were most prominent in Soviet war propaganda while the fighting was on the soil of great Russia, but after the Red Army pressed westwards and entered the Ukraine and Belorussia, Soviet and internationalist notes were sounded more frequently. In victory, Stalin rejected the Russification of the cause, and directed a Communist revival in the Soviet Union and internationally. Stalin demanded that historians recover the neglected perspective of class struggle. The leading historical periodical during the war, the Academy of Sciences' *Istoricheskii zhurnal*, was disbanded in 1945 in favour of a new journal, *Voprosy istorii* (Questions of History). The first issue of the periodical explained the change by noting the necessity to 'enhance the international authority of

Soviet socialist culture' in order to promote a 'scientific materialist understanding of history'. Marxist historiography had gone through several phases, the editors explained. In the early days after the Revolution, the emphasis had been on the sins of tsarist imperialism, which was held to have contributed little to international culture. The debunking of great Russian nationalism, they argued, had led to overemphasis on the national aspirations of lesser Soviet peoples (they specifically named the Kazaks, Tatars, and Bashkirs) and an endorsement of 'petty-bourgeois nationalism'. After that there developed, said the editors, a deviation of opposite character, on the side of 'great power chauvinism' and wholesale departures from class analysis. The present task was to lead a return to dialectical materialism.[4]

In order to accomplish this, it was now necessary, said the *Voprosy istorii* editors, to redouble efforts to master the theoretical contributions made by Stalin: his letter to *Proletarskaia revoliutsiia* in 1931; his 1934 conspectus on Soviet history, written with A. A. Zhdanov and S. M. Kirov; his letter to the Politburo of 19 July 1934 commenting on the publication of an article by Engels; the *History of the Communist Party of the Soviet Union (Bolshevik) – Short Course* of 1938; and his essay *On Dialectical and Historical Materialism*. In considering this list, one is struck by how meagre Stalin's theoretical contributions appear alongside not only those of Lenin, but of almost any of the 'enemies of the people', Trotsky, Bukharin, Zinoviev, Kamenev, and others, who had been liquidated in the purges of the thirties. The letter to *Proletarskaia revoliutsiia* is a criticism of the 'rotten liberalism' of a Soviet author whom Stalin suspected of smuggling in Trotskyism by his apparent approval of the tradition of Rosa Luxemburg in German and Polish Communism. The 1934 conspectus is a series of complaints about the excessive abuse of the tsarist tradition at the hand of historian M. N. Pokrovskii and others. Stalin's letter on the article of Engels has to do with Engels's antipathy for Russian tsardom and the support he seemed to lend to the foreign policy of Imperial Germany in the 1890s. The *Short Course*, of which *On Dialectical and Historical Materialism* is a chapter, tells the story of the Bolshevik party in terms of the bogus anti-Leninist plots and conspiracies presumably liquidated by the purges of 1936–38.

The theoretical output of Stalin, as judged by this list, may be said to comprise, aside from a bizarre rendering of the succession from Lenin to Stalin as a tale of mendacious and homicidal intrigues by Stalin's rivals, Stalin's own blend of Communist internationalism and Russian linguistic nationalism. Stalin would soon claim a leading role among the cultures of the civilized world for the Russian language, in an exercise of which we can scarcely imagine Lenin approving, at least not the Lenin who conducted the early meetings of the Communist International in German.

Yet Stalin was not at all bothered by his concessions to Russian nationalism. World Communism had to have a geographical centre in view of the fact that socialist revolution had first come in one country. Once one had said that, it was not a great step further to suggest that new additions to the family of socialism must be grouped around the movement's original home, Moscow. 'Through socialism to freedom', in the phrase of the pro-Soviet Menshevik Feodor Dan, by way of the 'Russian Idea'. Nevertheless, Stalin was not content to remain the *generalissimus* of the great Russian cause. He still aspired to be a tribune of the workers of the world.

Communist revival was therefore on the order of the day both in Russia and in countries where Communist parties did not exist. This applied to parties that had virtually been liquidated by the repressions of the 1930s, such as Hungary and Poland, as well as whose wartime self-liquidation had been pursued as a policy in the same conciliatory spirit that induced the dissolution of the Comintern. In April 1945, French Communist Jacques Duclos criticized Earl Browder and the leadership of the US Communist Party for their wartime liquidationism, an action that prompted the reconstitution of the American Communist Party a year later. The Duclos Letter is sometimes taken as an opening shot in the Cold War, but it was only part of the same process of rebuilding the Communist parties that was going on in east Europe. It did not mean going over to tactics anticipating an immediate fight for power, even in countries where the Red Army was dominant.

In fact Stalin's policy was quite different. He started from an assessment of traditional national interests of the powers. He did not assume that the United States had left isolationism behind to take up a permanent presence in Europe, nor did he judge that the old relationships among the former great powers of the interwar period could be ignored. Stalin's policy was based on the hope, which proved to be futile, that the United States would not object too strenuously to a 'normal' sharing out of lands won from the axis, in a division of spoils more or less corresponding to the disposition of the armed forces at the close of fighting. It was assumed that the wellsprings of American policy would continue to be what they had always been, Manifest Destiny and the Monroe Doctrine, alongside a continuing interest in the Far East. The American aims in Asia seemed clear enough from the formulations of the Cairo conference of November 1943: friendly relations with the Kuomingtang in China and retrocession by Japan of all that she had acquired since 1894–95, including recognition of Soviet rights in South Sakhalin. The Chinese nationalist Kuomintang might object to Soviet activity in Sinkiang and Outer Mongolia, but that could certainly be negotiated.

Stalin assumed that Anglo-Soviet relations would be more difficult, although they would be eased somewhat by his recognition of British interests in Greece. This was managed by the Stalin–Churchill 'percentages' agreement of October 1944, which also accepted Soviet influence in Bulgaria and Romania. Thus the Soviet Union refused to back the Bulgarian claim to West Thrace. Soviet naval power in the Baltic and Black Seas would have to be accepted by Britain in much the same spirit as the Soviets accepted British power in the Mediterranean. Tensions might result in the Middle East, the key to which was Iran, but a certain sharing out, along the lines of the Anglo-Russian entente of 1907, would surely be possible. Stalin thought he could round out Soviet Armenia, Soviet Georgia, and Soviet Azerbaijan by taking small bits of territory from Turkey and Iran. He thought that an old-fashioned strategic compromise of the nineteenth-century type would be satisfactory to Britain, a compromise that would leave the Soviets with their Arctic Sea passage and trans-Siberian railway, and the British with their long and short routes to India through Suez and around the Cape.

That would allow the Kremlin to settle European affairs in concert with the French, whose traditional European interests could be seen to harmonize with those of the Soviets. Khrushchev reports Stalin telling him that de Gaulle wanted to restore Franco-Russian relations 'as they had existed before world war one'.[5] That suggested a return to a Franco-Russian alliance against Germany. French policy did in fact centre on keeping Germany weak by means of alliance with the Soviet Union. De Gaulle promoted international control of the Ruhr, assuming that this would mean a Soviet share in its production. In return, he thought he could count on Soviet support for French claims to the Saar, although this hope was to be disappointed. De Gaulle wanted agreement with the Soviets as a balance against possible British designs on the French overseas possessions. Syria and Lebanon had been occupied by British troops in the summer of 1941, and the British subsequently encouraged local nationalism to the point that de Gaulle would accuse them in 1945 of 'betraying the Occident'. He regretted, he said, that he could not wage war over the matter.

The Americans could not appreciate his outrage over the unilateral British invasion of Madagascar in 1942, nor his fears about British meddling in Djibouti. On the other hand, he suspected that the Americans themselves would use a United Nations rationale for an Anglo-American occupation of Bizerte, Dakar, and perhaps New Caledonia.[6] De Gaulle did not at first expect the US to take a permanent interest in European affairs beyond a few proclamations in United Nations gatherings. He thought that relations with 'old and courageous' England could be improved as soon as she agreed on Germany and on some 'outdated rivalries' in some other parts of the world. But he counted most on the Soviets, in a way that he did not count on the others, to see eye to eye with him on the European settlement.

Stalin's policy meshed with these suppositions like the bones of a skull. In the occupied east European territories, the former French allies of the interwar period (Poland, Czechoslovakia, Romania, Yugoslavia) were given better borders than Hungary and Bulgaria, the former revisionist opponents of the Versailles and Neuilly treaties. Poland received Pomerania and Silesia at the expense of Germany, an arrangement for which French agreement was obtained before that of Britain and the United States. Czechoslovakia got back south Slovakia (which had been tossed to Hungary by Ciano and Göring in 1938) and the Sudetenland, losing to the Soviets only Sub-Carpathian Ruthenia. Romania got back northern Transylvania, Hitler's 'second Vienna award' of 1940, from Hungary. This compensated Romanian national sentiment for the loss of Bessarabia to the Soviet Union. Bulgaria had to renounce West Thrace, which she had administered with German approval.

The eastern European countries occupied by the Red Army were not immediately sovietized. Usually all-party coalitions were organized on the same model as the 'Tripartite' coalitions in western countries such as France, Italy, Belgium, Luxembourg, Norway, and Denmark – with the important difference that in East Europe Communist control of the armed forces and internal security posts was always the norm. In most cases land reform was imposed rather than collectivization. If a Soviet economic model could be said to be in evidence, one might say that it was the model of the New Economic Policy (NEP) of the 1920s which tolerated markets, rather than the model of Soviet Five-Year Plans. The all-party coalitions were not completely dissolved in favour of Communist rule until 1947–48, when Tripartism collapsed in the west. To Stalin the eastern coalitions were hostages that he would suspend in the intermediate status of 'people's democracy' as long as Communist parties were permitted to participate in the Italian and French governments. And, of course, the western Communist parties were expected to play a role in keeping their countries' foreign policies favourable to Soviet aims. When we judge the *perestroika*-era ideas of Communists about the shaping of multiparty systems of Eastern Europe after the revolutions of 1989, it is useful to keep in mind the experience of the Peoples' Democracies of the early post-war period.

A theoretical underpinning of pre-Cold War Stalin foreign policy was provided by the work of E. S. Varga, an economist with a reputation in the Soviet Union and the Comintern that went back to 1919. Varga saw the western world convulsed by an economic crisis of large proportions, a crisis that took its shape from the fact that the United States had suffered much less than Western Europe. The Americans, he judged, had much of the world's gold and much of its industrial production, but the dollar shortage in Europe meant that nothing could be bought from them except on credit.[7] The United States, moreover, engulfed as it was in a wave of post-war strikes, was itself in a major economic crisis.

American relations with Europe would be difficult, Varga thought, because of US designs on the British and French colonies. For this reason more than any other, the Anglo-American antagonism should be regarded as the primary one in world politics.[8] Varga was greatly impressed by the fact that during the war Churchill had wanted to create a second front in the Balkans in order to prevent Soviet influence from having a monopoly in the east, but that Roosevelt and Stalin had stopped him.[9] He reasoned from this that friction between the Anglo-Saxons would prevent their cooperation against the Soviet Union and other 'progressive' forces in Europe.

The 'democracies of the new type' would therefore be able to give a lead to political and economic reconstruction. Varga used this term primarily in reference to Yugoslavia, Bulgaria, Czechoslovakia, and Poland. This was at a time when Czechoslovakia was able to order its affairs on a basis of multi-party democracy without major interference from the Red Army. Varga thought that Yugoslavia was the most advanced of the New Democracies. 'In the person of Tito', he remarked, 'the movement has found a political and military organizer of the first rank, who has long been a national hero in the struggle and a recognized chief (*priznannym vozhdem*).'[10] Neither Hungary nor Romania, said Varga, had reached the same level as the other four countries, but they might advance in that direction under the right circumstances.

The Peoples' Democracies were no accidental feature of the Soviet occupation, according to Varga, but a more general outgrowth of the crisis of capitalism which had caused the war. Just as the wartime organization of Imperial Germany in the 1914–18 war had suggested the elements of socialist government to Lenin, the wartime economies of the western capitalist world had shown signs of incipient socialism. It was no coincidence that Labour had won power over Churchill and the Tories at the end of the War. Now the Labour government was taking steps on the path of peaceful transition to socialism. Varga took pains to counter any suggestion of a chasm between East and West based on the difference in social and economic systems. He stressed the diversity of the Peoples' Democracies, and invited comparison of the multi-party democracy of Czechoslovakia with those of France and Italy. 'The rise of the states of the New Democracy clearly shows', said Varga, 'that government by the toilers is possible even with the preservation of the exterior forms of parliamentary democracy.'[11]

Nor should the economies of the Peoples' Democracies be made to copy the Soviet model. On the contrary, Varga frankly called them 'state capitalist'. This not only invited comparison with the Soviet economy of the NEP period in the 1920s, but recalled the position of Lenin in 1919 and his description of NEP in 1921 as 'state capitalist'. Varga wanted to see the 'bourgeois nationalizations' then being carried out by Britain's Labour government as part of the same process that was transforming eastern Europe. No doubt the Tripartite regimes in the West, in which Communists participated, would over a period of time follow suit. Varga deliberately blurred the distinction between the dictatorship of the proletariat and capitalism; the Peoples' Democracies were neither, he said, but actually dictatorships of 'toilers of town and village' exercising 'only some functions' of the dictatorship of the proletariat. The old terms and distinctions should not be regarded as dogmatically binding. Everything was possible. Stalin himself had declared to Milovan Djilas that 'today socialism is possible even under an English king'.[12]

In historical accounts it is sometimes supposed that a 'soft' line of Varga's was under fire from its first appearance, with Stalin's heir-apparent Andrei Zhdanov, the most influential leader after Stalin himself, opposing it with the more hawkish analysis, according to which the world was divided into 'Two Camps'. This view is an interpolation from Zhdanov's speech at the founding conference of the Communist Information Bureau (Cominform) at Szklarska Poreba in September 1947, in which he did indeed warn of a division into two camps. By that time Varga's views had come under criticism from Soviet economists. The available evidence, however, does not support the thesis of a Varga–Zhdanov struggle over whether to start the Cold War. Zhdanov's most important public utterances in fact gave full support to the Varga line of forward policy. He put emphasis on the Soviet need for a period of postwar domestic reconstruction, (which he called *perestroika*), while at the same time emphasizing the global reach of Soviet policy: 'The authority and significance of the Soviet Union in the international arena has grown immeasurably. The USSR now plays a great role in deciding every international matter.'[13]

Before 1947 Zhdanov did not argue on the basis of 'Two Camps', but assumed the continuation of the wartime alliance with the Anglo-Saxons along with adherence to the Yalta and Potsdam decisions to build the United Nations.[14] He saw challenges from a different quarter: 'While it is true to say that fascism has been destroyed, its kernel nevertheless has not been liquidated. There still remain, among the ranks of the liberated nations, some reactionary elements.' By the 'kernel' of fascism Zhdanov was referring to resistance to the demands that the Soviets were making on Turkey and Iran, both of whom had been pro-German before the War. Stalin wanted bases at the Dardanelles and rectification of the Turkish border so as to bring the Kars and Ardahan districts into Soviet Armenia and Soviet Georgia. From Iran he wanted oil concessions comparable to those enjoyed by the British. After the Anglo-Soviet invasion in 1941, the Soviets had occupied Iranian Azerbaijan, Gilan, and Mazanderan on the southern coast of the Caspian Sea. In December 1945, the Communist Tudeh party had established an 'Autonomous Republic of Azerbaijan' under Jafar Pishevari, who had been a leader of the 'Gilan republic' established briefly in the same area after the First World War. Lenin and his colleagues had decided to abandon the Gilan Republic once they secured a treaty with the Persian monarchy with strong anti-British overtones. But now Stalin felt that a territorial division was in order, in line with the promises that Churchill and Roosevelt had made him during the War to the effect that Soviet interests would be taken seriously in both Turkey and Iran. In this, Stalin remembered the tsarist precedents from his youth. Just as the Far Eastern settlement at Yalta meant to him revenge for tsarist Russia's defeat by the Japanese in 1905, so now he expected a division of spheres in Iran comparable to that made by the Anglo-Russian Entente of 1907.

His speech of 9 February 1946, taken by some in the West as aggressive and even bellicose, nevertheless contained praise for the 'anti-fascist coalition of the Soviet Union, the United States, Great Britain, and other freedom-loving states'.[15] Stalin mused that the Soviet Union had once been compared to a less lucky multinational state, Austria-Hungary; however, his country was, he thought, no 'colossus with feet of clay', as the West liked to think, but a truly progressive and growing multinational state. Its umbrella could now cover Finns, east Prussians, Ruthenians, Bukovinans, Moldavians, Armenians, Azeris, Kurds, and others not previously protected by the Kremlin.

Stalin's policy, as seen in official statements and actions, would have made sense in terms of the traditions of the European nations. No nation now counted as much as the United States, and the policy could succeed only if the United States were to act as expected and return to a policy of isolationism. But the United States did not do so, and ultimately cast its vote in favour of the Cold War, rather than accept the steady expansion of Soviet influence in Europe and Asia. One can see the Soviet strategy line coming apart at each moment of disappointment. The United States took the Iranian case to the United Nations in January 1946. Churchill made his famous Fulton, Missouri speech in March 1946, with its call for a 'fraternal association' of the US and Britain, delivered at the same time as an Anglo-French agreement was reached on the Levant. Stalin could not ignore that, but even while noting that Churchill's call for a 'racial union' bore the threat of a new war, he also predicted that western peoples would not heed the voices of the 'reactionaries'.[16]

Zhdanov also refrained from attacking the prognosis of forward policy offered by Varga, despite his increasing concern about western policy. In November 1946, he still referred to 'freedom-loving states' in the West and restated the thesis that the anti-fascist war had resulted in a democratic peace.[17] He noted two tendencies in the United Nations, one struggling to preserve the unity and agreement of the great powers, and the other seeking to use its forum to make 'atomic propaganda' (a reference to the Baruch plan then under discussion). But he did not reckon with a permanently divided Europe at this point. Instead he cited the general pro-socialist trend in Europe and pointed proudly to the successes of 'brother Slav countries', Yugoslavia, Czechoslovakia, and Poland.

Zhdanov and Tito appeared to be close politically during the period before the purges of their followers that engulfed the eastern bloc after Tito's expulsion from the Cominform in June 1948. Like Varga, Zhdanov had envisioned Tito's Yugoslavia at the forefront of the new democracies. 'Anti-democratic' tendencies in the West were fighting to ruin Tito's policy of annexing Trieste, to which policy Zhdanov gave strong support.[18] Correspondence later released by the Yugoslavs showed that Stalin cautioned against supporting Tito on Trieste. Zhdanov similarly praised the work of Georgy Dimitrov, who had just taken power in Bulgaria, and who would soon be offering plans with Tito for a Balkan Union. Dimitrov's Fatherland Front was, he said, an example to democratic forces in postwar Europe, led as it was by the living symbol of resistance to the Nazis in the Reichstag fire trial and the champion of the Popular Front of the thirties. Zhdanov stressed that the victory of Labour in England and the Tripartite bloc in France showed a tremendous momentum toward the Left throughout Europe.

Soviet policy thus adhered to an analysis that promised multiform political gains in postwar Europe without incurring opposition from the United States. A turn from this line only occurred in the spring of 1947, when the Truman Doctrine promised aid to Greece and Turkey and the Marshall Plan proposed American aid to rebuild the western European economies. Western governments were given an alternative to Tripartite coalitions. They quickly accepted the aid offers and abandoned Tripartism. The Varga line had failed. This was the cause of the eruption of criticism by leading Soviet economists of Varga's ideas in May 1947.[19] 'Mistakes of a reformist character' were discovered in Varga's work. When sovietization got under way in the New Democracies, he was criticized for having described their economies as 'state capitalist'. After the Communist parties had left the French and Italian governments, it was discovered that he had been mistaken in the notion that 'right socialists' were fit partners for government coalitions in western countries, and that he had failed to recognize them as 'treacherous accomplices in reaction'.[20]

Varga had to recant. He admitted in 1948 that he had been misled by the experience of Communists in the French and Italian governments.[21] He admitted that he had been wrong to have compared wartime economy in bourgeois states to socialist planned economy. British Labour's nation-alizations, he said, were not really progress toward Democracy of a New Type:

> The state in England, where the Labourites have come to power, re-mains nevertheless a bourgeois state, a state of monopoly capital, the main function of which consists in securing the class rule of the big bourgeoisie. . . . the state of the Peoples Democracies, by contrast, is a state of toilers, exercising the functions of a dictatorship of the pro-letariat in the interest of all the toilers, against the exploiting classes.[22]

His previous designation of east European economies as state capital-ist no longer made sense as Communists were driving the non-Communist parties out of power. When Tito fell out with Stalin and was expelled from the Cominform, Varga had to reverse his earlier enthusiasm for the Yugoslav revolution. He now warned in his self-criticism of the danger of backsliding in the New Democracies, specifically citing the Yugoslav case. Soviet propaganda would be taking things still further by 1949, claiming that Yugoslavia had restored capitalism, and even become 'fascist'.

Soviet foreign policy between 1944 and 1947 had not been overtly ag-gressive in the military sense, if one excludes Yugoslavia's activities in Greece and its claims against Italy for Trieste and against Austria for Slovene Carinthia. These actions cannot, of course, be said to be com-pletely contrary to Moscow's wishes; Stalin's well-known doubts about the chances of winning in Greece would certainly have been suppressed as easily as his doubts about the Communist cause in China, had the campaign been a success. He did not like to dictate every move to his subordinates and comrades, but merely intervened when they failed or went too far. On the other hand, Stalin clearly wanted restraint in areas like Greece, China, and Indochina, lest he raise resistance to his plans in Europe. Even so, Stalin himself did not act with sufficient caution to make the Varga line work. The nuances that distinguished his all-party coalitions in eastern Europe from the sovietized regimes that came later were completely lost on western policy-makers, who saw only the mon-opoly held by Communists in the key internal security posts and noted the very Stalinist way in which political opponents were usually treated.

Stalin also aroused suspicion by his own claims against Iran and Turkey. Even while he had wartime assurances from Roosevelt and Churchill that Russian interests in these areas would be considered, he should have known that the United States, if it should reconsider, would only be further antagonized at the sight of Soviet policy treading in the path of traditional tsarist imperialism. His efforts only made sense on the assumption of an American return to isolationism, that is, to a primary interest in Asian and Pacific affairs. Stalin's America analysis was wooden and too much influenced by the pattern of the prewar years. In general it suffered from the thesis of an irrepressible Anglo-American antagonism that Stalin inherited from Lenin and his other Bolshevik forebears, a thesis that has never repaid the political and emotional sums wagered on it.

Varga's line was all the same an alternative to the Cold War. It was the only real model of a multiparty system in the Soviet bloc, so it would still have appeal for bloc Communists in the era of Gorbachev's *perestroika*. In 1989, Hungarian First Secretary Karoly Grosz spoke of returning to Varga's ideas. 'We are now participants', he said, 'in an undertaking of similar dimension and significance as in the period of the national democratic revolution between 1944 and 1948 . . .'[23] This was, of course, no real model of a multiparty democracy; Stalin's heavy hand at all times weighed down on the all-party coalitions and eventually liquidated them altogether. But Communists of the next generation who were fond of calling themselves democrats were nonetheless bound to find the myth of the Varga era irresistible.

And it is natural to ask whether this alternative, which could have been pursued by American policy-makers simply by not resorting to the Marshall Plan and Truman Doctrine, would have been preferable to what did ensue. In Europe, poverty and dislocation were already developing along the lines of 1918–24, when there were massive waves of strikes and civil disturbances, even an attempt by revolutionaries at seizing power in several countries. In Italy there were factory occupations in 1920 that led Lenin and Trotsky to think that the revolution was approaching. The German Communists did try to take power by insurrection in 1921 and 1923. Right-wing gangs found the atmosphere conducive to their rapid growth and schooling. In a post-1945 Europe recovering through its own efforts, scenes like this were definitely not excluded. To be sure, they might not have resulted in anarchy or a slide toward chaos. But it is fair to ask: how could the United States have become accustomed to watching Europe languish in poverty for the sake of staying on good terms with Soviet Russia?

II. *RECUEILLEMENT*: CLOSING OFF EASTERN EUROPE WITH COVERING ACTION IN ASIA, 1948–54

Two tasks had seemed to flow from the analysis of Varga and the policy offensive of 1944–47: to obtain a share in the production of the Ruhr, secured presumably by Soviet participation in its international control; and to probe areas on the southern flank from Greece to Iran for possible British weakness. If the Anglo-American antagonism were as profound as Varga thought, these rather modest pursuits might proceed without a major confrontation. But the United States and England did not fall out. The presumed benefits of their disunity over Germany never materialized. From the time of Secretary of State Byrnes's Stuttgart speech in September 1946, which looked ahead to a possible economic union of western zones, Soviet statements became more and more pessimistic. If the Anglo-American antagonism were to pay off anywhere, it should certainly be in Germany. The establishment of Bizonia and the announcement of the Marshall Plan ruined their fondest hopes.

Soviet policy therefore made a turn. The countries of the New Democracy were closed off from the West and underwent an internal process similar to Soviet Russia's when in 1928 she had deserted the cautious policy of the NEP for agricultural collectivization and the Five-Year Plan. The shining light of the New Democracies, Tito's Yugoslavia, had to be jettisoned lest the Soviets became involved with the United States in war over Greece. Czechoslovakia, in many ways living on Soviet forbearance from the time of Beneš's wartime agreements with Stalin, was brought behind the East–West line of division by a Soviet-sponsored coup d'état. Henceforth Czech industrial production would have to substitute for the lost share of the Ruhr, and Czech uranium mines would have to fuel the Soviet nuclear programme. The retreat in the east was covered by an attack on western access routes to Berlin, no doubt on the assumption that the United States was unlikely to find a way to defend the city against blockade.

Zhdanov's fall paralleled the turn. In his place Georgi Malenkov rose to prominence, backed closely by police expert Lavrentii Beria, since 1946 also in charge of the nuclear programme. At the founding meeting of the Cominform in September 1947, Zhdanov joined Malenkov and Suslov in denouncing the French and Italian Communists for 'tailing' their non-communist government coalition partners. He urged a general struggle against enslavement of western Europe under the Marshall Plan. Zhdanov described the world breaking down into 'Two Camps', a pronouncement that, as we have seen, has been taken to indicate that he was against the orientation of restraint that had been pursued for three years. In fact the speech at the Cominform meeting was Zhdanov's swan song. He was performing a last service to assist the policy turn by criticizing his own former views and supporters, in time-honoured fashion. In much the same way, Zinoviev had denounced his German ally Ruth Fischer in 1926, and Bukharin fought the Right Danger (views he had previously sponsored) in 1928. Zhdanov dropped from view a few months later and died in August 1948. Significantly, his decline and fall paralleled the purges of east European Communist leaders such as Gomulka, Rajk, and Kostov, as 'Titoists' and 'national deviationists'. These were accompanied by purges of Zhdanov's supporters in the 'Leningrad affair' of 1949. The foreign policy turn to the theory of the 'Two Camps' was certainly no victory for Zhdanov.

Western Kremlinologists speculated that Malenkov had never seen real possibilities for advance in Europe by political means but had instead envisioned military gains to be made in Asia. It was striking that during the winter months of 1948–49, while the attention of the West was riveted on the events attending the Berlin Blockade, the Chinese Communists launched a campaign that netted them crucial victories, at Mukden in November and Beijing in January 1949. This led to the intriguing thesis that Malenkov, as head of a Central Committee Far Eastern Commission, had redirected Soviet foreign policy from west to east, and used the Berlin Blockade to cover the Chinese Revolution.[24] It was said that Stalin had studied the geopolitics of Karl Haushofer, who had instructed Hitler in the twenties, and that the Soviet dictator had been impressed with Haushofer's conception of a Eurasian bloc of Russia, Germany, and Japan to do battle with the Anglo-Saxon imperialists. Stalin had followed the policy more consistently than Hitler, so the argument ran, making pacts with Germany in 1939 and Japan in 1941. According to the testimony of H. K. Konar, a pro-Chinese Indian Communist, and Wang Ming, a pro-Soviet Chinese Communist before his defection to the Soviet Union, Mao Zedong had clashed with Zhdanov repeatedly since the thirties and looked favourably on Malenkov, whose rise coincided with his own turn toward civil war against the Kuomintang. Khrushchev's later victory over Malenkov is given as a reason for estrangement between the Soviets and Chinese in the late fifties.[25]

The Berlin Blockade gave a signal of Soviet unhappiness with western moves toward the formation of a West German state, moves that were now backed by a newly anti-Soviet France, a France that had reversed alliances, in Moscow's eyes, in a craven attempt to receive Marshall Plan aid. Throughout the siege, the Soviets called for abandonment of western plans for a new West German state. Molotov insisted that despite everything progress could always be resumed with 'the implementation of the well-known plan of international control of the Ruhr'.[26] It was not to be. Instead of breaking the western camp, the blockade produced NATO.

One result of Gorchakov's *recueillement* of the nineteenth century had been that Russia, in turning away from Europe, was still able to expand in central Asia, absorbing Kazakhstan, and the southern khanates of Khiva,

Bokhara, and Kokand. Stalin too seemed to think that while turning his back on Europe, he could continue to advance in Asia. The Soviets tested their first atomic bomb in September 1949, breaking the American monopoly. A new burst of confidence fuelled Stalin's hope that the victory of the Chinese Revolution would not be the last blow to be struck in the east. According to Khrushchev, Stalin gave Kim Il-sung a green light, with Mao concurring, to attack South Korea in 1950. A rapid campaign would win this prize easily, thought the three, because it would trigger a mass rising against the South Korean government of Syngman Rhee. Stalin, on Khrushchev's account, was apparently satisfied that the United States would not intervene.[27] He thought that he was only rounding out the boundaries of a sphere of influence outside the essential defence perimeter of the United States, as recently defined by General MacArthur and Secretary of State Acheson.

The Korean War was only an episode in the process of turning away from Europe, but it ended by further inflaming the military confrontation in Europe. Secretary Acheson took it more seriously than Stalin could have foreseen, as he explained to a congressional committee:

> You cannot overestimate the seriousness of this Korean matter. It brings the danger of war very close. It means that we must act as though war might occur at any time, and we have got to think that the real pressure, when it comes, the critical pressure, is likely to be in Europe and not the Far East.[28]

Korea thus smoothed the way for German rearmament, a cruel and ironic turn of events for the French, especially for those like Jean Monnet and Robert Schuman, who hoped to make French adherence to NATO easier to accept by solving the German problem through economic integration. Korea also recast the discourse on Indochina. The West had previously considered it a problem of post-colonial adjustment, but now it became part of the global problem of resistance to Communist aggression.

The French were seemingly compensated for the defeat of their plans for keeping Germany weak by American sympathy for their interests in Asia. This became more evident when Monnet channelled the discussion into a project for a European Defence Community. Eisenhower and Dulles were already very much worried about a Vietminh victory in Indochina and even thought it was in some ways more dangerous, from the standpoint of fears of falling dominoes, than a Communist victory in Korea. Their desire to win the French over to the EDC prompted the hope that support for the French, 'picking up the check in Indochina', as Eisenhower's Chief of Staff Sherman Adams put it, would do the trick.[29]

On the other side, Soviet policy was just as firm in opposing the EDC.

Moscow's propaganda warned throughout the period in which the EDC was under consideration that it amounted to rearming the *Wehrmacht*, and thus represented a defeat for the 'national forces' of France. If they were to appeal to these forces, the Soviet leaders would have to take into their calculations the indisposition of French nationalists in Indo-China.

In his last theoretical document, *Economic Problems of Socialism*, published for discussion in spring 1952, Stalin stressed, as he had many times, the crucial importance to the Soviet Union of differences among the NATO allies. He reminded his countrymen, in the manner of Prince Gorchakov, that a 'united west' was not a historical tendency, and would not be held together simply by the unity of capitalists in the face of world socialism. Back in the twenties, Britain and the United States had originally helped Weimar Germany to recover (Stalin omitted reference to the help rendered Germany by the Red Army) with an eye toward making her a bulwark against Soviet Russia. But, in the end, Germany directed her armed forces against the West, and when she attacked Russia, the western bloc did not help her but joined Soviet Russia in opposition. The lesson that Stalin was trying to teach was that the 'Cold War' of the twenties and thirties was not primary, but antagonisms between the imperialists were. Britain and France were maritime powers who jealously guarded overseas interests from American interference and disliked penetration of the European economies by American investment. West Germany and Japan could also be counted on to try to 'break from American bondage' before long.[30] Indeed, while Stalin argued that wars were inevitable between capitalist countries, he insisted that war was *not* inevitable between socialism and capitalism. The western peace movement could serve as a powerful moral support of the Soviet bloc in the effort to counter aggressive acts by the West.

Stalin's foreign and domestic policies in the last years of his life suggest that he sensed the limits of his reach in Asia, but also that he was resigned, in the absence of some immediate break in the front of the western powers, to a more or less prolonged period of war in the east. He seems to have feared that the momentum of the criticism of Varga would establish the case for consumer goods production as against heavy industry. The economist Ostrovityanov, for example, who had criticized Varga's 'reformism', nevertheless wrote expansively of the Soviet state 'using the law of value as an instrument for the planned management of the economy'.[31] Even in those pre-*perestroika* days, Soviet economists could not resist the idea that the market might make their system more rational. Stalin suspected that a new edition of the pro-market ideas of his old rival, Nikolai Bukharin, might reappear. In *Economic Problems* he reminded the comrades of previous suggestions that profitability be used as a criterion for allocating resources and warned that following this path would put an end to expansion of the economy. The law of value (the logic of the market) could not be used to regulate the economy. It was a mistake to make a fetish of economic laws.

Stalin wanted to keep the Soviet economy in a permanent state of siege and to wear down the western armies in far-flung engagements all over the globe. He was encouraged by the thought that he was at least abreast of the United States in the race for the hydrogen bomb. He judged that his grip on things was stronger and his nerves steadier than the states-men he was up against. Just a few months before his death, he was pre-paring to clear the decks of potential opposition by a purge staged to follow 'revelations' about a Doctors' Plot, with charges of assassinations by Kremlin physicians. There is reason to think that Molotov, Kaganovich, Voroshilov, and many others from among his old associates would have been carried off in this whirlwind. But his death in March 1953 cut the campaign short.

As would later be demonstrated in the transition from Brezhnev to Gorbachev, a foreign policy of confrontation relying on sheer will, even confrontation far from the main European theatre of conflict, is simply not capable of surviving a succession of leadership. New leaders want to make a fresh start. So it was natural that, when Stalin died, Malenkov and his associates reached the not very aggressive conclusion that their Asian policy was exhausted. They made peace in Korea. And after the French defeat at Dien Bien Phu, they apparently induced Chou En-lai to persuade Ho Chi Minh to be content with control over only the north of Vietnam. The Soviets wanted the votes in the French Assembly of the Gaullists and Mendès-France's Radicals against the EDC. It was obvious to the western diplomats at the Geneva peace conference ending the Indochina war that a deal 'in the margins' had been struck between the Soviets and Mendès-France to limit Ho in return for Mendès's agree-ment to torpedo the EDC.[32]

The failure of the EDC could be seen as a victory for the Soviet policy of keeping Germany the 'apple of discord' among the western powers. But the United States did not give up so easily. Dulles warned of an 'agonizing re-appraisal' of US commitments in the wake of its failure. As a result of this pressure, the German rearmament that the supporters of the EDC had originally sought to contain within the framework of a European multilateralism happened all the same. German rearmament had been forced on a recalcitrant France by the United States and Britain. In re-sponse, Mendès-France's cabinet took the decision, in December 1954, to build nuclear weapons, the first to be tested within five years. At the same time, the British, having helped forge this arrangement, themselves benefited from the relaxation of the McMahon Act (legislation enacted in 1946 that had restricted exchange of nuclear information), a relaxation which permitted American aid to their nuclear programme. The persistent feeling harboured by the French, that they were never permitted to have the same relations with the United States as the British, was now abetted by the developing 'special relationship' between the Anglo-Saxons.

The Soviets had expected and hoped that the Western bloc would be rent with differences, something to which their policy was supposed to contribute. That the Western powers were quarrelling was thus a kind of success. But the period of *recueillement* had ended with a peculiar sort of increased friction among the Western powers that only produced multiplied threats, often nuclear ones, against the Soviet Union. In 1954–55 with the post-Stalin succession struggle bringing Khrushchev to the top, the shortcomings of the previous policy period were increasingly evident. Some new way, perhaps by political means, would have to be found to resume the advance.

III. POLICY OFFENSIVE: KHRUSHCHEV COURTS THE NON-ALIGNED, 1955–63

While they wanted a new beginning in foreign policy, Stalin's heirs also knew that they could profit from the existing differences among the Western powers. The French were not happy about the Indochina arrangement settled at Geneva. The French Right was chagrined to see a prized part of the empire lost without even a residue of neutral states with whom some kind of relations could be maintained. Instead, they saw a Communist state and a string of states dependent on the United States. Nor was the situation better in other parts of the French empire. In the Maghreb, the United States seemed to be encouraging nationalist forces in Morocco and Tunisia. The British were less uneasy about the United States, but they were also concerned about their Middle Eastern interests. The American action in overthrowing the Mossadeq government in Iran in 1953 put US interests on an equal footing with those of Britain, in an area where the British had previously had no competitor. As a result, there was some foreboding in Egypt that a wounded Britain would try to compensate by 'doing a Mossadeq' against Nasser.[33] For their part, the French also saw Nasser as a source of their trouble with Algeria, and suspected that the Americans enjoyed their discomfiture. From the Soviet standpoint, there was reason to remember Stalin's prediction that the Western states could not preserve their unity just on the basis of anti-Communism. One point that Varga, in his 1947 self-criticism, had refused to recant was his thesis on the conflict among the interests of Britain, France, and the United States.[34]

Malenkov and Molotov had ended the wars in Asia, and Malenkov spoke as if he wanted to ease tensions in the West and blunt the edge of a nuclear NATO. To this end he made his famous statement to the effect that the atomic bomb does not obey the class principle, that is, that a nuclear war would be as damaging to Communism as to capitalism. Khrushchev quickly corrected him, asserting that capitalism would suffer worse losses. Khrushchev was attempting to court the military people who had helped him arrest Beria in 1953, in order to finish the job against Malenkov. With his criticisms of Malenkov's ideas about increasing consumer goods production at the expense of heavy industry, Khrushchev's positions gave the appearance of a genuinely Stalinist programme, in the spirit of *Economic Problems of Socialism*, one that must have impressed Molotov and the rest of those who wanted to continue in the footsteps of the dead dictator. Khrushchev was aided by the opinion of the Soviet generals to the effect that, if they permitted the economy to go over to the production of consumer goods, they would be condemning their country to permanent inferiority in the arms race with the United States. The Soviet Union had experienced the first of three fateful national security debates; the next would occur in the sixties after Khrushchev's fall; the last would be the fight over Gorbachev's *perestroika*.

Khrushchev needed a broad front among the political and military leaders against Malenkov, but once Malenkov was defeated, he quickly decided that he did not want to be a mere first among equals in a collective leadership with them. A general secretary, as would be forcefully demonstrated with the case of Gorbachev, is in a position to introduce policies that give him a lever against opponents. By 1955 a Khrushchev foreign policy having this function had begun to take shape, one that was bound to create friction with those who upheld Stalinist orthodoxy. Khrushchev proposed in effect to resume the line of march that had been followed in the days of Zhdanov's ascendancy and abandoned at the time of the rift with Tito. The Soviets had ended up disavowing the Yugoslav probes into the British sphere in the eastern Mediterranean in 1948 because of the excessive danger. But Khrushchev now wanted to seek reconciliation with Tito and take advantage of the latter's influence among the bloc of neutrals that seemed to be taking shape around Nasser, Nehru, Sukarno, U Nu and others at the Bandung Conference in May 1955. Khrushchev wanted better relations with all the Asian and African neutrals, all the more in view of the fact that the American Secretary of State John Foster Dulles regarded neutralism as apostasy. But he especially wanted to act in the Middle East to break up the Baghdad Pact being formed by Britain at this time.

Reaching out to the Third World (the 'peace bloc') required continued harmony with China. Already in 1954 Khrushchev had promised Mao

Zedong a Soviet withdrawal from Port Arthur, Sinkiang, and Manchuria, and agreed to an exchange of technical information on nuclear energy. The Soviets transferred to China control over their joint-stock companies mining uranium and thorium in Sinkiang since 1949.[35] They wanted to construct more active positions in Asia. They agreed with the Indonesian Communist D. N. Aidit's slogan for a coalition of Communist and national forces. They offered aid to both Indonesia and Burma. They cast aside all fears about the 'national bourgeoisie' in Asia and dubbed it a progressive force.

In Europe, they decided on a treaty to neutralize Austria, an idea that had a certain appeal for Tito. And, against the opposition of Molotov and Suslov, Khrushchev met with Tito and went so far as to admit a share of the fault for the estrangement between Yugoslavia and the USSR. When, in January 1956, he told the Twentieth Party Congress of the errors of the Stalin period, he was not only putting the reconciliation with Tito on an ideological footing, but also sharpening up his dispute with Molotov, who defended not only the break with Tito but the entire record of Stalin. In thus breaking with certain tenets of the Stalinist heritage, Khrushchev undermined the positions of those east European leaders who had taken power as a result of Stalin's purges of 'Titoists' and 'national deviationists'. Like Gorbachev's *perestroika*, Khrushchev's revolution in foreign policy entailed a revolution in ideology and, as it turned out, an east European revolution as well.

It is unlikely that Khrushchev ever thought the whole policy through. He was an impetuous and bold man, quite unlike Stalin in that respect. Many of his most important foreign policy decisions were made under the pressure of his political struggles with opponents, in an attempt to strengthen his hand against them. The deal whereby Czech arms were sent to Egypt in September 1955 was intended to bolster Egypt's claim to leadership of the Arab world, as against Iraq, at first the only Arab state in the Baghdad Pact. But it also sharpened up the confrontation with Khrushchev's Politburo critics, 'those narrow-minded skunks', as he called them in his memoirs, who tried to defeat his Middle East policy.[36]

Khrushchev's east European policy was not simply a reflection of the general thaw, but was in some respects a correction of it. This was the case with regard to the experiment of Imre Nagy's team in Hungary, which rose up simultaneously with the Malenkov leadership in the Soviet Union, and which was ended in 1955 on the initiative of Moscow's pronouncement in favour of heavy industry production. Patching things up with Tito in effect caused a re-thaw. But even that exploded in the revolutionary events of 1956 in Poland and Hungary.

That Imre Nagy, recalled to power by the Hungarian Revolution, should then contemplate withdrawal from the Soviet bloc seemed to some a natural consequence of Austria's neutralization. As they would in 1989, the former Habsburg empire nations were reaching for national policies reminiscent of the pre-1914 Austro-Marxism of Otto Bauer and Karl Renner, against which Lenin had once cast fire. Gorbachev would return to these tempting ideas of international liaison in the centre of Europe. In 1956 a wing of the Yugoslav leadership also looked with favour on this developing central European bloc. Not surprisingly, its members were from Croatia and Slovenia, and they hoped that the developing autonomy of central European forces might tilt Yugoslavia away from excessively 'Balkan' (that is, Serbian) influences.[37]

But Hungary's break with the Warsaw Pact was too much for Khrushchev, and even for Tito, who endorsed the Soviet intervention. The lesson presumably was that Tito's non-alignment was to be considered a special case and not to be emulated in principle; or perhaps Tito simply sensed that a collapse of the Soviet bloc, despite what it had done to him, would be a disaster for the cause. Perhaps he had the same response as Trotsky in the thirties, defending Soviet Russia despite its anathemas against him.

The policy offensive fared better in the Middle East. The Anglo-French–Israeli Suez intervention of 1956 was turned back by the withdrawal of US support. After this had become clear, the Soviets made veiled threats of a nuclear attack against Britain and France. This was the first of Khrushchev's horrible attempts to mimic the nuclear diplomacy of John Foster Dulles (or rather the 'brinkmanship' popularized in magazines). Anwar Sadat relates in his memoirs that at the time he thought the Soviet 'threat' to be nonsense but that it succeeded in impressing Nasser, perhaps because that version of events better tended to vindicate Nasser's foreign policy.[38] In any case, the humiliation of Britain and France was real enough. The Duncan Sandys Defence White Paper of 1957 reflected that in its emphasis on the nuclear side of British defence, at the expense of the conventional. The French nuclear programme also won more support as a result of the Suez fiasco.

As with the crisis over German rearmament in 1954, once again disunity and failure in the Western alliance seemed ironically to multiply nuclear threats against the USSR. Nevertheless, Khrushchev could claim that his foreign policy since 1955 had been largely successful. The American line of opposition to neutralism had been broken with Eisenhower's puncturing of British and French efforts to pursue an independent course. The talk about 'rollback' of Soviet power in eastern Europe had proven to be nothing more than posturing; it was clear that the United States left that area to the Soviets as their sphere. The Baghdad Pact was unhinged by the overthrow of the Iraqi monarchy in 1958. The demonstration of ICBM capability in 1957 raised Soviet international influence to the next rank, casting doubt both on the efficacy of the American notions of Massive

Retaliation and on the willingness of America to stand for Europe when its own cities were under threat.

Khrushchev's internal opposition did not view the matter so favourably. Molotov, Malenkov, and Kaganovich rallied under the banner of resistance to Khrushchev's departures from Stalinist orthodoxy. It could be argued against Khrushchev that it was only his vain pursuit of Tito that had led him to criticize Stalin, and that this had weakened the position of the entire bloc in 1956. The Anti-Party Group, as they would later be called, used these arguments to try to force a showdown in summer 1957. They were defeated by Khrushchev in a tense confrontation in the central committee, in which Khrushchev got the support of the army. Later in the year, however, Khrushchev was forced to dismiss his ally in this struggle, Defence Minister Zhukov, an act which put him for the remaining years of his rule under the restraint of the small Politburo group that had supported him against the Anti-Party Group, with ideologue Mikhail Suslov at its head. Molotov was removed from the leadership but he continued to exert pressure in various ways. He and Suslov had participated in Stalin's campaign against Tito, and they stood by their record. Yugoslavia was henceforth criticized as Revisionist and strong objection was voiced to the outrageous idea of Tito being a guide for the east European states (that had been exactly Varga's opinion). Yugoslav replies to these criticisms left no doubt that they considered Suslov to be their main source.[39] Khrushchev, trying to find a position in the centre, was forced to warn against both Revisionism (Tito) and Dogmatism (Molotov).

The restraints on Khrushchev were not obvious at the time in the West. It was widely believed that the defeat of the Anti-Party Group had ended the period of collective leadership and left Khrushchev unchallenged. The influence of Suslov was in most cases not taken into account. And since the Chinese leaders had expressed opinions and given advice to Khrushchev during the tumultuous events of 1956, they assumed a role as vocal contributors to Soviet foreign policy, at first friendly, but increasingly abrasive toward the end of Khrushchev's time in power. They thought the Soviet ICBM heralded a new era for the Soviet bloc and were disappointed not to receive more active support on Quemoy and Matsu in 1958. According to later Chinese statements, they had been promised help for their nuclear programme in 1957 but had then been denied it in 1959. Khrushchev's own account of this decision in his memoirs paints Chinese motives in the darkest tones. His other testimony on Mao's views of nuclear weapons alleges that Mao looked with equanimity on a nuclear war between the two superpowers from which presumably China would emerge as at least a survivor. In the polemics between the Chinese and Soviet parties in the sixties, Soviet writers offered many quoted statements of Mao's that sent a chill up the spine. At the same time Khrushchev was competing for influence in the Third World and could not permit himself to appear to have weak nerves, especially since his denial of nuclear technology to China meant that his was the only nuclear power in the socialist camp.

At the same time, this complex man showed a realization that without some measure of detente with the west he could not continue to take advantage of the various changes taking place in Asia and Africa. For Khrushchev, as for Zhdanov before him, and Suslov and Gorbachev after him, Detente was a premise of forward policy. After 1960, the Cuban Revolution also came under the wing of Soviet power. Khrushchev talked stridently about the Monroe Doctrine being a 'rotting corpse'. Yet he dared not abandon the idea of coexistence with the West. In his dualistic policy, the West would have to be placated and kept at bay while 'class and national struggles' throughout the world steadily undermined its power. He thought he could blunt with gentle entreaties the formidable nuclear threats made by the Western proponents of Massive Retaliation. He did not assume it to be an easy task. The Chinese made everything more difficult by insisting that *sputnik* had given the Communist bloc the upper hand in the nuclear field, which to them meant the upper hand in general. A Chinese endorsement of Soviet Russia as the centre of the bloc, in the effusive speeches at the 1957 meeting of Communist parties, had the corollary that the Soviets must use their nuclear prowess to foster gains for world Communism.

Yet *sputnik* and the Soviet ICBM had accelerated Western plans to reorient NATO toward a tactical nuclear defence of Europe. With Britain and France licking their wounds after Suez, that meant redoubled reliance on Adenauer's Germany, which in those days was often described both in the East and West as America's first ally. *Sputnik* had provoked a debate in the United States that entailed a reconsideration of the idea of Massive Retaliation. The Rockefeller and Gaither Committees' reports were benchmarks in the turn toward to what later became known as Flexible Response. At the same time there seemed to be emerging a ring of American IRBM installations around the entire Soviet bloc. The Soviets did not see the IRBM strategy as it was advertised, that is, as establishing firebreaks to control a nuclear war, but rather as an attempt to threaten a surprise counterforce strike against their own weapons.

Ironically, Khrushchev thought he had learned to live in the world of Massive Retaliation, confident that he had the wherewithal to match vague nuclear threats. We now know that Soviet Russia did not have the missile arsenal predicted for it at the time. Khrushchev enjoyed the psychological edge of western assumptions about a missile gap in the Soviet favour, but he was in fact relying on a minimum deterrent. He was not at all intimidated by the idea. In fact, in submitting to the Supreme Soviet a proposal for the reduction of Soviet forces by 1.2 million men in 1959, he assured that body that the possession of nuclear weapons would keep the Soviet Union impregnable. Khrushchev's Russia was really only the latest nuclear country to adopt the perspective of Massive Retaliation.

Thus 'armed', Khrushchev plunged relentlessly into the series of confrontations, from the Berlin Crises to the Cuban Missile Crisis of October 1962, that brought the world, including his own military and party colleagues, to the edge of the cliff. Reeling from these withering tests of nerve, Suslov and his Central Committee colleagues managed to remove him from power in 1964. After that date, Khrushchev's policies were bitterly renounced and described retrospectively in the press as 'voluntarism, subjectivism, and hare-brained schemes', a verdict that sensible people all over the world could endorse. Khrushchev was clearly the most dangerous man ever to direct Soviet foreign policy, perhaps the most dangerous man in world politics since Hitler. Those who overthrew him could congratulate themselves on a service rendered to all humankind. That Khrushchev was later so blithely admired by many prominent Soviet intellectuals who had Gorbachev's ear tells us a great deal about their judgement, and about the mind of Gorbachev. The years following the Test Ban Treaty of 1963 and Khrushchev's ouster were blessed by a lessening of tensions. However, the lesson ultimately drawn by the Suslov–Brezhnev leadership was an equally dangerous one: that Khrushchev's actions had only failed for being premature and that, after a period of *recueillement*, arms building, and another attempt at reconciliation with the Chinese, the course on which Khrushchev had embarked might again be resumed.

V. *RECUEILLEMENT*: WITHDRAWING FROM EXCESS COMMITMENTS, 1964–73

The impression gathered by outsiders was that Soviet Russia sank into herself after the removal of Khrushchev and began to act, as de Gaulle often put it, as a second-rank power. This made her available for a new connection with Europe. 'The rise of a powerful and militant China on their eastern flank', said the General, 'causes the Soviets to inject a note of sincerity into the couplets that they periodically devote to peaceful co-existence.' Western literature gradually became convinced that this was so, and offered various explanations of the decline in the vigour of Soviet Communism, sometimes attributed to behaviour in accordance with a law of deradicalization of revolutionary regimes. This sort of analysis was pushed along by western New Left ideas about the irrelevance of the Soviet model and the usurpation by a bureaucratic elite of the heritage of the Russian

Revolution. It joined a stream of Chinese invective about the betrayals of 'Soviet Revisionism'. Despite the fact that the consensus shared among these disparate groups 'saved the appearances', as was once said about Ptolemy's astronomy, it was not built for the ages. Soviet Communism had not been domesticated. In fact the Soviet Union had embarked on a vast programme of arms building designed to make good the deficiencies that Khrushchev had attempted to conceal by a policy of bluff and brinkmanship.

The deficiencies were formidable ones. There was a yawning gap between Soviet arms and Soviet purposes. According to the rhetoric of the Statement of the Eighty-One Communist Parties at their Moscow meeting in 1960, the coming trend in world politics was to be one of National Democracy, not a 'bourgeois' democracy, but something like Varga's 'Peoples' Democracy'. The National Democratic regimes, of which Nasser's Egypt and Nkrumah's Ghana were examples, were thought to be necessarily transitory, facing tasks that could only be successfully carried out by Marxist-Leninist parties. During the sixties the hope for a guided transition of Third World dictatorships along a 'non-capitalist path' to a form of Communism was to be more often disappointed than fulfilled. Its outstanding test case, Cuba, demonstrated its central problem, for Cuba lay within a zone of American naval power that had proved impenetrable at the time of the Caribbean crisis of 1962. Moreover, it continued to require Soviet aid while it owed its existence, in large part, to fortune. Cuba thus demonstrated the need to build a Soviet navy capable of worldwide power projection, in much the same way that Berlin required from the West a nuclear doctrine stretched to the maximum. Soviet support for Cuba, dating from 1960, could not in any sense be described as a vital Soviet national security interest. But Soviet ideological ambitions had led to the necessity of global reach.

At the same time, the Soviets had to struggle to achieve missile parity with the United States, which they did by 1969. The settlement of the Cuban Missile Crisis had prompted the Partial Test Ban Treaty, signed in summer 1963. This pointed in the direction of an eventual arms control regime with the United States. But it also meant to the Chinese leaders a terrible threat of US–Soviet nuclear condominium:

> Recently, while fraternizing with U.S. Imperialism on the most intimate terms, the Soviet leaders and the Soviet press have been gnashing their teeth in their bitter hatred towards socialist China. They use the same language as U.S. Imperialism in abusing China. This is a U.S.–Soviet Alliance against China pure and simple.[40]

The Chinese leaders who, as recently as 1957, had been praising Yugoslav movement on the path to socialism, now called Yugoslavia a 'fascist' state.

This was a pointer to the future. Fascism is on the other side of the class line, and not a charge to be used by Communists in fraternal quarrels. By the precedent of the Second World War, it justifies an alliance with a bourgeois state. Accordingly, by 1971, the Chinese would be calling the Soviet Union itself fascist.

The perceived threat of US–Soviet condominium had a result more favourable to the Soviets in Europe. Like Mao, de Gaulle also feared what he called a 'super-Yalta', an East–West deal at the expense of Europe. He judged that Soviet policy after the Cuban Missile Crisis was recoiling and weak, and thus more fitting for an approach that might provide new options for France. French displeasure with being eclipsed in Indochina and unhappiness with a presumed lack of US support for French interests in North Africa has already been noted. The United States certainly showed no sympathy for French work to build their own nuclear deterrent. But even in this, French efforts were driven by the bitterness of their ex-imperial status. Foreign Minister Couve de Murville traced his own attitude on the gulf between French interests and those of the Anglo-Saxons to the landing of US Marines in Lebanon in 1958 – alongside the British troops in Jordan. This was an assertion of American and British power, even at the cost of a nuclear crisis with the Soviets, in an area where French interests had once prevailed. French views seemed more in tune with those of Khrushchev than with their NATO partners.[41] Dulles told David Schoenbron of a meeting with de Gaulle at the time in which he informed the General that, in the American view, France was only a regional power with no interest in the Levant.[42]

De Gaulle did not see the unity of the Occident against the Soviet Union as the primary desideratum. In fact he judged that the Cold War was over. In a world that was politically multipolar, the Anglo-Saxons were dominant on the seas and cooperating in the nuclear field, in both senses excluding France. China was in an analogous position *vis-à-vis* the Soviet Union. Therefore solidarity among the continental powers, France and Russia, must replace the relationships developed between 1948 and 1963, relationships that reflected the division of Europe. A centre of gravity must be created around a Europe from the Atlantic to the Urals. For de Gaulle the outcome of the Berlin and Cuban crises had demonstrated Soviet strategic inferiority so clearly as to make a wounded Soviet Union a natural partner for France.

That the Americans should be increasingly bogged down in Vietnam was unfortunate for them, but perhaps an understandable result of their failure to make use of French good offices to arrange a neutralist solution. Most important, France could not permit herself to be forced to finance the Vietnam folly. The dollar balances accumulating in the European central banks as a result of American balance of payments deficits were a symbol of the recovery of Europe and the end of its tutelage and division as dictated by Yalta. The United States could not pretend that these dollar balances meant nothing or that their backing could be made by treasury bills, or indeed anything but gold. French policy here reverted to an attitude that it had assumed in the 1920s. With the repatriation of capital from Algeria after the end of the French war there, the French position in opposition to the dollar, or rather, to the gold exchange standard, was strengthened.[43] The French call for gold in 1965 triggered a rise in interest rates and inflation in the United States and opened another front in the Franco-American contest in which US policy would be shown to be vulnerable.

For its part, the Johnson administration was not thinking of foreign policy in the Cold War sense in which it had been conceived in 1948–63. It judged that the Soviet Union had been tamed by the events of 1961–62, and that the same lesson would have to be administered to the Chinese Communists and to their allies all over the world. It considered the Vietnam War to be an exercise in containing China in which the Soviet Union was seen to be playing only an auxiliary role. The US was making a general demonstration of the cost to be borne by those inflicting insurgencies on Third World countries. This was a little like the anti-Communist policy applied by the Western powers in previous decades, but Containment in its original form had been blurred to the point of obscurity. The US was fighting China and the radicals of the world; France was fighting the US and drawing closer to Russia; China was sympathizing with initiatives of European independence, quarrelling with the Soviets on tactics, and preparing for the worst in its struggle against the United States. The confrontation between the superpowers was not at centre stage. It appeared very much to be a multipolar world of political conflict.

The events of 1968 exploded this picture and returned world politics to a concentration on the Soviet–American relationship. The Tet Offensive launched by the North Vietnamese broke the resolve of US war policy. It triggered a run on the dollar in Europe that would ultimately dislodge gold from the $35/ounce price established at Bretton Woods in 1944. Europeans could regard this as a kind of revenge for the US sales of sterling at the time of the Anglo-French attack on Suez in 1956. Tet also convinced a meeting of American notables that the war policy as pursued up to then was a dead end. And Tet encouraged Democratic party opposition to the leadership of Lyndon Johnson. Much the same impact on de Gaulle's policy was effected by the Parisian general strike of May–June 1968. The flight of capital into France that had made possible his intransigence toward the dollar now became a flight out of the franc. Before a year had passed, de Gaulle was out of office and his eastern initiatives were succeeded by the *Ostpolitik* of Willy Brandt. In the summer, the Soviets ended the war of nerves that they had been conducting with Dubcek's Czechoslovakia and invaded the country.

The comfortable myths of the sixties were ground under. It had been assumed that the middle rank powers would be able to lead the way in a post-Cold War world. Yet, while the Chinese were seemingly strengthened by Tet, French policy was in tatters. It had also been assumed that the superpowers were no longer capable of dominating traditional spheres of interest. Yet, while the United States appeared mortally wounded in Vietnam, Soviet policy was more robust and wilful than ever.

The closure of an era could be seen more clearly in the following year. The ripening of the Sino-Soviet antagonism was visibly demonstrated in March by armed clashes on the Ussuri River border between the two countries. The Soviets threatened the Chinese nuclear facilities at Lop Nor in Sinkiang.[44] Following on this crisis, a reversal of alliances by the Chinese would develop over the next two years, during which Henry Kissinger would refer to China 'only half-jokingly' as a new member of NATO. Perhaps some idea of the Chinese assessment of the superpower balance can be got by comparing the Nixon Doctrine, announced on Guam in July 1969, with what the Chinese called the 'Brezhnev Doctrine', in the form of Soviet statements on the Czech invasion. Nixon had in effect changed John F. Kennedy's 'we will go anywhere, we will pay any price to preserve freedom' to the equivalent of 'we will go to certain places, we will occasionally assist others up to a point . . .'. This meant that the American troops were coming home. On the other side, Brezhnev's 'doctrine of limited sovereignty' unabashedly proclaimed Soviet willingness to defend the gains of socialism anywhere. To be sure, this is not a difference between the limited and the unlimited. But it does show that the American exit from Vietnam, a precondition of normalization between China and the United States, had reduced American influence to the point where China could consider it a fit partner for an anti-Soviet combination. The USSR by contrast, had emerged from its *recueillement* as the primary pretender to hegemony.

VI. POLICY OFFENSIVE: SEEKING REVOLUTIONARY ADVANCES, 1973–80

Just as the winners of the two world wars had fallen out after victory, the Sino-Soviet bloc, already under severe strain throughout the sixties, came apart completely after American forces were defeated in Vietnam. During the Vietnam War, the Soviets and Chinese had encouraged an international revolt against Containment while they quarrelled angrily about tactics. But in victory one of them would have to break away and combine with the United States, which they both thought to be entering a chronic state of decline. Both Russia and China wanted a weakened US as an ally against the other. The Nixon–Kissinger Detente was designed to manage the competition between them in order to permit the US a small measure of recovery.

If the Chinese were encouraged by the apparent local weakness of the United States, calling the US a 'wounded animal', the Soviet leaders were no less so. The Soviets shared in the heritage of the 1963 Partial Test Ban Treaty, part of which was the hope of taming the other superpower by means of arms control negotiations. American failure in Southeast Asia would, they thought, aid the process by promoting a sense of 'realism' (that is, defeatism) in US foreign policy.[45] The spectre haunting the Soviets in the sixties was not the United States but China. Soviet literature on the activities of Western radicals and American blacks demonstrates vividly a primal fear of the influence of old ideological enemies, anarchism, Trotskyism, and other 'petty-bourgeois radicalism', and their perceived threat to make common cause with Maoism. This was combined with the horrifying spectacle of Chinese cultural revolutionaries in Beijing toppling a statue of Chernyshevsky, the Russian radical democrat of the 1860s who was revered by the Soviet Communists. At the time of the skirmishes on the Ussuri River in 1969, the Soviets had wanted to gain US approval for action against Chinese nuclear facilities. After failing to obtain that, they had to face the prospect of developing cooperation between the US and China. The reversal of alliances proceeded. That the Chinese press, in the summer of 1971, began to call the Soviets 'fascists', showed that for China the Soviet Union, and not the US, was now the principal enemy.

Increasing rancour in Sino-Soviet relations prodded the Soviets in their pursuit of Detente with the West, as de Gaulle had already foreseen. But the Soviets had for some time maintained that Detente was inevitable, in view of their achievement of missile parity. The same Suslov–Brezhnev leadership that had made the last turn to *recueillement* was now turning to a new forward policy. They had to exercise forbearance in adjusting to the negotiating style of Nixon and Kissinger, themselves adjusting America to the loss of the war in Vietnam. The American architects of the Detente policy would have been less than human had they not looked back wistfully to the *tour de force* of John Foster Dulles in 1954, the last time the Communists had won a war in Southeast Asia. At that time, the Soviets had restrained the Viet Minh in Indochina because of their interest in stopping German rearmament and their need for French support to defeat the EDC. The Viet Minh were limited to North Vietnam, and German rearmament proceeded nonetheless. Nixon and Kissinger would have liked to perform a similar feat, using the Soviet desire for arms control to save the day in South Vietnam. But the Soviets never accepted the American position that US–Soviet relations were connected to American fortunes in Southeast Asia. They could see that Nixon and Kissinger needed to restore the trajectory of US policy toward Detente and to divert attention from the Vietnam War, which had seemed to devour American foreign policy in the 1960s. The American opening to China was supposed to help with this. It not only pushed Detente along but, as Kissinger was to write later, 'Its drama eased for the American people the pain that would inevitably accompany our withdrawal from southeast Asia.'[46] But there would be no repetition of 1954. In return for following the American etiquette, which they did without yielding any geopolitical position, the Soviets would soon have SALT, the ABM treaty and, in the Helsinki Treaty, a European settlement denied them since 1945.

While Detente with the United States was unfolding, the economic climate in the West was changing rapidly. Claims on gold, mostly from Europe, that threatened to exhaust the American holdings were only halted by the closing of the US gold window in August 1971. The Soviet press noted this break in the seemingly inexorable ascent of Western postwar prosperity, but did not ascribe to it the importance that it would just a few years later. Perhaps this was because the American recession could still be seen as a continuation of the movement of the sixties, that is, as one result of a European assertion against the United States that had begun with Gaullism. For the Soviets European conditions were always primary. By contrast, when the European economies were themselves shaken by the first oil shock that accompanied war in the Middle East in 1973, Soviet analysts began to envision a crisis of formidable proportions in the capitalist west as a whole.

This change of analysis was to mark the end of the period of *recueillement* and the start of a new policy offensive. Suslov now saw west European capitalism engulfed in 'crisis and monetary chaos'. In the past, he said, recessions had been overcome, but this time there would be no 'second wind' (*vtoroe dykhanie*).[47] Today, he argued, 'the entire world of capital is being shaken by class battles, and the workers' assault on the positions of the monopolies is constantly building ... Their demands are increasingly aimed at the very foundations of capitalist society.'[48] The policy of Detente was a reflection of 'newly propitious conditions' (*blagopriatnyi klimat*) provided by the emerging 'collective security' system in Asia. The policy of the Western hard-liners had been broken. The overthrow of dictatorships in Portugal and Greece, thought Suslov, should provide Communists with confidence in their opportunities. Another Soviet writer called the situation the third of the twentieth century's general crises of capitalism, the two world wars having been the climax to the other two. This was an analysis in keeping with Stalin's postwar dicta.[49] The west

European masses were thought to be more radical than they had been in decades, especially in Portugal, Greece, France, and Italy, and the Communists more active.

These Soviet writers did not see the same opportunities in the United States as they saw in Europe. It seemed that they had been frightened by the American turmoil of the sixties and were grateful to see it subside while the radical tide rose in Europe. In any case, they judged that the United States could no longer be a motor for European prosperity. The economist A. Mileikovsky claimed to see 'pathological changes' in the US economy in 1969–71, marking the exhaustion of the Keynesian solutions to the cyclical problems of capitalism. As a result of the Vietnam War, he claimed, the 'Keynes cult' had been overpowered by 'galloping inflation' (*galopiruiushchuiu inflatsiiu*).[50] In the past, unemployment had been regarded by the capitalists as the first enemy, but now inflation was centre stage. Thus did the Soviet analyst quaintly describe the rise of monetarist theory in the United States. The gloating was transparent; the return of the West to classical deflationary ideas clearly pleased the Soviet economist, being more in keeping with his own classically Marxist analysis.

There was, however, nothing classical about the tactics that were prescribed. The leftist military dictatorship in Peru and the cooperation in Portugal between the Communists and the Armed Forces Movement prompted a more creative view of the revolutionary role of the armed forces, a view differing from the opinions found in all the writings of Lenin and the other Bolsheviks. 'As can be seen in Portugal and Peru', wrote T. Timofeyev, the director of the Academy of Sciences' Institute of the International Workers' Movement, 'relations between the proletarian vanguard and the armed forces represent a unique form of alliance'.[51] A praetorian strategy was deemed more useful in Portugal than a broad democratic front. Alvaro Cunhal of the Portuguese party was quoted approvingly as saying that 'voting does not entirely reflect the strength of the Communist party, its influence, its ability to mobilize the masses or its role in the Portuguese revolution'.[52]

The old alliances, with the leftists or social democrats, were perhaps inappropriate to the current circumstances, wrote a leading authority on the western New Left of the sixties. The radicals of that era had played into the hands of the bourgeoisie with their 'anarchism', while the social democrats had rejected Marxism contemptuously, as they had since 1951.[53] Timofeyev accused the Italian *Manifesto* group, the Trotskyists, and the Maoists of a mistaken view of the 'automatic relation of economic crisis and revolution' in contrast to the Communist parties' correct 'strategy of the offensive'.[54]

Suslov's point of view may not have been gospel in the Politburo at this time. There was some evidence that he was being criticized in 1974. The rehabilitation of N.A. Voznesensky, the economist and critic of Varga whose work Suslov had attacked in 1952 after Voznesensky had been shot, seemed to show that someone still wanted to embarrass Suslov.[55] Yet his views seemed to be seconded by Brezhnev at the 25th Party Congress in 1976. Brezhnev said that the crisis of capitalism could only be compared with that of the 1930s; it was a 'general crisis that continues to deepen'.[56] He pointed with pride to the advance of Communism in Vietnam, Laos, Cambodia, Angola, Mozambique, Guinea-Bissau, and Somalia. The Chinese, by contrast, had put themselves at the disposal of 'the world's worst reactionaries'. After listing the other diplomatic gains of the recent period he addressed those who asked whether the advent of Detente had not 'frozen' the revolution: 'Suffice it to recall', he said, 'the great revolutionary changes that have taken place in recent years.'[57]

And there were more to come. The collapse of the Portuguese empire in 1974 provided the encouragement for Marxist regimes in the former colonies (Angola, Mozambique, Guinea-Bissau, Cape Verde and Sao Tome). It also lent intensity to the campaign of SWAPO in Namibia. In 1978 a coup installed the Khalq (Peoples') Communist Party in Afghanistan. Zaire's mineral-rich Shaba province was invaded by former Katangese gendarmes, thought to be assisted by Cuban and Angolan troops. Cuban and Soviet forces helped a pro-Soviet coup in South Yemen. By 1981 this regime would be part of a Soviet-sponsored defence pact with Ethiopia and Libya. An agreement between Vietnam and the Soviet Union served as a prelude to the invasion of Cambodia by Vietnam. The Chinese answered this by invading Vietnam in January 1979, the same month in which the regime of the Shah of Iran collapsed, to the great excitement of the Iranian Tudeh party. In the summer came the victory of the Sandinistas in Nicaragua, and in December the Soviets invaded Afghanistan.

It appeared that the positions laboriously established over decades by the western policy of Containment were all going to collapse in a heap. Soviet policy had been for some time demonstrably successful in the Third World, but the strategists from the Central Committee's international department were even more enthusiastic about radical changes sweeping the countries between Greece and Portugal. Most important was the Portuguese revolution, which Washington correctly saw as having far-reaching implications for Europe in general.[58] The Soviet prognosis was for a wave of revolutions bringing Moscow-inspired Communists to power. Nevertheless, they misread the wave: the movement was in the direction of democratic governments in Europe and, in subsequent years, other parts of the globe. Many of the European Communists seemed to understand this better than Moscow. Increasingly as Italian, Spanish and other parties saw opportunities for inclusion in bourgeois governments, they took their distances from the Moscow-approved tactics of Cunhal and proclaimed themselves Eurocommunists, in order to make their peace with the democratic alternance of the West. On his side, Suslov left no doubt that he considered Eurocommunism the worst kind of betrayal. In the end, however, it probably demonstrated a better understanding of the constitutional realities of western Europe than the bizarre tactics of Suslov and Cunhal.

And in fact, while Brezhnev and his colleagues had every reason to look back on their policies of the last decades and feel rather full of themselves, the curve they were describing was at its apogee and was going to take them back to earth. There may have been opposition in the Politburo to the ambitious schemes and actions of Suslov and Brezhnev. The impulsive decision to invade Afghanistan, coupled with the decision to use the Polish military to suppress the *Solidarity* movement caused some disquiet in the leadership and seemed to have prompted Andropov and his disparate allies to attack the Brezhnev system from within in December 1981. This action began the reform of what later came to be misleadingly called the regime of 'stagnation' and a shift in external policy back to *recueillement*, changes we associate with the rise of Gorbachev.

VII. *RECUEILLEMENT*: GAINS OF THE SEVENTIES UNDER ATTACK, 1980–85

The year 1980 marked a dramatic turn of fortune for Soviet international policy. Positions seemingly gained in the seventies began to unravel in various ways throughout the eighties. While the Soviets were wrestling with the difficult problem of succession, unpleasant news from the outside world continued to come in. The emergence of *Solidarity* in 1980, and the failure of the December 1981 coup fully to suppress it, meant a permanent threat of its imitation in eastern Europe and even in European Russia. The Tudeh party in Iran was ruthlessly crushed as Khomeini's revolution threatened Soviet Central Asia. The Gulf war between Iran and Iraq showed that countries in the region were developing a missile capability that could put Soviet targets under fire. Soviet tanks were destroyed in large numbers in the fighting in Lebanon in 1982. The use of Exocet missiles in the Falklands War of the same year did not promise better for the burgeoning Soviet blue-water navy. The Soviet-supported Ethiopian Dergue proved incapable of suppressing the Eritrean rebels. In Afghanistan, 'vertical envelopment' with attack helicopters and heliborne infantry did not turn out to be the counter-insurgency marvel that Soviet military literature had expected it to be. The attack helicopters themselves, on which much of the Soviets' hopes depended, proved to be vulnerable to American *stinger* missiles. The Reagan administration showed itself capable of arming and training insurgencies against Soviet-supported regimes, so that many of the gains of the seventies now seemed reversible and, at the least, extremely expensive to maintain. The nuclear threats that had been made against Europe with Soviet SS-20s were answered by American Pershing IIs and ground-launch cruise missiles which posed a shocking threat to Soviet command and control. The programme of American ballistic missile defence through Reagan's Strategic Defense Initiative presented a challenge of seemingly limitless proportions to Soviet budgetary resources and technology. In the broadest sense, the costs of empire seemed capable of exhausting and overpowering the Soviet economy.

These were the conditions prevailing at the time that Gorbachev was being groomed to become general secretary. No one then would have taken him for a critic of Brezhnevism. In 1988, in the midst of the *glasnost* campaign against the regime of 'stagnation', he had to caution his most militant supporters to keep in mind his own political debts to the past. He angrily reminded *Ogonyok*'s Vitaly Korotich, that 'we all kissed Brezhnev's ass! All of us!'[59] On his way up, Gorbachev gave the Brezhnev Politburo every sign of pleasure in Soviet international achievements. He shared the assumption of Gromyko and the others that Detente with the West must be revived. Gromyko thought in terms of the *recueillement* that had resulted in the SALT and ABM treaties of 1972. At the least, *recueillement* implied a renunciation of some of the most ambitious foreign projects and the unpleasant necessity of settling up by means of 'national reconciliation' in Afghanistan, Angola, and Cambodia. Yet this was not the first renunciation of its kind in Soviet post-war policy. After the first period of forward policy, the Soviets had to renounce the hope of guiding western Europe to Peoples' Democracy. The second period of forward policy was based on the settlement of wars in Asia. The second *recueillement* required renunciation of the nuclear diplomacy used by Khrushchev to ignite the Berlin and Cuban crises. Gromyko's policy now necessarily entailed political concessions in the Third World, in eastern Europe, and even in Soviet domestic affairs. But there was always the hope that eventually, at some future date, the advance might be resumed. *Recueillement* had always made possible a new forward policy.

This was a highly sensible line with obvious attractions. Yet in order to overcome Gromyko's influence and rise above the collective leadership, Gorbachev had to find policies that went beyond it. He had to revive the precedent of the periods of forward policy. Most of all he looked to the first period, 1944–47, when the Varga line had prevailed. Even after his resignation in December 1991, he made this clear in his first syndicated newspaper column:

The real opportunity was missed by both east and west after the victory over Nazism and fascism. There were various causes and motivations for that historic error: one side feared democratization in its own country and heightened dialogue with its ally in victory, while the other side feared the spread of Soviet influence. The result was the beginning of the cold war. Europe and the world lost a very great opportunity between 1945 and 1947.[60]

Gorbachev's views were in keeping with a recurrent theme in Soviet political history, what might be called Advance by Revival. Khrushchev had sought to revive the ideas of the previous period of forward strategy, when Zhdanov and Varga had seen Yugoslavia as an exemplary Peoples' Democracy. This had helped to sharpen Khrushchev's differences with Suslov and Molotov, who had played a prominent role in the campaign to anathematize Tito. In Gorbachev's 1989 gambits on the neutralization of Hungary one could see a revival of the Khrushchev forward policy with its neutralization of Austria and proposals for 'disengagement' of the two military blocs. Gorbachev's east European supporters would revive the whole Varga line of multi-party coalitions, and *glasnost* writers would try to revive precedents from the anti-Stalin oppositions of the twenties. The stimulus to these efforts was the much resented attempt of the 'conservatives' to participate in the process of reform, to stay abreast of it and to keep Gorbachev in the position of first among equals. Gorbachev's response was to ransack the past for ideological leverage against them, and to advance increasingly radical measures.

Gorbachev thought he could master world politics 'by political means', that is by daring political initiatives that did not require the use of force but only a willingness to swim with the currents of history. He was confident that he understood history better than others. The slightest Politburo opposition, sometimes expressed only in murmurs of reluctance, made him press his ideas more vehemently. By 1988–89, his projects had become so vast and ambitious that they no longer bore any semblance of a coherent design. Like Captain Ahab, he had thrown the quadrant overboard. He could not know that at the end of his voyage there lay, not just a haven from the Cold War, but the shipwreck of the entire Soviet enterprise.

3 Hero of the Harvest

> The one coming to power with the aid of the great finds it more diffi-
> cult to maintain himself, for he is a ruler surrounded by many who
> think themselves his peers, so that he cannot manipulate or command
> them as he would like.
>
> Niccolo Machiavelli

He certainly did not find the idea attractive. Even had he been of such a
mind, Gorbachev could not simply have stepped into Brezhnev's shoes.
The Brezhnev system of rule was *sui generis*, a collective leadership seemingly headed by a prominent and often, at least before the end, vigorous
figurehead. As exalted a personality as Brezhnev appeared to be, however, he was never strong enough to dispense with the group of leaders
who brought him to power. He was never in fact any more than first
among Politburo equals. He had to content himself with official praise
and honours, which he devoured with relish. They tossed him a fleet of
western automobiles: as Gogol says at the end of *Dead Souls*, what Russian doesn't love fast driving? Despite his having served in the War as an
ordinary political commissar, he was allowed to assume the title of Marshal of the Soviet Union. He made sure that accounts of his military
career inflated his importance and portrayed him as a great commander,
in order that he might be decorated with the Order of Victory. He arranged to have himself awarded the Lenin Prize for Literature for his
ghost-written memoirs. At ceremonies near the end of his reign, he seemed
to be collapsing under the weight of his medals and honours. But in fact
he was never more than the creature of the Politburo collective, which
exercised the real power.

Its true linchpin, the 'ideologist in chief' in a country where ideology
was everything, was Mikhail A. Suslov, the only survivor from the top
leadership of the Stalin era. Suslov had direct control over the complex
of institutions dealing with education and culture, guiding the work of
the Main Political Administration of the Armed Forces and the Central
Committee Information Department. He oversaw the press, radio, television, cinema, and publishing. He published nothing in his own name,
but his word on anything touching Marxism-Leninism was authoritative
and final. This meant that on larger issues, especially those bearing on
foreign policy, he led the small informal groupings that set the agenda
for policy. He was as well a balancer of personal ambitions among the
top leaders. It was he who managed and choreographed their order of

appearance at all the important state ceremonies and celebrations, such as Lenin's birthday, Armed Forces Day, or the anniversary of the formation of the Soviet Union. In the newsreels, when a group of leaders went to the airport to see Brezhnev off on a foreign trip or to welcome him back from one, on assembling for a photo, they would glance over to Suslov to get his quiet cues as to where and in what order they were to line up. Tall and fit, modest and polite, with the long thin fingers of a pianist, he was a genteel cardinal among the other apparatus men who looked to him for a lead in ideological and political matters.

In the thirties he had been a 'Red Professor', teaching political economy at the University of Moscow and at the party's Industrial Academy, where the young Nikita Khrushchev was his student. He came into the Central Committee Secretariat in 1947, and played a role in defending Stalin's views in the course of the leader's last political campaigns, which featured some pronouncements on economic theory. After the death of the leader in 1953, Suslov guarded Stalin's legacy, as expressed in a work he probably helped Stalin to write, *Economic Problems of Socialism* of 1952. He defended the primacy of heavy industry in the debate of 1953–54 against Malenkov's attempt to shift the emphasis of the economy over to production of consumer goods. He bitterly opposed Khrushchev's 1955 reconciliation with Yugoslavia, which in the end he was unable to prevent. But he recovered the lost ground and had Khrushchev criticizing Tito intermittently after 1957. By 1964 he emerged triumphant over Khrushchev himself, having arranged the combination that deposed him at a Central Committee plenum in the fall of that year. The Brezhnev–Kosygin leadership group, therefore, was from the first his creation. After surviving some challenges, he emerged as its ideological mentor and the guardian of its collective leadership principle against any possible usurpation by Brezhnev.

Some have likened the Brezhnev regime to the famous one-hoss shay in the poem by Oliver Wendell Holmes:

> You see, of course, if you're not a dunce,
> How it went to pieces all at once,–
> All at once, and nothing first,–
> Just as bubbles do when they burst.

Actually it only began to unwind with the death of Suslov in January 1982, just nine months before Brezhnev's death in November. The ideological secretary had by that time reached agreement with Brezhnev on the designation of Chernenko as successor rather than the more widely respected Andropov. Had Brezhnev died in January and Suslov in November, the succession would no doubt have turned out differently and Andropov would have been prevented from getting to the top. But the death of Suslov was the beginning of the end for the Brezhnev system. Since he was not personally ambitious, as number two man he had been the perfect one to keep the limits on the General Secretary, and the ideal protector of collective leadership and 'stabilization of cadres'. After Suslov's death, however, no less than three candidates appointed to his ideological post (Andropov, Chernenko, and Gorbachev) would themselves use it as a stepping stone to the top job of general secretary. There could be no surer proof that the unique Suslov–Brezhnev system was at an end.

It was unrealistic, therefore, for the Politburo to suppose that it could elect Gorbachev and still continue with collective leadership in the old way. Suslov had not bequeathed them a constitutional limit on the power of the general secretary. Suslov's system had depended on his own unique authority, accumulated over some thirty years, from the time of his central role in the regime of Stalin. Who then could be the new Suslov? What would become of collective leadership? Gorbachev would not be as easily reined in as the mediocre and, as it turned out, terminally ill Chernenko. He would be bound to want changes and could hardly be denied a new leadership team. Most of the Politburo leaders realized as well that changes would have to be made. But how could the line be drawn between retiring the old Brezhnevists and establishing Gorbachev's absolute personal rule? That was the question of questions. In a dictatorship, didn't one have to have a dictator? For his part, Gorbachev had before him only two precedents outside the Suslov–Brezhnev system for his struggle to become something more than first among equals, which for him meant something more than a new Brezhnev. There was Stalin's way, the gradual accumulation of police power, exercised finally by terror over the party and the entire society. Or, there was Khrushchev's way, the path of reform, one that had ended in defeat at the hands of a vigilant Politburo and Central Committee. Gorbachev had to devise a new method of making Khrushchev's way succeed. The course that he was eventually to set was full of contradictions: it demanded the erection of constitutional foundations for an office of President and the transformation of the Soviet polity by the introduction of what amounted to a new legal system, one moreover that would have had to endorse Gorbachev voluntarily as the leader of a one-party state. And alongside the new leaders and the new form of leadership, Gorbachev would need a whole new Soviet ideology.

None of this was foreshadowed in Gorbachev's early works or in the speeches he made after winning the general secretaryship in 1985. His boldest measures would be in response to the pressure of the Politburo opposition that arose in January 1987. Gorbachev, who began by extolling

Brezhnev's achievements and promising continuity, thus turned into the resolute enemy of the 'stagnation' (*zastoi*) regime of Brezhnev, against which he came to propose a 'revolutionary' reconstruction (*perestroika*). As we will see, he who began his time in office by praising Stalin's wartime leadership and rehabilitating the arch-Stalinist Molotov would eventually incite an explosion of revelations in the Soviet press about the crimes of the Stalin era. In a country which had always arrogantly boasted of a more humane and rational social system than that of the outside world, he would proclaim that the Soviet Union 'lacks political culture', and must therefore 'learn democracy' anew. Of this the patrons who brought him to power in 1985 had no inkling. Like the Politburo supporters of Stalin in 1925, they liked their candidate best because they thought him least likely to produce unpleasant surprises.

VICTORY ON THE GRAIN FRONT

Although we are still sorting out the details of Gorbachev's youth and early life, an account can be pieced together from his own remarks and the material made available in official biographical sketches published when he first came to power. A great deal of basic information dribbled out in often conflicting versions. It took years to determine, for example, where and in what circumstances Gorbachev's father died. An official biography related that he fought in the Second World War, but gave no other details. Then Gorbachev himself mentioned 'losing' his father in the Carpathians, an area where his unit had apparently fought. Since the elder Gorbachev had in the thirties been involved in the leadership of a collective farm in the Stavropol *krai*, where the campaigns for collectivization of agriculture were most furious and prolonged, the question emerged whether he might have been caught up in the initial collectivization campaign or in later purges. A television documentary in 1989 apparently settled the question with a shot of the tombstone in his village bearing the date 1976.[1]

Gorbachev was born in 1931 in the village of Privolnoe in the Krasnogvardeisk district in Stavropol *krai*, in the North Caucasus, an area populated mainly by Russians and Ukrainians, with small minorities of Circassians and Turkic-speaking Karachais. Wheat and sunflowers grow there in abundance, and cattle and sheep are raised. The area is well-known for its mineral water spas, to which most of the Soviet leaders repaired at least occasionally. At the time of Gorbachev's birth the entire area was convulsed by Stalin's campaign to collectivize agriculture by force. Gorbachev's grandfather was an activist in the campaign, which resulted in the arrest and deportation of a great many peasant families who resisted.

The north Caucasus was also hard hit by the famine of 1932–33, which may have had an impact on Gorbachev's family. The official accounts did not say. We are especially in the dark about the activities of his father. A classmate whom I interviewed who was part of his circle of friends at school in Stavropol described Gorbachev as a very bright, slightly temperamental boy who was often boisterous but not particularly rebellious. He tells of an incident when he was ten, during a lunch period in the school courtyard. One of the boys had a sandwich made with white bread, which immediately provoked excited questions from the others: where had he got it? Did he have to wait in line? The boy with the sandwich, who happened to be the son of a local Communist party official, replied: 'We have our own stores. We don't have to wait in line.' Young Gorbachev was bitterly incensed at this answer.[2]

The atmosphere in the school was typical of Soviet education in that period. Hymns and poems of praise were offered up to Stalin on a regular basis. The students heard him daily described as Father of the Peoples, Great Friend of Children, Transformer of Nature, Genius of Mankind. The great hero for Soviet youth of that time was Pavlik Morozov, a boy of 14 who, during the collectivization, had, in a burst of patriotic zeal, turned in his own father to the authorities of his village for concealing grain, after which the man had been shot. Warnings were regularly issued to the students about internal enemies of the regime against whom vigilance must be constant. Those who knew Gorbachev at this time report that, without being particularly militant, he tended to accept things as did the others, and did not raise objections. A normal Soviet youth of the period.

During the war years, at least one of his friends was deported to Siberia with his whole family. And German troops conquered and occupied Stavropol *krai* for five months in 1942. They promised to dissolve the collective farms and to reopen mosques and churches, and urged the Moslem and other non-Russian nationalities to declare autonomy and resist Moscow. And they met with some success. 'National Committees' were formed and volunteer squadrons of horsemen were organized to fight alongside the German forces. After the Russian armies managed to recapture the area in 1943, Stalin ordered mass deportations. At the end of the year, when Gorbachev was 12, the entire population of 80,000 Karachai were deported to Central Asia from an area only a few miles from his home village. Gorbachev must have known of these and other deportations. Some of the deportees were later permitted to return by the order of Khrushchev in the late 1950s, by which time Gorbachev was a Komsomol official in Stavropol. He helped in relocating some of the families and admired the decision of Khrushchev to reverse Stalin's edict against them.

Near the end of the war, Gorbachev went to work as a combine op-
erator at a Machine Tractor Station, according to official biographies.[3]
No mention is made of a break in his secondary education; the employ-
ment, therefore, must only have been during the summer months. In 1986,
he told an audience of his work experience and the 'tempering' he had
got from it, and noted, however, that he had only done this job 'periodically'.[4] The Soviet convention in official biographies was to downplay
the role of formal education in the moulding of character and to stress
work experience and a common background of life among toilers.
Gorbachev was awarded at the age of 18 for his labours, the Order of
the Red Banner, a decoration that he displayed with pride when he went
to Moscow for studies at the Lomonosov State University. The distinction may have been instrumental in gaining him admission.

At Moscow his field of concentration was law, which would give him a
training and outlook that would set him apart from his future party colleagues, most of whom had been trained as engineers and factory managers and had earned higher degrees only later in their careers, usually
by correspondence. In contrast to future colleagues, he got a thorough
training in public speaking, and he learned to argue a point persuasively
in public. To complement his deep voice and his warm smile, he developed a talent for oratory. On the other hand, legal expertise was not
worth much in Soviet politics. Gorbachev even expressed regret that his
training had not given him 'a real speciality' and had failed to make him
more than a talented generalist. He revealed a nagging fear of being a
bit glib and shallow.

In 1950 Soviet law was not thought of as protection for the unjustly
accused, but rather as a sword of righteousness wielded against the
ubiquitous internal enemy. The leading figure in Soviet jurisprudence at
the time was Andrei Vyshinsky, prosecutor in the Moscow trials of the
thirties, and later foreign minister, a former Menshevik in pre-revolutionary days, but nonetheless a rabid tormenter of Stalin's real and perceived enemies, who argued shamelessly that the confession of the accused
was perfectly adequate to prove guilt, even in the absence of other evidence. One could see this attitude in action daily. The mood of the period
was shaped by the Leningrad Affair of 1949, in which thousands perished
as a result of the discovery of a 'plot' among the followers of Leningrad
party boss Andrei Zhdanov. This was the result of a change of line, about
which we will say more later, and the whisperings of Zhdanov's rivals,
Georgy Malenkov and police expert Lavrenti Beria, who incited Stalin
against the Zhdanovites. By the time Gorbachev had joined the party in
1952, the wind had shifted again, and a potentially more sweeping purge
was threatening to engulf the old Stalinists and supporters of Beria. This
was the affair of the 'Doctors' Plot'. The press told amazing stories about

the activities of the 'doctor-poisoners', who were said to be intent on assassinating Stalin and other leaders. The campaign associated with the 'plot' was closed down immediately after Stalin's death in March 1953. Ironically, Gorbachev was immersed in the study of the principles of 'social-ist legality' at a time when the regime came close to returning to the lawlessness and oceanic terror of Stalin's purges of the thirties.

Gorbachev did not burn with the desire to emulate great prosecutors like Andrei Vyshinsky. In fact, his classmate Zdenek Mlynar, a Czech Communist who later supported the Prague Spring in 1968 and suffered persecution and exile, tells us that he was critical of the atmosphere of vigilance and repression. 'Lenin did not arrest Martov (a leading Menshevik and trenchant critic of Bolshevism); he let him emigrate from the coun-try,' Gorbachev is reported to have told Mlynar.[5] One can interpret this as an expression of sympathy for the idea of moderation toward political opponents, perhaps also as a judgement about the Stalin purges, one to the effect that, even if Stalin's opponents had been wrong – and Gorbachev, it should be stressed, never gave any hint of doubting this in his entire career – it had nevertheless been a mistake and an unnecessary excess, after having defeated them politically, to shoot them. Mlynar's Gorbachev may have been anticipating the position that Khrushchev took in his famous de-Stalinization speech to the Twentieth Party Congress in 1956.

Other reports offer a different view. 'He was one of the boys with his colleagues', reports another classmate, Lev Yudovich, of the US Army Russian Institute in West Germany, 'and he played up incredibly to those in authority . . . It was obvious he realized that his party political activity would get him further than his studies would.'[6] Komsomol work in the hysterical atmosphere of 1950–53 can only have been grim fare, in view of the press's constant calls for vigilance and ruthlessness in rooting out Stalin's hidden enemies. Nevertheless, we have testimony from classmates that he was fiercely dedicated to his party work and passed every test of militancy.

Gorbachev also experienced the thaw after Stalin's death, which Mlynar tells us, made a profound impression on him. Mlynar's Gorbachev was apparently a sensitive personality, philosophically inclined, profoundly affected by his reading of Hegel, a student who loved to quote the Ger-man philosopher's aphorism to the effect that 'the truth is always con-crete'. In the pages that follow we will find the mature Gorbachev frequently reconciling himself to an unpleasant reality by recalling this phrase.

It was at this time that his life was profoundly changed by his relation-ship with the beauteous Raisa Maksimovna Titorenko. She was a philo-sophy student with an interest in political questions, even ideological ones. Originally from Stavropol herself, she had made herself aware of the cultural life of the capital, and her good looks made her a centre of attraction with many admirers. Gorbachev met her at a class in ballroom dancing. She was impressed with his 'lack of vulgarity', as Mlynar put it, and his appreciation for the various concerts, plays, and other events that they attended together. Gorbachev was reading widely in the western intellectual tradition, that is, the part that leads through the French Enlightenment to German philosophy, English political economy, French socialism, and eventually Karl Marx. In a later recollection, he described himself in this period as a spiritual seeker, and Raisa as a companion with whom he could share his most intimate and perplexing thoughts.

In the eyes of his friends and associates, they were an idyllic pair. Raisa has spoken of her luck in having such a good friend for a husband. In some ways the other party leaders could only envy Gorbachev for having such a splendid asset in dealing with the often difficult social situations that party leaders at the upper levels had to face. On the other hand, Raisa was so unlike the other less lustrous wives of leaders that she was often as much the target of attention as he. This was not always to work to maximum advantage when he attained the post of General Secretary. His critics were to raise the question of Raisa having undue influence over him, a notion inevitably prompted by his famous statement in 1985 that he 'discussed everything' with her. In 1987 a mysterious video of one of her shopping sprees in the West was to surface in Moscow, perhaps, it was speculated, a KGB production designed to use her interest in fashion to discredit him. At the Nineteenth Party Conference in 1988, a delegate was to ask from the floor about her presumed hold over him. This sort of thing was not usually taken seriously. Yet she certainly had her influence, reinforcing his genteel side and his distaste for the common style of work of many Soviet managers and politicians, men who liked to have it said about them that when they walked into a plant or an office, 'the earth shook beneath them', and who all the same never failed to flatter and extol the genius of their superiors, or even to shower them with gifts. She had some part in forming his attitude toward the inadequacies of the traditional Russian political culture. One would not insist that this was a key to his personality, but its importance in Gorbachev's conception of the goals of reform should not be ignored. Once he had reflected that the USSR lacked political culture, something he said openly in 1987, he was on the way to denying that Communism had any claim to be an advance on the West.

They had a Komsomol wedding, sharing a wedding night in a dormitory left vacant by prior agreement with its occupants. They were not able to live together until married student housing became available a few months before graduation in June 1955. After that, he returned with Raisa to Stavropol, where their only child, a daughter, whom they called Irina, was born the following year. He took up a position in the Komsomol apparatus under Vsevolod Murakhovsky, the Komsomol first secretary. He did well enough that, when Murakhovsky went on to work in the Stavropol *krai* committee the following year, he recommended Gorbachev as his replacement. The two men would have a long association, with Gorbachev, in a reversal of roles, later actually a patron of Murakhovsky, who by 1987 was appointed First Deputy Chairman of the Council of Ministers. Gorbachev had become First Secretary of the Stavropol district Komsomol in 1958. Then he left the apparatus of the Young Communists briefly in order to become an organizer of one of the rural agriculture units in 1962. In leaving his Komsomol work at this time, he was probably dodging the problem of choosing sides in a struggle then being waged by KGB chief Aleksandr Shelepin, who wanted to strengthen his hand in Moscow by winning support in the Komsomol apparatus.[7] Shelepin was to figure prominently in the removal of Khrushchev, but his influence was to wane in the Brezhnev years. A year later, Fyodor Kulakov, First Secretary of the Stavropol *kraikom*, who was to become an important sponsor of Gorbachev's, appointed him to head the department of Party Organs for the district.

The fall of Khrushchev in October 1964 benefited Gorbachev. His patron Kulakov was brought to Moscow to head the agriculture department of the Central Committee. Kulakov had been involved in the cabal against Khrushchev that was organized by Suslov. During the late sixties and early seventies, Kulakov would be among a select group in Suslov's Secretariat who in effect dominated the Politburo as a caucus. Gorbachev had in him a powerful future supporter if he hoped to become head of the Stavropol district, a post that would promise him a seat on the Central Committee. To improve his chances he added to his résumé by obtaining a second degree in agronomy through correspondence. While many politicians did this in the Brezhnev years, the degree gave to Gorbachev a noteworthy dual competence that was bound to set him apart. It helped to qualify him for the post of Second Secretary of Stavropol *kraikom*, to which he was appointed in 1968.

That was a difficult year for Soviet foreign policy, already reeling from the internal crisis at the time of the war in the Middle East in 1967. There were mass demonstrations in the Polish cities in March and then the experiment of the Dubček government in Czechoslovakia. Mlynar, who had returned home, became assistant to the General Procurator during the Prague Spring, which only ended with the invasion of Czechoslovakia at the end of the summer. Mlynar kept in touch with Gorbachev through this period and reports him as being sympathetic to Dubček's liberalizing aims, and even to the general idea of different countries proceeding on their own specific path of development.[8] Perhaps Gorbachev would not have gone that far, and certainly not in public. But many Communists had misgivings about the decision to crush Dubček's regime. Suslov himself was extremely reluctant, for the sake of the world Communist movement, to use force against the Czech Communists. At any rate, the invasion did not hurt Gorbachev; by 1970, he became *kraikom* First Secretary, and the following year a full member of the Central Committee.

As head of Stavropol *krai* Gorbachev was in a position to greet and attend the various important leaders who came to the region's spas and sanatoria to take the waters. This was how he became acquainted with Andropov, who was born in a village near the popular spa of Kislovodsk and returned there frequently. He and Raisa showed him around, made him comfortable, and in general made a favourable impression on Andropov. Gorbachev in this way became known to a number of other Kremlin personalities. He impressed all with his intelligence and tact, his straightforwardness and lack of pretence. Gorbachev's various patrons might often have been at odds with each other, but Gorbachev seems somehow not to have accumulated enemies, not even the enemies of his friends.

His main responsibility was for the district's agricultural production. In this Gorbachev employed two very different approaches. First he experimented with a system of brigades under contract that worked without quotas and were paid according to their results. Initial successes led to a broadening of this system to the whole district in 1976. Then, in 1977, he made an abrupt turn, resorting instead to what became known as the Ipatovsky method, so called after an area in the district where it was first used. It involved large fleets of harvesters and trucks rather than the small brigades. At the time, the Ipatovsky method produced good results and was praised in the party press. *Pravda* went so far as to call Gorbachev a 'hero' of the Ipatovsky harvest.

Suslov visited Stavropol in the spring of 1978 to award Gorbachev the Order of the October Revolution for his work in agriculture. Suslov himself had formerly been First Secretary in Stavropol, at a time when he could scarcely have noticed Gorbachev. It is said, however, that first secretaries are keenly interested in those who succeed them as they move up, not knowing what sort of incriminating documents they may have left behind or what sort of 'historical' information may surface in the future. That Gorbachev rose to the top in Stavropol is one indication that he was not feared or suspected by Suslov. Gorbachev and Raisa took him on a tour of the district and made a good impression. Using the Ipatovsky method, Stavropol and the other regions produced a record harvest in 1978. Gorbachev was of course delighted with the result, but no more than his patron Kulakov, who was responsible for agriculture in the secretariat and got credit for this extraordinary 'victory on the grain front'. Kulakov was becoming a major star in Kremlin political circles, rumoured by some to be in line for much higher things and, in view of Brezhnev's failing health, perhaps one day even the top job. In July, however, Kulakov died in mysterious circumstances, perhaps from an overdose of alcohol at a time when he was recovering from stomach surgery.[9] Andropov, and perhaps also Suslov, immediately suggested the Hero of the Harvest, Mikhail Gorbachev, as Kulakov's replacement in Moscow.

The appointment, however, had to be cleared by Brezhnev himself, who always insisted on participating in personnel matters, despite his increasingly enfeebled condition. He was working a drastically shortened week and had to be assisted in climbing stairs or in other physical activity at ceremonies. He was duly brought to meet Gorbachev in September 1978, on his way to a ceremonial appearance at Baku. The two converged at the spa of Mineralnye Vody. Gorbachev was apparently not Brezhnev's first choice, in view of his closeness to Suslov and Andropov, but Brezhnev had to cut the others some slack in order to have his own protégé and designated heir apparent, Konstantin Chernenko, brought into the Politburo. So the appointment of Gorbachev was part of a compromise, so typical of the civilized manoeuvres that characterized the era of Stabilization of Cadres.

Later, during Andropov's rule, the Ipatovsky method that had brought Gorbachev much fame and fortune was described as so much eyewash typical of the Brezhnev period. In looking back on the experience after he came to power, Gorbachev duly concluded that the Ipatovsky method had been a detour from the more productive 'brigades under contract', a system he said should be employed more widely. This was cited in 1985 as evidence of Gorbachev's liberalism. Which views were Gorbachev's own? Was the turn in 1977 simply a result of a decision made in Moscow to which Gorbachev only reluctantly assented? Material is not available to provide a full answer. One can at least conclude that Gorbachev was not so in love with any particular agronomic idea as to make a political false step in defence of it. In traditional Soviet politics one does not rise to the top by means of spirited stands based on principle. And indeed how can the harvest of wheat be a matter of principle?

Gorbachev's excellent standing with a variety of patrons thus resulted in his election to the post of Central Committee Secretary for Agriculture. Amazingly, the loss of his closest patron had not ruined Gorbachev's chances for this position. He had been lucky enough, at this sensitive moment, to be able to transfer from the 'tail' of Kulakov to that of Andropov. This must have been what Raisa meant when she reflected later on their debt to Andropov, that 'We owe everything to Yuri Vladimirovich.'

GOOD WILL TO ALL

Gorbachev's rise was hardly meteoric. He had spent 22 years in his home district before making it back to Moscow. Still he was outwardly in every sense a product of the Brezhnev era, a partisan of its ascendancy over the hasty improvisations and 'hare-brained schemes' of the Khrushchev period. And, one would have thought in 1985, living evidence of the proposition that 'stability of cadres' could in its own slow and deliberate way bring forward the best that Soviet society had to offer. We have no reason to doubt his sincerity in speaking with pride of Soviet economic achievements, of the rebuilding of the armed forces and the attainment of missile parity at the end of the 1960s. This must also apply to the gains made by Soviet foreign policy in the seventies, the achievement of detente with the United States, the triumph of national liberation struggles in the Third World, and other Soviet gains in world politics, including the downfall of some of the most entrenched anti-Soviet personalities, such as the Shah of Iran. There is no real evidence of any distinctive or critical Gorbachev idea about foreign or domestic policy, nor even of a distinctive approach in his special field of expertise, agriculture.

Nor was he a partisan of one or another party in the Politburo quarrels at the end of the Brezhnev era, such as the reported differences over the decision to invade Afghanistan. Nor could he be a factor in discussions about the succession to Brezhnev. Yet these matters affected him directly. At the end of the seventies, the overwhelming consensus among Western sovietologists, and perhaps among the Politburo members that they studied, was that, with the death of Kulakov, the heir-apparent was Andrei Kirilenko, an old friend of Brezhnev's, and a man of roughly the same age. Almost none of the observers of Soviet politics would have guessed that Andropov, Gorbachev's closest associate in the Politburo, would succeed Brezhnev. His position at the head of the KGB was judged, correctly, to be a liability to Andropov. Brezhnev had seen to it that his own close associates G. K. Tsinev and V. M. Chebrikov, and his brother-in-law (who had married the sister of Brezhnev's wife) S. K. Tsvigun, were installed as Andropov's deputies. With Andropov thus hemmed in by Brezhnev men, he was thought unlikely to be able to use the KGB to intervene in Politburo matters.

These plans were upset in 1979. In the spring Kirilenko suffered some sort of medical setback that would cause his health to decline in the following years.[10] This paralleled a decline in his political fortunes. Instead there arose to prominence Brezhnev's chief aide, Chernenko, a man widely viewed as intellectually limited. There is reason to think, however, that his candidacy may have appealed to Suslov for exactly this reason, as he would be a figure unlikely to upset the collective leadership and its vision of continuity. With Brezhnev's workload cut down because of his health, Chernenko was called upon to chair sessions of the Politburo, offering its members a preview of what life would be like under his leadership. Apparently they disliked what they saw. This was especially the case with a 'national security' group, including Defence Minister Dimitri Ustinov and Foreign Minister Andrei Gromyko, both of whom, it was subsequently revealed, had worked closely in the Supreme Defence Council with Andropov and thought highly of him. Now the consensus was broken, with Suslov on the one side promoting an orderly procession of elders in the collective leadership, and on the other, a group who felt that the responsibilities of leading a nuclear superpower must be put into the hands of the most capable man.

Even after it became clear from press accounts that the knives were out, at the beginning of 1982, Western observers, oddly enough, paid little attention to this aspect of the conflict, preferring instead a mechanical analysis of warring factions grouped on the basis of competing bureaucracies and patronage loyalties. Overlooked was the fact that Andropov was widely recognized by his peers as the most able man. Andropov was certainly feared, but he was also respected. That made it possible for him to break out of the containment that had been imposed upon him. Leadership of the KGB was by no means a claim to leadership of the party. Everyone, including the Western observers, remembered the case of Beria after the death of Stalin in 1953. He too had represented a 'national security' candidacy, with his connections to the foreign intelligence apparatus, the East European Communist leaders, and the nuclear programme. Yet none of those who combined to arrest Beria really thought that in doing so, they were depriving the Soviet Union of its best politician and its most able Marxist theorist. Moreover, Andropov had something that Beria had never had, an opportunity to influence matters *before* the death of his predecessor.

This was partly a result of his own efforts and partly a result of the conjuncture of events. At the end of 1981 it was clear that Chernenko was the heir-designate. Soviet foreign policy, which had passed from success to success up to the invasion of Afghanistan, then seemed to be encountering stormy weather. The first ground operations in Afghanistan were not immediately successful, and the Soviet forces were answering with more intense aerial bombing.[11] In Poland, where the movement of *Solidarity* had risen up the previous year, Suslov's calls on the Communist Party to 'reverse the course of events' had been met by a refractory party congress in the summer of 1981 that elected a new leadership, an event that convinced him that the Polish party itself was now no longer reliable. By December a *coup* by the Polish armed forces and security troops would effectively, at least for the moment, crush the *Solidarity* movement. Brezhnev applauded what he called 'timely measures'. Still, it appeared that instead of having matters under control, the Soviet party leadership was getting in increasingly over its head.

At this point Andropov made his move in what came to be known as the Affair of Boris the Gypsy. This concerned a certain Boris Buriatia, a colourful character who had been a star of the Moscow Gypsy Theatre, a singer at the Bolshoi Theatre, an alleged dealer, along with some state officials, in an illicit trade in diamonds and foreign currency, and the lover of Brezhnev's daughter Galina. The activities of Boris were well known, but no one was eager to expose a scandal that would embarrass the Brezhnev clan. This was at least the case until the end of 1981, when most of the principals were suddenly arrested, undoubtedly on the initiative of Andropov's KGB. As the affair unfolded, a strange 'satirical' article appeared in the December 1981 issue of the Leningrad literary magazine *Avrora*. It was carried on page 75 (Brezhnev was celebrating his 75th birthday) and described a concerned public's watch over the anticipated death of a 'great writer' (Brezhnev had received a Lenin prize for his memoirs), a death that never seemed to come, to the disappointment of his admirers. 'But I think that he will not keep us waiting very long', wrote its author, 'He will not disappoint us. We all believe in him. We want him to finish those labours that he has not yet finished, and hasten to gladden our hearts.' This ominous and leaden attack on Brezhnev must have had the approval of Leningrad party chief G. V. Romanov, like Gorbachev an ambitious and bold man who had reason to hope for an end to the precession of the elders.[12] Its concurrence with the first arrests in the affair of Boris the Gypsy leads to the conclusion that Romanov and Andropov were in a bloc. After coming to power in 1983, Andropov would reward Romanov by bringing him to Moscow to work in the Secretariat.

KGB First Deputy Chairman General Semyon Tsvigun may also have been part of the campaign, probably by giving the order for the arrests that opened it, despite his relationship by marriage to Brezhnev. On 19 January he apparently killed himself. Yet Brezhnev, surprisingly, did not sign Tsvigun's obituary notice, as was customary. According to an account given the *Los Angeles Times*'s Robert Gillette, by a source 'proved reliable in the past', Tsvigun had engaged in an intense argument with Suslov over the arrests, which Suslov had tried unsuccessfully to prevent. It ended with Suslov telling Tsvigun that he had betrayed the party and 'you've nothing left but to shoot yourself'. Tsvigun then took the advice. The confrontation so shook Suslov that he himself had a stroke or heart attack, and died on 25 January. At one blow, the two most prominent obstacles to the rise of Andropov were fortuitously removed from his path.

The affair also suggested the distinct outlines of a conflict between the KGB and the Ministry of Internal Affairs (MVD). Reports indicated that the KGB had been trying for months to assume direction over a general campaign against corruption, against the resistance of the MVD. The latter was a stronghold of Brezhnev men. If the defecting KGB major Vladimir Kuzichkin is to be believed, the KGB had warned Brezhnev repeatedly against the invasion of Afghanistan.[13] The *coup* that accompanied the invasion, carried out under the guise of shaping up the personal security arrangements for Afghan dictator Hafizullah Amin, was directed by MVD deputy chief Viktor Paputin. Brezhnev had bypassed the KGB, depending instead on the other security service, in the hands of supporters and cronies who had attached themselves to him when he was party chief in Moldavia in the early fifties. One of these was Interior Minister Shchelokov. Another was Chernenko. The *coup* resulted in the death of Paputin, after which Yuri Churbanov, the husband of Galina Brezhnev, was promoted to First Deputy Chairman.[14] Further indication of the Moldavian–MVD connection is given by the fact that one of Andropov's first acts on succeeding Brezhnev was to replace Shchelokov as Interior Minister with a KGB professional, Vitaly Fedorchuk. Tsvigun, however, was also part of the Moldavian group. Viktor Chebrikov, one of the Brezhnev men from the time of his tenure in Dnepropetrovsk, also must have turned against Brezhnev at this time, to judge by the appointment he received once Andropov had power. But Chebrikov had as well some ties to Chernenko, alongside his years of working with Andropov. Despite the desperate nature of the conflict, it is doubtful that the bureaucratic antagonisms were personified in long-standing blocs waiting for a chance to have it out. More likely, these men, who had worked together for a long while, preferred to continue to do so as long as things went well. The eruption of the crisis caused by Brezhnev's choice of Chernenko as his successor had forced them into the unpleasant necessity of choosing sides.

Gorbachev was among those reluctant to desert a position of goodwill to all. Brezhnev, Suslov, and Andropov had been his most important patrons and tutors. Part of his attraction to them was his seeming capacity to steer clear of blind factional commitment. He did not have to take a side definitely until the May 1982 meeting that elevated Andropov to the position of Suslov's replacement. Along with the other Soviet leaders, Gorbachev must have been sobered by the passing of Suslov, the architect of the regime of post-1964 who had given the leadership such extraordinary security and stability in a different transitional period. And it seemed that, in the *absence* of Suslov, the Politburo was trying to act as a *collective* Suslov. It called to mind the period after the death of Stalin when the reflex of the leadership was to revive the party against the ambitions of the security forces and the state machinery. So now the leaders seemed to want to revive Suslovism. This complex and in many ways confused urge may be understood in terms of two motives: first, to preserve collective leadership by retaining the device of a 'second secretary' responsible for ideology, and second, to resist the rise of a new Khrushchev who might abuse the collective and reintroduce the turmoil of the 1953–64 period.

These seemed compatible with the election of Andropov, whose intellectual and political abilities helped them overcome fear of his KGB connections. Knowing what they did about Kremlin politics, they simply had to overlook the KGB campaign against Brezhnev and the death of Suslov. In April, they chose Andropov to give the Lenin Day speech, a ceremonial occasion for the celebration of Soviet Marxism, usually delivered by one who is, if not the number two man himself, highly regarded in the field of ideology. At the May Central Committee plenum Andropov was made Suslov's successor in the Secretariat, again in a close vote with Brezhnev and Chernenko probably opposing. Andropov's support came mainly from Ustinov, who nominated him, Romanov, who continued on the course set at the time of the affair of Boris the Gypsy, and perhaps Vladimir Shcherbitsky, Victor Grishin, and Arvid Pelshe. Gorbachev probably cast his vote for Andropov, although some in the West actually thought that he had voted against.[15]

ANDROPOV'S *TOUR DE FORCE*

Gorbachev still had responsibility for agriculture and made public pronouncements only about matters such as the Food Programme on which Brezhnev had staked so much. However, things did not go as well as expected in his bailiwick. Poor harvests had prevented the 1976–80 Five-Year Plan from reaching target. Indeed, agriculture had in general done poorly since Gorbachev assumed charge, even in his Stavropol base. Gorbachev must have hoped that the cup could pass from him. But in fact he was not punished for failures in agriculture, because he was now in a position to be considered as a generalist, something that happens to Soviet politicians when they reach Politburo level, and because he impressed all as a leader who should be prepared for future responsibilities.

On Brezhnev's death in November, the general secretary's post passed, as expected, to Andropov. Again the impulse to behave like a collective Suslov impelled the Politburo to make Chernenko the number two man. This could not have been out of recognition for Chernenko's capacities with ideology. The Politburo was simply saying that it was not giving the Soviet Union to Andropov as a personal fief; he had to be limited, and the best way was to limit him with the rival of the moment. The leaders thus desperately invested in the idea of continuity, while they sensed that a wholesale generational change was looming.

Andropov did not share their reverence for continuity. Before he had been in office a year, there were widespread changes in every sphere of state and party life. Ten of 25 Central Committee secretaries were replaced. Fully one-quarter of the *oblast* and *krai* secretaries were removed for new men. In June, Andropov took on the chairmanship of the Presidium of the Supreme Soviet, becoming in effect the Soviet President and combining in the two posts, General Secretary and President, something that Suslov and the Politburo had denied Brezhnev until 1977. Chernenko was paid little attention at first. It was not he but Gorbachev who gave the Lenin Day speech in April 1983. Andropov had gathered around him a group of younger men, some of them former associates of Kirilenko, in a kind of unofficial *orgburo*, or perhaps more precisely, a caucus, under whose direction he put Gorbachev. He wanted to establish Gorbachev as a kind of counter-second secretary, in somewhat the same way Khrushchev had once tried to set up the ideologists Pospelov and Ilychev as a counter to Suslov.

Andropov wasted no time setting his foreign policy agenda, centring on efforts to regain some of the momentum of Detente lost in the past few years. Andropov cited many times the great achievement of the Brezhnev years in reaching missile parity with the United States. But he also stressed that this was an achievement that had been ensured by the Detente process that had reached a climax in the Helsinki Final Act of 1975. Indeed, Andropov insisted, against Politburo and military doubters, that it was precisely Detente that had made possible the political gains of the late seventies. It was nonsense, he thought, to maintain the simple-minded proposition that the economic superiority of the Soviet defence industry and a policy of toughness were responsible for changes in the world balance. The gains of Soviet policy had been primarily the result of political efforts. And a new political effort was required to maintain them. The most immediate challenge was the possible deployment, according to the NATO decision of 1979, of Pershing II and ground-launched cruise missiles (in response to the earlier Soviet deployment of SS-20s). Andropov let some of his first European visitors know that reopening dialogue on the INF question was his first order of business. They sensed that he might be able to change the atmosphere of confrontation. German foreign minister Hans-Dietrich Genscher came away from one interview greatly impressed with Andropov's grasp of the issues, down to the numbers and characteristics of the weapons systems.

Andropov had to find a way to respond to the idea of a 'zero option', first suggested by Helmut Schmidt's German Social Democratic party. If the Soviets would remove their missiles, the West would refrain from deploying theirs. The idea at first seemed ridiculous and utopian in Washington, an airy product of the European peace movement. But in November 1981 Reagan decided to offer the zero option to the Soviets. No one thought it a real possibility. But Reagan loved to beard the Soviets with broad challenges that he knew they would not meet. Andropov, who had been deeply involved in shaping arms control strategy prior to coming to power,[16] offered a number of ingenious initiatives and formulae over the next few months before he was incapacitated by illness. Their essential animus was the desire to stop the NATO deployment, while at the same time matching the British and French long-range missiles and keeping the Soviet deployments against China out of the negotiations. In view of the positions taken in Washington, Andropov had become locked into a situation in which he was compelled to match the United States, NATO, and China, while the United States reserved the right to match the Soviets. It looked like an insoluble dilemma guaranteed to ratchet up the totals on both sides and ensure an indefinite race for superiority. Andropov, despite his considerable resourcefulness, was forced to stick to hard bargaining with a seemingly implacable adversary. Moreover, the zero option, toward which Gorbachev would eventually find his way, was closed off to Andropov, so unacceptable would it have been to the Soviet strategic rocket forces. They, along with the air force, the air defence forces, and the electronic warfare command, were said to be strongly supporting Chernenko rather than Andropov.[17]

Andropov hoped that the contrast between his policy and President Reagan's, in which Reagan appeared not to be willing to move an inch, would favourably impress European opinion. He hoped especially that German voters would be influenced by the anti-Euromissile movement to remove Kohl from office in the elections of March 1983. But this hope was dashed with Kohl's re-election. The United States remained committed to the zero option, at least as declaratory policy, while offering an 'interim solution' with equal ceilings at any level, provided that Asia-based forces would be counted. Andropov found himself boxed in. And in the same month, quantity, as the Soviets like to say, turned into quality, when Reagan announced his Strategic Defense Initiative. Soviet military people, highly concious of their inferiority in the latest computer-related military technology, were alarmed, not to say panicked, by the amazing schemes of missile defence floated by the supporters of the SDI. Suddenly, halting the SDI became the first requirement of Soviet foreign policy.

Chernenko supporters found new life with Andropov's failures, and especially when Andropov's health began to falter. By summer he was unable to appear in public at all. And the centrepiece of his foreign policy collapsed in the autumn, when the first of the Euromissiles were deployed.

But Andropov, even operating from a hospital bed, was still formidable. At the December plenum, he brought a number of new allies into the leadership, full Politburo members Vitali Vorotnikov and Mikhail Solomentsev, and candidates Egor Ligachev and Viktor Chebrikov. These were men who had either languished in obscurity under Brezhnev, as in the case of Ligachev, who had spent 17 years in Tomsk, or former associates of Kirilenko who had no particular affection for Chernenko. One would assume that Andropov, given another three or four years, would have disposed of Chernenko's challenge in a convincing manner, and might even have found his own way to some of the strokes that Gorbachev made later.

As it turned out, he had only a few months to live. His kidney failure forced him to depend on a dialysis machine, ironically a piece of medical technology in which the Soviet Union was far behind the West. Yet, even in his weakened condition, he hinted at a new orientation and provided Gorbachev with a basic direction. In discussing a draft of an article with speech writers in the autumn, he emphasized the importance of abandoning Brezhnev era assertions that the Soviet Union had reached a mature socialism, and only required 'further perfecting'. This, he said, was ludicrous in view of the relative economic backwardness of the Soviet system, the complacent dogmatism of its managers, and its generalized indiscipline.[18] Andropov was making an important point. It is not a long step from this proposition to the idea that the socialist countries do not belong on a separate and parallel track alongside the rest of the international economy, but should instead join it, and even 'go to school' to it, as the first advocates of Russian Marxism, the 'Legal Marxists' of the 1890s, had once urged for Tsarist Russia.

MOSCOW OR LENINGRAD

Andropov did not have time to develop his inklings on these themes, much less translate them into policy. His energies were absorbed in the fight to establish a regime to supplant the leaders of the Brezhnev era. As it turned out, even this was beyond his physical capacities. On 9 February 1984, the nation got the news that Andropov had died. Faced with the prospect of another succession, the old guard of Brezhnev-era leaders was forced again into the role of kingmaker. They were required, as they saw it, to act collectively as Suslov would have acted, yet they were most influenced by the remaining members of the 'national security bloc', Ustinov and Gromyko.[19]

At the end of Andropov's life, the only Politburo members with seats on the Secretariat, aside from Andropov and Chernenko, were Gorbachev and Romanov. That seemed to say that these two were the alternatives for the position of general secretary in the long run. In addition to this ultimate choice, however, there remained the problem of Second Secretary Chernenko. At the time he was set up as Secretary for Ideology, the hope was that Andropov would hold sway until the entire old guard, including Chernenko, had gracefully and gradually passed from the scene. Now Chernenko was an immediate problem. But there was reason to think that he might be of service in sorting out the struggle among the next generation's leaders.

Romanov and Gorbachev were a different story. They personified a traditional division in the Soviet leadership since Lenin's time. Leaders having Leningrad at their base had traditionally represented a powerful left alternative in policy struggles. The city of the Revolution, the second city of the Soviet Union, had usually been mobilized behind programmes emphasizing strong party (as opposed to state) input, the primacy of heavy industry and centralized planning and command, a striving for autarky in the national economy, and a highly ideological foreign policy of often anti-western character. Grigorii Zinoviev had represented this orientation at the Fourteenth Party Congress in 1925, then Andrei Zhdanov in the thirties and forties, and Frol Kozlov during the Khrushchev period. Romanov undoubtedly struck the senior comrades as one who, intentionally or not, called up these echoes.

On the other side, Gorbachev, if for no other reason than that he was Romanov's rival, stood in the historical place where the Moscow programme had stood, an outlook favouring solicitude toward the peasant, legality and economic rationality in the industrial field, a willingness to use market mechanisms up to a point, and a foreign policy open to more economic and political ties to the West. Nikolai Bukharin had defended this outlook against Zinoviev and the Leningraders in 1925–27. Georgy Malenkov had stood essentially for the Moscow line in the course of his struggles with Zhdanov and Khrushchev. Aleksei Kosygin in the sixties had suggested an economic programme in the same spirit.

In the past, the two policy orientations had in effect taken turns in an alternance that marks out the periodization of Soviet domestic and foreign policy history. Turns from one line to another were always directed by a centrist who tilted his weight from one side to the other: a Stalin, a Khrushchev, or a Suslov. Of these the kindest and gentlest of the great centrists was Suslov, and it was his example that the old guard now sought to follow. Despite the quarrels of the last few years they tended to regard Chernenko as one of their number, perhaps because they hoped to use him as a prophylactic against personnel changes so sweeping that they themselves might be washed away. At the same time, Gromyko and Ustinov would have had to be superhuman not to be tempted by the opportunity to enhance their own standing and assume virtual control over foreign and defence policy in a transitional Chernenko regime. But they remained conscious of their responsibility to choose between the real candidates of the future, Romanov and Gorbachev.

They could not fail to be impressed with the difference between them. They remembered some unpleasant precedents: how, for example, despite all calculation, Khrushchev had emerged ahead of Molotov and Malenkov in the struggle to succeed Stalin. Khrushchev had stood on the same policy positions as Romanov, renewed emphasis on heavy industry and aggressive ideas about foreign policy. As spiritual heirs of Suslov, their greatest fear was a new Khrushchev. So everything suggested the same compromise solution: make Chernenko the general secretary in return for a consensus behind Gorbachev as his second secretary and presumed heir. Some objections were raised. Tikhonov suggested that Gorbachev was not ready for the job. But Ustinov and Chernenko himself insisted, no doubt because of prior agreement. Gromyko suavely moved the arrangement along.[20] Assuming that Chernenko's health was not robust, this amounted to anointing Gorbachev and at the same time attaching to him a governor (a 'braking mechanism' as Gorbachev would later put it) in the form of a gradual transition under Chernenko.

In the eyes of the Soviet public, this was not a very impressive vote of confidence in Chernenko, and subsequent pronouncements by Soviet leaders indicated that their confidence was in fact not abundant. Whereas they had spoken of Andropov as a leader with a flair for politics and a grasp of theory, a man who gave profound speeches, made 'contributions to foreign policy', and offered 'solutions to topical problems of socialist construction', they described Chernenko condescendingly as a 'talented organizer of the masses', an 'ardent propagandist for Marxism-Leninism', a 'staunch fighter' for the implementation of party policy. The message was not lost on careful observers, especially those in the military. During the Chernenko interregnum the television public was treated to extraordinary demonstrations of the independence of high military officers. It was shocking for foreign journalists who had never seen officers giving their own press conferences without party people, officers who sometimes ignored or turned their backs on the leader during an address. This gave the impression that the military opinion of Chernenko was nothing short of contemptuous.

Actually military policy under Chernenko seemed to return to the grooves in which it had moved under Brezhnev. The war in Afghanistan had not been pursued vigorously under Andropov. But Chernenko made a last attempt at a breakthrough against the Moslem rebels. In March and April offensives were taken up in several directions, toward three cities, and in the Panjshir valley, a stronghold of the *mujaheddin*. Five thousand troops participated in the operation, along with heavy artillery and, for the first time, high altitude heavy bombers flying missions from bases in the Soviet Union. Great losses were inflicted, especially on the local village population, but the rebels were not defeated. Indeed, the massive refugee problem that the Soviets deliberately increased in these horrific bombings would soon make their problems all the greater. People who had no real national consciousness would develop it in the exile camps in Pakistan, largely out of hatred for the Russians. The analogy with the US effort in Vietnam is particularly striking. Soviet reliance, in the fifth year of the war, on aerial bombing in non-urban areas may be compared with the American strategy after 1965.

The same impotent rage was evident in the Soviet response to the deployment of Pershing II and ground-launched cruise missiles in Europe. This had begun in November 1983. The Soviets walked out of the INF and START talks. They announced the intention to station their nuclear submarines closer to the American east coast and to prepare them to fire on a 'depressed trajectory' at American cities. This was done, one Soviet official claimed, in order to match the flight time to Moscow of the Pershing II and by getting Soviet forces 'under ten minutes'. These bizarre measures reflected broad Soviet anxieties about their strategic position *vis-à-vis* the United States, a seemingly demonic adversary facing them with an array of threats. The D-5 missile for the Trident II nuclear submarine, due to be deployed before 1989, menaced their land-based missiles. The SDI threatened to stop these missiles if launched. What was then called 'emerging technology' weapons (special long-range sensors and terminally guided munitions, some of which would be used in the Gulf War in 1991) threatened crippling deep strikes against their European armies. Against these the Soviets had to contemplate expensive new programmes for ballistic missile defence and smart weapons for the European front. They had to consider ways to respond to the increasing vulnerability of their land-based strategic missiles, perhaps even a costly new programme of submarine-launched ballistic missiles in order to get more of their deterrent out to sea.

Gromyko and Ustinov now had virtual control over foreign and defence policy. This was, at any rate, the impression gathered by foreign visitors, including the redoubtable German Foreign Minister Hans-Dietrich Genscher. Yet there seemed to be a bellicosity of Soviet policy that led many in the West to ask if the Soviet military did not have a voice of its own. Chief of Staff Nikolai Ogarkov gave an interview in May to the military paper, *Krasnaya zvezda*, in which he stressed the broad spectrum of threats to Soviet security, the most important of which, he said, was that of the 'emerging technology' weapons: 'automated search and destroy complexes, long-range high accuracy terminally-guided combat systems, unmanned flying machines, qualitatively new electronic control systems.' These, he said, were responsible for a 'revolutionary turn' in warfare with a 'global effect', especially when combined with 'weapons based on new physical principles'.

Ogarkov had distinguished himself for years as the most hawkish among the various spokesmen for Soviet foreign policy. In 1981, he had claimed openly that the Soviet Union could fight and win a nuclear war in the event that deterrence failed. More recently he had been associated with the position that the constant upgrading of nuclear forces actually *reduced* the danger of a surprise nuclear attack, while the technological improvement of conventional forces presented a more pressing threat. His speeches stressed the contrast between the instability of nuclear war as against the stability of conventional war.[21] He championed high levels of spending for technological improvements, even if these would have to be secured for the military budget by means of reducing consumption, an option even Chernenko was loath to contemplate.

Ogarkov's reputed close relations with Romanov, who spoke of the international situation being 'white hot', gave the looming Romanov–Gorbachev rivalry an added urgency, as it seemed to foreshadow a third great debate about national security. The first had been in 1954 over Malenkov's consumer goods programme; the second over the Kosygin reforms of 1965. In both cases the heavy industry–arms production position had proved victorious. Now the whole orientation of the Soviet economy was again in question in a struggle for the succession. But the struggle was suddenly cut short by the news in September 1984 that Ogarkov had been demoted, an action that was accompanied by the suggestion by diplomatic officials that this had been because of his 'unpartylike tendencies'. The hawkish argument that Ogarkov had been carrying on in the military journals must have intersected with a related argument in the Politburo. The timing of his demotion gives further indications of the reasons for the move. Romanov was in Ethiopia, on a mission of long-standing importance to Moscow, to influence the Dergue (the leftist Ethiopian military leadership) to declare itself a Marxist-Leninist party. Romanov was overseeing this process, and attacking the idea that Washington and Moscow could be brought to agree on arms control measures as something that could be believed only by 'naive people', when Ogarkov was removed. It was said that Ustinov, who would die in December, had feared the prospect of Ogarkov succeeding him as Defence Minister, and thus helped in engineering Ogarkov's fall as a last service to the party.

Romanov dropped from sight for almost a month. Then he reappeared to announce that Ogarkov had been reassigned to the command of the largest part of Soviet western area forces. Romanov also spoke in a more subdued voice about the importance of resuming talks with Washington.

He was scurrying to fall in line with a return in Soviet policy to negotiations which, Gromyko claimed, would include all the existing nuclear forces – 'absolutely new talks' – as he put it. The change of line was undoubtedly the result of a struggle having as its centrepiece the contrast between Gorbachev and Romanov. There was reason to believe that it was conducted along the lines of the classical Moscow–Leningrad confrontation of right and left programmes. Those who defeated Romanov showed that they wanted Gorbachev and the softer policy line, and also perhaps that they feared that the tense international atmosphere might produce a rally of military and hard-line Politburo forces capable of defeating Gorbachev as a similar lineup had defeated Malenkov in 1953–54. Romanov–Orgarkov constituted a threatening reminder of Khrushchev–Zhukov. The last sudden removal of a senior military man in such a dramatic fashion had been the fall of Zhukov in 1957.

At the same time, the action was a kind of defence of Gorbachev, perhaps on terms that were agreed when Chernenko had first taken power. It was assumed that there was no love lost between Chernenko and Romanov. The latter had initiated the campaign against Brezhnev at the time of the affair of Boris the Gypsy, in December 1981. The campaign against Brezhnev was essentially a campaign against Chernenko's candidacy. Andropov had rewarded Romanov by bringing him to Moscow to work in the Secretariat in 1983. The arrangement that brought Chernenko to power in 1984 was probably designed to promote Gorbachev and weaken Romanov, so that the demotion of Ogarkov could be seen as the fruit of that agreement. Shortly after it was announced, *Pravda* editor Viktor Afanasyev permitted himself to go along with a Japanese journalist's suggestion that Gorbachev was the 'Second General Secretary.' The post does not exist, but such slips (Afanasyev retracted it afterward) showed that Gorbachev's status as heir apparent had been bolstered.

THE IDEOLOGY POST

The core of the responsibility of a second secretary, or number two man in the Politburo, is ideology. Gorbachev had broad responsibilities under Chernenko, not confined to agriculture, as is suggested by his virtual invisibility at a plenum on agriculture held in October, 1984. Yet Gorbachev's imprint on ideology under Chernenko remains a puzzle. His statements did not break new ground, nor was there the slightest suggestion of upbraiding or reining in Chernenko on ideological matters.

In fact, ideology, when Gorbachev was responsible for it, was markedly retrograde. The most profoundly symbolic act undertaken was the rehabilitation, in July, of V. M. Molotov, probably the single closest collaborator that Stalin ever had. Molotov, at the age of 94, was living in Moscow as a pensioner. It was revealed that he had been expelled from the party after the Twenty-second Congress in 1961, after Khrushchev gave the fullest description to date of his activities in the 'anti-Party Group' that had tried to oust him in 1957. Khrushchev had taken the opportunity to redouble his call for a break with Stalin's ideological tradition. Indeed one could argue that rivalry with Molotov was the main motive for all Khrushchev's criticisms of Stalin. To rehabilitate him in 1984 was thus a deeply meaningful act for the current leadership.

Who was Molotov? A Bolshevik since 1906, he worked on the staff of *Pravda* in 1907, and became a member of the Petrograd Bureau in 1917. He was on the *Pravda* editorial board, the Central Committee, and the executive committee of the Petrograd Soviet. He was not prominent among the leaders of the October Revolution, but filled a number of positions before being brought into the Secretariat in 1921 in the purge of secretaries who had disagreed in various ways with Lenin in the debate that climaxed at the Tenth Party Congress. He was probably already by this time tied to Stalin, whom he supported in the series of conflicts with Trotsky, Zinoviev, Bukharin, and others after Lenin's death.[22] He took a position on the extreme left during the collectivization of agriculture and made himself a spokesman and defender of the 'Third Period' Comintern line that kept the German Communists from making common cause with other parties against Hitler. This was also his orientation after Stalin turned to a more moderate policy after 1933, a fact which caused his star to eclipse until the mid-point in the great purges of 1936–38, when he returned with a vengeance as advocate of a pact between the Soviet Union and Nazi Germany. The pact would be considered henceforth by the leaders around Stalin, with only partial justice, as his brainchild. With the triumph of his line he became Foreign Minister in place of his rival of the preceding years, Maksim Litvinov.

Molotov was the first spokesman for foreign policy during the war years and in the first years of the Cold War, but lost his post to Andrei Vyshinsky in 1949. On most accounts, Molotov was out of favour with Stalin in the dictator's last years, despite his willingness to perform every service. He had abstained on the Politburo vote that sent his own wife to prison as a spy.[23] He was himself no doubt slated for liquidation in the 'Doctors' Plot' purge that was cut short by Stalin's death. Regaining the Foreign Ministry in 1953, he supported Khrushchev's hard line against Malenkov in 1954, but then, with Suslov, fought Khrushchev, because of the latter's intention to patch up the split with Tito. Molotov was removed from the foreign ministry in 1956, then from the Politburo after his attempted *coup* against Khrushchev, a stroke that, had it succeeded, would probably have placed Molotov at the head of the party. After its failure, he did not disappear but remained a critic, giving succour to the elements in the leadership who sympathized with Mao's attacks on the Soviet leadership. When Khrushchev was in the United States in 1959 he told reporters that the West should not want to see an end put to the one-party system in the Soviet Union because, he said, if there were another party, it would be pro-Chinese and be headed, no doubt, by Molotov.

As a political symbol in Soviet history, Molotov personified the left in Stalinism, in the sense that someone like Bukharin or Malenkov personified the right. Molotov's increased influence had usually accompanied the violent turns to the left: toward the collectivization policy in 1928, toward the reorientation of Soviet foreign policy and the intensification of the purges in 1937–39 (a turn accompanied by renewal of collectivization drives in some of the constituent republics), toward reliance on heavy industry in 1954, and later toward a position of sympathy with international Maoism. His rehabilitation could be interpreted in various ways, each worth considering and perhaps not exclusive of one another. It was a rebuke to the memory of Khrushchev and a statement of sympathy with Suslov's massive redirection of affairs after 1964. It was also a way of answering some Soviet intellectuals critical of the Stalin era. Roy Medvedev's *All Stalin's Men* had appeared in the West in 1983, a book that contained painfully accurate portraits of Molotov, Suslov and other Stalin cronies. The rehabilitation was a way of saying to Medvedev, who was isolated at the time in Moscow by round-the-clock police watch and followed ostentatiously,[24] that the de-Stalinization process initiated by Khrushchev (whom both Roy and his brother Zhores were known to admire) would not be resumed.

In addition, it was a landmark in Gromyko's campaign for political continuity. Molotov had, of course, had deep roots and genuine influence in the Foreign Ministry since 1939. Gromyko had been among the *vydvizhentsy*, (the 'promoted ones') who profited from the purges in the Soviet institutions at the end of the thirties. In the spring of 1939, when he was a researcher at the Academy of Sciences' Institute of Economics in Moscow, Molotov had chosen him for a diplomatic career. His and Molotov's attitudes toward Litvinov and their other predecessors were probably similar.[25] Molotov was actually Gromyko's first chief, who shared with the young diplomat the experience of attendance at the wartime and postwar conferences with the world's great leaders. Gromyko would have been an unusual man if he did not retain some pleasant memories of the heyday of Soviet–American cooperation and, even afterward, of the greatly enhanced prestige of Soviet power in international counsels. Vyshinsky's postwar displacement of Molotov had hurt Gromyko, causing him to lose his post as Soviet representative to the United Nations. He had been sent to be Ambassador to Great Britain. He only recovered politically with Stalin's death, as was the case with Molotov. He finally attained the position of Foreign Minister with the fall of the man who replaced Molotov in 1956, Dmitry Shepilov. So Gromyko had more ties to Molotov than to any other leader. Although he is usually thought of as a diplomat and exponent of *realpolitik*, he has always made his views compatible with those who, like Molotov, were deeply ideological Stalinists.

Another aspect of the Stalinism of the year 1984 is the revival of the 'Anti-Zionist Committee of Soviet Public Opinion', a shadowy organization designed to combat 'Zionist' ideological tendencies. One of the first political assignments of the Russian nationalist politician Vladimir Zhirinovsky was to infiltrate Jewish organizations at the behest of the Committee.[26] Its propaganda argued that Zionists and Nazis had collaborated in the thirties, giving as evidence material referred to, in a much different context, by Hannah Arendt in her famous book, *Eichmann in Jerusalem*. This material concerned the Jewish Councils that had helped under extreme compulsion with the movement of Jews to transit camps. Not everything the committee mentioned was obviously anti-semitic, but the anti-Israel tone was clear. The committee ceased its meetings within a year, but the episode again caused questions about the nature of Soviet ideology under Gorbachev. Was the committee's activity designed to serve the aim of improving links to conservative Arab regimes? In January, for example, a major arms deal supplied weapons to Jordan that the United States had previously refused to supply. Perhaps this was a kind of revenge for the defeat of Soviet-equipped forces in Lebanon in 1982. Or, was the Chernenko–Gromyko leadership underlining the stance Suslov had taken in favour of Russian nationalists in 1980–81? This odd episode does not permit an easy interpretation.

Nor was Gorbachev's precise role in the formation of Soviet ideology very clear. One can say that Soviet ideology in the period of Chernenko's rule, when Gorbachev was in charge of it, was a distillation of the purest and most unreconstructed (perhaps the best word) Stalinism. The indications are strong that Gromyko, who had more control over foreign policy and perhaps more political influence than at any time in his long career, was the architect and defender of the outlook exhibited by the regime. That was one based on *realpolitik*, desirous of a renewal of negotiation with the United States and confident that Gromyko, the student of Prince Gorchakov, would be a natural partner for Western students of Bismarck, such as Henry Kissinger. Return to detente was not to be accomplished by dramatic political strokes but by hard bargaining of the type that had created the arms control agreements of the Nixon–Kissinger era. And it was to be compatible with a broader range of international activities in the Third World.

Not much of this outlook survived into the period after Gorbachev came to power in March 1985. For that reason, it is difficult for us to imagine Gorbachev as the source of the thinking of the Chernenko inter-regnum. Yet there he was, a confident Second Secretary, pronouncing on various matters outside his traditional bailiwick of agriculture. Up to the moment he came to power, he satisfied all the expectations of all the Brezhnev-era elders, who regarded him as the most able man they could choose. His sponsors must have thought that one day Soviet posterity would recognize their achievement in finding this man and guiding him to power through the minefield of Soviet succession. On the other hand – a sceptic might have pointed out that their man was a lawyer who had never practised law, who had spent a long career in agriculture, who knew nothing of foreign affairs, who had got the attention of his superi-ors because he was First Secretary in a resort area, whose qualifications were rather like those of Prince Rainier of Monaco or the mayor of Las Vegas. In choosing him, their intention was to preserve the continuity of the methods of the Brezhnev–Suslov system. If, however, Lenin had been there looking over their shoulders, he might have suggested a note of caution, as he once did about Stalin, to the effect that 'this cook will prepare peppery dishes'.

4 'Acceleration of the Perfection', 1985–87

As I see it, no one else in the world had, or has, more power than I had in 1985.

Mikhail S. Gorbachev, 1992

When Gorbachev came to power in 1985, the question Western observers of Soviet politics were required to answer was: How long would it be before the new General Secretary had consolidated his power? We had to make policy with this man in mind – if, that is, he was truly in charge. In 1953 we had thought it was Malenkov and it turned out to be Khrushchev. Was it really Gorbachev now? When would we know? In truth, there was no adequate answer to these questions, if by the consolidation of power was meant a point at which the leadership of the new General Secretary was no longer challenged. There was, of course, no constitutional mandate, so Gorbachev, like every Soviet leader before him, had to fight for power by meeting two challenges: to find the policy that most served the needs of the nation and the cause on which it was founded, and to build his political credit and patronage against the opposition which could be expected to mount if his policies should miss their mark.

Gorbachev's election was the apparent result of a narrow Politburo vote. 'Those were anxious days', Yegor Ligachev told the Nineteenth Party Conference three years later, but the choice of Gorbachev was ensured 'thanks to the firm position taken by comrades Chebrikov, Solomentsev, and Gromyko and a large group of province first secretaries'.[1] Gromyko had spoken shortly after Gorbachev's election of how it was that, when Chernenko's health had faded, Gorbachev had chaired Politburo sessions 'in a brilliant manner'. By reason of his own responsibilities for foreign policy, said Gromyko, he had understood, 'perhaps more clearly than certain other comrades' Gorbachev's fitness to guide the party in the international arena.[2]

The rest of the rally to Gorbachev may be attributed to the legacy of Andropov, who had brought Solomentsev and Vorotnikov into the Politburo, with Chebrikov and Ligachev as candidate members, in December 1983. This group of Andropov appointees would be a solid source of support for Gorbachev until January 1987 when their enthusiasm began to waver. They were never part of a Gorbachev 'tail' but a loose bloc having in common a grudge against Suslov's regime of 'stabilization of cadres', and a desire to effect a general change of leadership by the whole-sale retirement of Brezhnevists. It was a perversity of the Soviet system that the power was not simply granted to the leader for a stated period but had continually to be shored up by these informal means, that is, by political struggle. Often this meant cultivating the enemies of one's rivals. This had been the case with Stalin on his way up as much as with Andropov.

The class of December 1983 could be counted on to retire the Brezhnevist old guard with a certain sense of mission, but at Gorbachev's election, they were not an overwhelming majority. Grishin, Tikhonov, Kunayev and probably Shcherbitsky (who was on a visit to San Francisco at the time of the election) were regarded as their opponents. Romanov probably supported Grishin against Gorbachev. According to Ligachev's account, in the Politburo meeting held on the night of Chernenko's death, Gorbachev's opponents balked at appointing him chairman of the funeral commission, an act that would have smoothed the way for his appointment as general secretary. The meeting broke up without a decision and Gorbachev, as Second Secretary, had to assume the responsibilities for the ceremonies by default. But the issue of the succession was at that point far from clear. The next day, Gromyko told Ligachev that he had decided to nominate Gorbachev. Ligachev regarded Gromyko's support as decisive.[3] When the foreign minister made his nomination at the meeting held that evening, all opposition melted away.

Gromyko actively sought Gorbachev's election in view of the alternatives. In fact, however, he only stood to lose in influence if there were too rapid a turnover of Politburo personnel. In a broader sense, he had grown in stature as a leader in the course of guiding the difficult succession from Brezhnev to Gorbachev. He was no longer seen by the top party men simply as a striped-trousers type, but rather as a trusted veteran whose ties to the era of Molotov had been sharply underlined in 1984. He was, that is, the very personification of both the Stalinist ideological tradition and the gains made by Soviet foreign policy since Stalin.

As had been demonstrated since the demotion of Ogarkov in late 1984, Gromyko was prepared to resume efforts to revive SALT-era diplomacy with the United States. He was trying not to appear too abject. Reagan had joked in August about abolishing the Soviet Union ('the bombing will start in five minutes') and Gromyko had rejected the conciliatory gestures made by American officials in the wake of the incident. Nevertheless he went to Washington in October and agreed to reopen talks on arms control at Geneva.

Gromyko wanted to get back to the measured bargaining over weapons systems, the 'bean-counting', the calculated trade-offs between 'offsetting asymmetries' that he remembered from the days of detente. He hoped that the election of Gorbachev would not upset his work of the previous year in reopening negotiations. It would certainly take a while for the full transition to a new Politburo team to be achieved and for the initial bargaining with the United States to ripen. Gromyko wanted continuity in both regime and policy matters. As it turned out, however, Gorbachev could not live within these limits. As the next years were to show, the Soviet Union was pressed on many fronts to make more dramatic changes. The 1984 campaigns in Afghanistan were a failure and the war looked like a dead end. The Reagan Doctrine, with its promise of aid to anti-communist insurgencies, seemed to threaten 'two, three, many Afghanistans'. Cooperation of China in these efforts would only firm up the 'strategic relationship', as Senator Percy and General Haig had called it, between Washington and Beijing. Responses to the SDI, to the threat of new weapons for NATO, to the D-5 warhead for the Trident II submarine, to an array of challenges to the Soviet military budget, would have to be found.

The answer indicated by the party of Stalinist continuity, dogged sym-metrical response in the form of redoubling military efforts in Afghani-stan and building, whatever the cost, the weapons necessary to meet all threats, cannot have been attractive to Gorbachev. It was advocated most strongly in Ogarkov's book, *History Teaches Vigilance*, which was pub-lished in 1985. But Gorbachev must have envisioned himself following this path to perdition as titular leader of a Politburo collective; he may have considered by analogy the collapse of American leadership under similar burdens at the end of the 1960s.

Not wishing to imitate this example, Gorbachev had to break with Stalinist continuity, both in regime and policy. To be sure, he only came to this realization by stages, after exhausting half-way alternatives. His first con-ception, for example, was to build on the achievements of the preceding regimes by introducing 'acceleration' (*uskorenie*) into Soviet economic life. But this slogan passed out of currency within the first two years of his leadership. Instead, he came to the position that the previous regime had been backward in virtually all areas and that Soviet Communism required a break with 'Stalin–Brezhnev stagnation'. This seemed to sug-gest something at first unthinkable, a reconsideration of the legacy of Khrushchev. In fact some observers of Gorbachev noted very early that when his name was mentioned alongside that of Khrushchev, he did nothing to discourage the comparison.[4]

Disabusing himself of the notion that he could represent continuity with political traditions of his own past was not something that Gorbachev did painlessly. But there was no real choice: he had to establish his own prerogative in the Politburo, and he had to attack the accumulating prob-lems. Within two years of his election, therefore, Gorbachev was proclaiming the need for a new ideology of Soviet socialism, one that broke decisively with the views of the past. In particular, he would call it a mistake to have thought that the Socialist bloc of countries could develop on a separate track from the rest of the world. This idea was being precisely articulated by 1988. Andropov had given indications that he was thinking the same thing in 1983. In terms of Soviet history, it was a revival of the conceptions debated in the Politburo of Lenin in 1921–22, when the Bolsheviks had sought to pursue British Prime Minister Lloyd George's initiatives for the integration of Soviet Russia into Europe. It was, moreover, a rejection of the theory of the 'Two Camps' as elaborated by Soviet theorists at the beginning of the Cold War in 1947. As such it seemed to recall the forward strategy of 1944–47, as described by Varga, a strategy that relied heavily on the presumed advantages of Soviet initiatives toward western Europe and the dream that eastern and western Europe might find a new accommodation.

ALIEV VERSUS LIGACHEV

However, a break with Gromyko's conception of continuity was not visible in the spring of 1985. The Lenin Day speech was given by Gaidar Aliev, as a reward for having supported Gorbachev. He stressed that the election of Gorbachev had been in accordance with the 'strategic line of the party' as outlined by the Twenty-sixth Congress (1982) and subsequent Central Committee plenums. That line called for 'acceleration (*uskorenie*) of the social-economic growth of the country and perfection (*sovershenstvovanie*) of all sides of its life'.[5] Aliev spoke of the socialist countries as constituting a self-sufficient 'world system'. He underlined the importance of opposing imperialist aggression in Afghanistan, Lebanon, South Africa, and Central America. Monopoly capital, plunging into a 'deepening general crisis', was still baring its fangs. The party must remember Lenin's advice that 'the imperialists would not dare to encroach on our security if they know that such an attempt will end in their defeat'.[6] With this in mind the party must be mindful of the role of the armed forces in the defence of the motherland.

Aliev would take part in important meetings with Nicaragua's Daniel Ortega at the end of the month. There was reason to think that he had been chosen for the ideological post by the electors of Gorbachev. However, one day after the Lenin Day speech, the Central Committee appointed Chebrikov, Ligachev, and Ryzhkov to full membership. Ligachev would emerge in the following months as the 'second secretary' in charge of ideology. At that time he was thought to be Gorbachev's alter ego, a man whose thinking was perfectly in tune with that of the General Secretary.

As could be expected, a certain tension remained between Ligachev and Aliev that was only alleviated by the retirement of Aliev in the autumn of 1987. This tension may also help explain the encouragement given to the Armenian claims against Azerbaijan in the spring of 1987, since Azerbaijan was Aliev's old bailiwick and the dispute afforded an occasion to criticize him.

Chebrikov, who controlled the KGB, was a veteran of decades in security work. In June he published an article in the party's theoretical organ, *Kommunist*, hailing the election of Gorbachev as an expression of the 'strategic line of the party for the acceleration of the social-economic growth of the country and the perfection of all sides of the life of society', thus echoing Aliev. He paid tribute to the work of Andropov in the service of state security for 15 years, following this with praise for the foreign policy of detente. He credited Andropov's contribution to efforts to contain the danger of nuclear war, which efforts, he said, were in the 'interest of all of humanity'.[7] Thus he sounded a note that would be more prevalent in 1987–88, the reference to the fight against nuclear war as a desideratum of foreign policy that transcended the class struggle.

In another voice, Chebrikov warned obsessively about the tireless activity of foreign intelligence services and 'anti-Soviet centres', with their impressive output of anti-Soviet literature. Imperialist centres, maddened by 'the deepening general crisis of capitalism', never abandoned their hope of an eventual 'ideological erosion' of Communism.[8] Thus the KGB had to persist in vigilance against multiform 'ideological diversions' and interventions in Soviet domestic life. Chebrikov was sounding a theme he would repeat at intervals in subsequent years, especially after Gorbachev's efforts to promote Soviet political culture resulted in the formation of numerous informal political groupings.

OGARKOV AGAINST THE WESTERN 'EXHAUSTION STRATEGY'

Fortified by the new appointments, Gorbachev began to speak to the world, yet he did not seem eager to break with tradition. On the 40th Soviet anniversary of V-E Day, he praised the work of the Soviet people and armed forces in defeating fascism. 'The gigantic efforts at the front and in the rear were guided', he said, 'by the party, its Central Committee, by the State Defence Committee headed by Joseph Stalin, General Secretary of the Central Committee of the All-Union Communist Party (Bolsheviks)'.[9] Western press reports described an outburst of applause at the mention of Stalin's name, with a show of enthusiasm from the military people present. It is noteworthy, of course, but Gorbachev may only have been responding in kind to the recent visit of Reagan to the Bitburg cemetery, where some Nazi SS men are buried. Gorbachev mentioned the visit, calling it an insult to the memory of the victims of the SS. However, in the same speech he spoke of 'genuine opportunities' to curb the spread of militarism and the arms race and stressed that the outcome of the competition between the two world systems could not be decided by military means.[10]

Gorbachev would restate this idea at intervals in the coming years. It has such a pleasant and hopeful ring that it has been tempting to view it as a unique contribution of Gorbachev's to Soviet New Thinking with all of its impressive initiatives. It would be more accurate to consider it a reassertion of a very old idea debated in Soviet elite circles since the 1950s, that the nuclear weapon does not obey the class principle, that is, does not discriminate between communists and capitalists, and that a nuclear war cannot provide a victory for communism, as the Chinese were frequently accused of thinking. Suggested by Malenkov in 1954 and by Khrushchev at the end of the fifties, the idea had taken hold as Soviet strategic capacities grew. Alongside it, however, was the persistent notion that a defence to nuclear weapons might one day be found.

Western analysts have debated as well the meaning of the growth in Soviet confidence in their ability to deter nuclear war. With greater faith in deterrence has grown a search for military alternatives to nuclear war.[11] The nodal point in this development is put by one analyst in 1966.[12] Others cite the famous 'Tula line' laid out by Brezhnev in a 1977 speech which denied the possibility of gaining nuclear 'superiority', in effect asserting the permanence of deterrence and providing a premise for a conventional war option.[13] The prominence of General Ogarkov was taken as a sign of Soviet interest in preparing for the challenge of a possible protracted conventional conflict in several theatres, one which, however, might not escalate to nuclear war.

In the early sixties, when the Defense Department of Robert MacNamara was developing the idea of Flexible Response, the Soviets had reacted with alarm. As the American strategists then saw it, the outbreak of war should not be a signal for nuclear attacks against the Russian cities. Instead, the US should calculate a series of lower levels of response, from conventional to tactical nuclear operations; and even at the end of the ladder of escalation, they should aim their attacks at Soviet nuclear weapons without touching Russian cities. The Soviets charged that the 'no cities' idea could not have a retaliatory character but must be part of a surprise attack, even a limited attack hoping to avoid escalation to all-out war. They insisted instead that any use of nuclear weapons would inevitably escalate to all-out war. They reasoned this way well into the eighties when NATO began to consider new strategies such as Air–Land Battle and Follow-on Forces Attack, making use of what were then called 'ET' (emerging technology) weapons (long-range sensors and smart munitions of various types) to blunt a possible Soviet attack and thus make conventional operations without nuclear escalation more feasible. But some Soviet military men, General Ogarkov chief among them, also noted the revolutionary character of these new weapons which, they argued, lent a certain stability to conventional operations, as contrasted with the extreme instability of any nuclear operations. Thus Ogarkov and his co-thinkers saw a new era in which the Soviets would need their own capacity for 'flexible response'.

At the same time the Soviets continued the modernization of their theatre nuclear option with medium-range SS-20 missiles that they had begun in 1977. They posed the threat to west Europeans of a Soviet conventional and theatre nuclear superiority that would be all the more overwhelming, or so charged the critics of SALT 2, alongside a super-power arms control regime that would in effect 'de-couple' US and European security. It was a new way of raising the old question: Would the US automatically risk its cities to defend European ones? The case of Soviet threats of missile attacks against European cities in 1956 might be cited, with claims, erroneous ones to be sure, that these broke western unity in the Suez crisis and deterred western action against Soviet suppression of the Hungarian revolution. The Euromissiles promised by the 1979 decision of NATO were an attempt to restore this coupling by increasing the 'automaticity' of the US nuclear response to a conventional Soviet attack. But, in Soviet eyes, they posed new threats. The short flight-time of the Pershing II missile menaced Soviet command and control centres and raised the spectre of the 'decapitation' strike that had already been discussed among some Washington analysts. Soviet threats against American command and control, by moving nuclear submarines closer to the American east coast in order to 'get within ten minutes', as one Soviet spokesman put it, could not hide the fact that the SS-20s had been a huge mistake.

Gorbachev would later say simply that the maintenance of nuclear parity at ever greater levels does not provide security, but its opposite. Nevertheless military people had to look at it differently, being charged with providing the national defence, including the defence of interests abroad, against every conceivable threat. Soviet military writers, with Ogarkov in the lead, found an array of threats in statements made by Reagan administration officials and Washington defence intellectuals. In an interview a few months prior to his demotion in 1984, Ogarkov described a new 'crusade (*krestovyi pokhod*) against socialism' with the United States at its head. The traditional supporter of neo-fascist organizations in the Federal Republic, the United States had always encouraged a campaign for a greater Germany. Now Germany was a vanguard of American 'preparations for a new world war', readied by a military budget in excess of three hundred billion dollars.[14] The US was active on many fronts, in Grenada, Lebanon, Angola, Nicaragua. The Soviet Union must remember Engels's admonition that wars are usually won through superior equipment. It must not succumb to the traditional Russian disease of *Manilovshchina*.[15]

Ogarkov's continued presence in the Defence Ministry after his demotion suggested that it was easier to reduce his political influence than to ignore his warnings.[16] Ogarkov put primary emphasis on ET weapons in conventional operations. This did not change the fact that since Reagan's announcement of the Strategic Defense Initiative during the time of Andropov's rule, facing this threat had been the first desideratum of Soviet policy. By the time Gorbachev assumed power, NATO's Nuclear Planning Committee had declared its support for the SDI and expressed a wish that the Allies participate in the programme at various levels. In July 1985, the Eureka plan was announced, organizing west European high-technology cooperation in information sciences, robotics, biotechnology, and a number of related fields. Some of the technologies were crucial to problems described by Ogarkov, for example Anti-Tactical Ballistic Missile systems, thought to be effective against both short-range and cruise missiles. The Soviets were not much in arrears on these – NATO had already taken note of the capabilities of Soviet SA-10 and SA-12 missiles.

Yet the technical arms race that was shaping up looked deadly for the Soviet budget. Moreover, Soviet military writers, in contemplating operations on the electronic battlefield, with strikes up to 100 km behind the point forces, put increasing emphasis on the factor of time. Pre-emptive attacks would have to be made against mobilizing forces. It made sense to recall that it had been the Russian mobilization of August 1914 that prompted from Imperial Germany, not mobilization but war, owing to the demands of the military timetable.[17]

The author of these threats, the America of Ronald Reagan, with European and Japanese allies and the active cooperation of China, was described by Ogarkov as having re-created the Berlin–Rome–Tokyo Axis of the 1930s. By 1985, Soviet commentators were also warning about French efforts to bolster nuclear cooperation with West Germany. Jacques Chirac, in a speech in Bonn in October 1983 had spoken of the absurdity of Britain and France trying to guarantee the security of western Europe without the involvement of West Germany. A French Socialist party document published in July 1985 suggested extending French nuclear protection to Germany.[18] A Soviet comment of the previous year shows how this trend was received:

> France ... has proposed the idea of European defense under which the FRG could lay its hands on French nuclear weapons. What explains

this behavior? The well-worn lie of a threat from the East is being used, reckoning on the short memory of the West European peoples. At the same time, efforts are being made to forget the times when Hitler speeded up the militarization of Germany to ward off the Bolshevik danger, and the ruling circles in Britain and France encouraged his actions, calculating that they would be able to channel Germany's attack to the East.[19]

The liberties taken by the Soviet commentator in describing the situation of the thirties will not obscure the gravity of his analogy or the rawness of nerves which called it forth. It was the same reaction the Soviets had showed to the formation of the Federal Republic, its rearmament and, at the end of the fifties, another perceived threat of its nuclear armament.

GORBACHEV'S ANSWER: *USKORENIE*

Gorbachev was determined to prevent tensions from building any further. Like Andropov, he thought he saw the possibility of a political response, and his first initiatives did not go beyond the parameters of Andropov's policy in 1982–83. He would not try anything radically new until he accepted the zero option in 1987, under circumstances we will describe. Andropov had made serious efforts at INF negotiations, without however departing from an ironclad position: that the Soviet Union would destroy all its SS-20s save only those required to match 150 British and French weapons. He had thought his position so attractive in West Germany as to assist the Social Democrats in their opposition to deployment of the NATO missiles. In effect he tried to intervene in the West German elections in March 1983. But he failed. The elections gave an unequivocal mandate to Chancellor Kohl who favoured the deployment. In the same month he was faced, in addition, with the American SDI. After Andropov's death, Gromyko had tried to reopen talks on all strategic weapons, including the 14 American aircraft carriers. Gorbachev, undaunted, was coming back for another try to realize Andropov's vision.

At a Central Committee meeting on science and technology in June, Gorbachev called for 'acceleration' (*uskorenie*) as the pressing task of the Soviet economy. With all the achievements of the postwar period, he said, one could not help but note a slowness in transferring to 'intensive factors' of economic growth. The slowing of growth in the last Brezhnev years could be attributed to a failure to make the transition. The achievement of strategic parity had been a product of this ironmongering, he argued, but now 'in the face of the aggressive policies and the menace of imperialism we must strengthen our country's defence capacity and deny our rival superiority over us'.[20] Far from embracing the idea of economic interdependence that had bounced around in the research institutes for years, Gorbachev stressed 'total independence' of other countries, 'especially in the strategic areas'. This must be bolstered by increasing investments in machine tools, micro-electronics, computer and informational technology.

Gorbachev urged that the average Soviet citizen work harder and better. In keeping with these demands, Soviet television produced a report on Aleksei Stakhanov, a coal-miner who in 1935 had mined 102 tons in six hours, establishing a standard for 'socialist emulation'. Stakhanovites had been the leading shock-workers in every industry and the vanguard of the Victorian labour regime of Stalinism, about which Gorbachev did not seem to be embarrassed. Nor did he hesitate to take measures against consumption of alcohol, something that would deal a serious blow to state revenues. Clearing the way, Gorbachev made sweeping changes in the economic ministries. Tikhonov resigned from his post as premier, to be replaced by N. I. Ryzhkov. Baibakov was removed at Gosplan; Patolichev, in the top leadership since 1946 and Minister of Trade since 1958, was replaced. Similar personnel changes were made in Latvia, Lithuania and Belorussia. Campaigns in Uzbekistan, Azerbaijan, and Kirghizia were taken up against embezzlement, nepotism, and other local corruptions.

It was not easy to determine what these measures meant. Were they moves in the direction of a massive campaign from above such as Stalin's first five-year plan? The shaking-up of the cadres seem to suggest it. Or were they, as some European observers suggested, only a prelude to a relaxation of state initiatives and a loosening of controls? It was rumoured that Gorbachev had charged the Institute of World Economy and International Relations with three research tasks: first, investigation of Lenin's economic ideas at the time of the proclamation of the New Economic Policy in 1921. This had provided the policy infrastructure for the mixed economy of the Soviet twenties. Second, study of Andropov's economic ideas, as raised in discussions with East European economists such as the Hungarian Janos Kornai. Third, historical review of the pre-revolutionary reforms of Stolypin, who turned over state and commune land to smallholders, a 'wager on the strong' intended to provide a prop for the monarchy on the countryside.[21] This would provide a glimpse of the Reformer Gorbachev that came more freely into view later. Stolypin was sometimes cited as a forerunner of the Communist theorist of the twenties, N. I. Bukharin, who advocated policies designed to permit freedom to the individual peasant. Could the revival of Bukharin's name in 1987 have been connected to Gorbachev's researches on Stolypin?

NEW FACES

There was no way of knowing where Gorbachev's policies would go in 1985–86. It was clear, however, that Gorbachev was moving quickly to form his own leadership team. Romanov, his main rival, disappeared in May and was dropped by the plenum in July. Boris Yeltsin and Lev Zaikov were brought into the secretariat. The Supreme Soviet elected Gromyko President (Chairman of its Presidium) and Edvard Shevardnadze took his place as Foreign Minister. A Western Sovietologist had already suggested this in April: 'Perhaps the simplest way to ease Gromyko out would be for Gorbachev to forgo for the time being his own election as president of the Supreme Soviet's praesidium and saddle Gromyko with this largely symbolic position.'[22] Gorbachev had begun to get control over the conduct of foreign policy. In the following year over 40 per cent of the ambassadors were recalled and changes took place at lower levels as well. Anatoly Dobrinin was brought home from the Washington embassy to replace Boris Ponomarev as secretary in charge of the Secretariat's International Department. Yuly Vorontsov was brought from Paris to be First Deputy Foreign Minister.

The changes bespoke a new outlook for international policy. The International Department was taken from a man who had worked in the Communist International in the thirties and given to one who had provided a back channel to Moscow for Henry Kissinger in the seventies. The power of Gromyko over foreign policy was broken, or at least bent, but his prestige in the Politburo remained and he was to be a force for stability and continuity for four more years.

Gorbachev took control of the Defence Council in August. He removed General Zaitsev from command of Soviet forces in East Germany. The head of their Main Political Administration (the political arm of the party in the armed forces), General Lizichev, was brought home to be overall head of the MPA, replacing General Yepishev, a strong Brezhnev supporter and opponent of liberal tendencies. General Tolubko of the strategic rocket forces was retired. He had helped to invest billions of rubles in the SS-20, with its solid fuel, its three warheads, reload capacity, and mobility (all of which would be traded away by 1987). Admiral Gorshkov, the father of the Soviet navy, the theorist of amphibious invasion by 'vertical envelopment' with heliborne naval infantry, was retired. The associates of Suslov in the secretariat, Kapitonov, Rusakov, and Zimyanin, were removed. The head of the Moscow committee, Viktor Grishin, was replaced by Boris Yeltsin.

Gorbachev had cleared away the most outstanding figures associated with the defence buildup of the Brezhnev period. Only Gromyko remained as an elder statesman in the Politburo. Correspondingly, while Gorbachev spoke of a political offensive to solve the most pressing national security problems, he also stressed ideological continuity with the past. This compromise position reflected the balance of Politburo forces, with a group of Andropov appointees (Aliev, Vorotnikov, Ligachev, Solomentsev, and Chebrikov) allied with the Gorbachev appointees (Zaikov, Ryzhkov, Shevardnadze) in order to encroach on the positions of the holdovers of the old guard (Kunaev, Shcherbitsky). As the backgrounds and the subsequent actions of the individuals in the first two groups would indicate, Gorbachev would have made a mistake to assume too much about their loyalties in the heat of action. This may explain why Gorbachev tried so hard to satisfy the demands of both continuity and renewal.

AGAINST NEO-GLOBALISM: DE-IDEOLOGIZATION OF FOREIGN POLICY

The lineup of forces described above held, for the most part, from the Twenty-seventh Party Congress, February 1986, to the January 1987 Central Committee plenum. During this time Gorbachev's foreign policy offensive broached new themes. Already, in a speech in October 1985 to a group of French parliamentarians he spoke of the need 'not to emulate medieval fanatics and not to extend ideological differences to inter-state relations'.[23] This bore on the oldest problem in Soviet foreign relations from the earliest days of Soviet power: how to distinguish between the requirements of the foreign policy of the Soviet state and the attitude of the Bolshevik party toward the Communist International. The notion of de-ideologizing Soviet foreign policy had appeared in the earliest days of Soviet power, for example, when Nikolai Bukharin spoke in 1922 of the possibility, in the event of a new war, of the Soviet Union allying with a bourgeois state (it would have been Weimar Germany) against another bourgeois state. Then there were pacts with France and Czechoslovakia in the thirties, and the experience of the Second World War, not to mention the many contacts and agreements of the postwar period. After the dissolution of the Comintern in 1943, there had remained the problem of coordinating a broad field of international activity by Communist parties and other groups, a problem that eventually became the responsibility of the international department of the Central Committee. In the first decade of the Sino-Soviet dispute, the Soviets spoke of containing the quarrel to party relations to keep it from affecting state relations, that is, to keep it in the socialist bloc.

Strictly speaking then, the idea of 'de-ideologizing foreign policy' was nothing new. Yet the phrase grated on some party and military people. The generals were expected to demonstrate a high degree of party-mindedness while at the same time assessing the military needs of the state from their political and ideological analysis of the changing world. So an assessment of threats made by the 'imperialists' would not be possible without 'class analysis'. The problem was compounded by the fact that the analysis had traditionally been made *entirely* by military people, and only recently had a number of civilian specialists been permitted to contribute. Military advisers were not sure of their role. General Ogarkov had got lost in this maze and overstepped the proper boundaries at the time of his demotion in 1984.

At the same time, Gorbachev was careful not to suggest that foreign policy exhausted the possibilities of Soviet action in international life. Even with the extraordinary effect of the formula of de-ideologization, Gorbachev had to elaborate, as will be described, the distinction between 'interstate' (*mezhgosudarstvennye*) relations, which were to be de-ideologized, and 'international' (*mezhdunarodnye*) relations, which were not.[24] This would seem to show that the international department was not to be subsumed in the foreign ministry, and was to retain its functions.

Gorbachev raised the idea of de-ideologizing interstate relations in the context of discussing relations between the superpowers in the nuclear field. He denied any desire to pursue what he called a 'Metternich type of balance-of-power policy, setting up new blocs and counterblocs.' He wanted, he stressed, a policy of 'worldwide detente'. The orthodox could read this as a broadening of Gromyko's line, while others might sense an implicit rejection of SALT-era diplomacy and the clumsy schemes of Brezhnev to forge a US–Soviet combination against China. Gorbachev wanted to be seen as absolutely orthodox and at the same time as transcending the orthodox line with broader and bolder political initiatives.

These would have had to come in the area of regional conflicts, which Gorbachev and his supporters seemed to tacitly recognize as the cause of the breakdown of detente at the end of the seventies. Reagan had made his view clear enough from his inauguration. His very first televised interview emphasized the need to restore 'linkage' in foreign policy, by which he meant that peace was indivisible and that regional conflicts must have a bearing on superpower relations in every field, especially the strategic field. The Soviets could not make mischief in the Third World and still have the arms control agreements that they wanted. That was in 1981. The Soviet position had been that there could be no such linkage, that arms control did not imply a counter-revolutionary bloc with the US to stop just wars for national liberation. On 16 January 1986, Gorbachev reiterated this position, saying that American affection for linkage reflected a 'reluctance to disarm and a desire to impose an alien will on sovereign nations'.[25]

Yet when Gorbachev met Reagan at the Geneva summit in November

1985, he agreed to conduct regular talks on regional conflicts. Gorbachev told a press conference at Geneva that a great deal in US–Soviet relations depended on how they saw the rest of the world. It made no sense, he argued, to maintain that the various conflicts and wars going on in the world were the product of the rivalry between east and west. Indeed, such a view, he said, was 'not only erroneous but extremely dangerous'.[26] Gorbachev's position was little different from the most strident positions taken by Soviet publicists at the end of the seventies when the world revolution seemed to them to be inexorably on the march in the Third World.

But the picture at this point was not so clear. Movements for which Moscow had expressed sympathy were not necessarily winning ground, but not losing it either. The prospects were brightest in Nicaragua where the Contra forces equipped by the Reagan administration had been driven into Honduras and could not claim to hold a single Nicaraguan town. The Sandinistas, aided by Soviet M1-24 helicopter gunships, long hailed as the miracle-worker against rural insurgencies, seemed to be in control. Contra leaders told disappointed interviewers in the United States that they had no immediate offensive plans of any kind. The leftist rebels in El Salvador, by contrast, were defying all attempts to root them out of the countryside and were carrying the battle in limited ways into the cities and towns. In Angola, also discussed specifically at Geneva, government forces, aided by Soviet equipment and Cuban cadres, made an attempt to wipe out the UNITA rebels of Jonas Savimbi in August–September 1985, only to have South Africa and the US prevent it by timely aid. Shortly after the Geneva summit, Reagan promised to send *stinger* anti-aircraft missiles to Savimbi. These, along with TOWs, 106 mm recoilless rifles, and other anti-tank weapons would take a heavy toll of Soviet tanks when the offensive was again resumed in September 1987.

Most confused was the situation in Afghanistan, where the Soviets seemed to be settling in for a long struggle. Before Gorbachev came to power the Chernenko–Gromyko regime began a programme of sending 6000 Afghan children per year to the USSR for ten years of schooling. In March 1985, as Gorbachev was being elected, the education system in Afghanistan was being revamped from top to bottom, with emphasis on the teaching of the Russian language and the rewriting of Afghan history to stress a historical friendship with Russia. The Soviets had pursued policies similar to these in their own Islamic republics in central Asia, policies that were judged, despite the fact that they had taken a generation to work, to be quite effective. This sovietization plan in Afghanistan bore all the earmarks of the thinking of Aliev, who had long argued that Soviet experience in the Moslem republics could provide advantages in dealing with the neighbouring states, including Islamic Iran.

Gorbachev's policy seemed to be searching for an alternative. At the Soviet Twenty-seventh Party Congress in February, Gorbachev had referred to Afghanistan as 'a bleeding wound' and promised to begin the withdrawal of Soviet troops. The quagmire in Afghanistan had been the work of the aggressive forces of imperialism, he said, which as usual sought to halt the course of history by the use of force. Socialism, by contrast, would not need to pin its future prospects on the military solution of international problems. He cited as a precedent Lenin's criticism of the 'war party' in the Bolshevik leadership of 1918 who had wanted to carry the Russian revolution westward on the ends of bayonets.[27] An interesting citation. The occasion of Lenin's criticism was the debate on the treaty of Brest-Litovsk, which ceded to Imperial Germany two-thirds of tsarist Russia's European territories. Despite the humiliation of thus caving in to German power, Lenin had advocated signing it. Gorbachev had reached for the classical case of a 'Leninist' reconciliation with an unpleasant reality.[28]

He had also implied that Generals Ogarkov, Sokolov and others who warned constantly of threats that must be met by Soviet conventional capabilities were not on a precisely Leninist path. Lenin was, to be sure, invoked selectively. The generals were at a disadvantage not being able to cite on their own behalf cases of the successful use of force. No one wanted to boast about the interventions in Hungary and Czechoslovakia.

Gorbachev's remark about Afghanistan being a 'bleeding wound' was a benchmark for a change of policy and leadership in Afghanistan. One of the Afghan leaders, Najibullah, attended the Soviet Twenty-seventh Party Congress and was fêted in a way that overshadowed the then head of state, Babrak Karmal, who would be forced out by May. Najibullah's regime would conduct a vast party purge later in the year, while he strove to broaden the party's base by holding local provincial elections and meeting with tribal leaders. This was done to the accompaniment of plans for 'national reconciliation' and a 'normalization' of affairs in the country. The Soviets made their own contribution by promising to remove six regiments. This was accomplished by October.

Gorbachev insisted that the restructuring in Afghanistan did not weaken Soviet resolve to defend the gains of the April 1978 revolution that had brought the Afghan Communists to power. He must have also considered, in line with the general idea of *recueillement*, that it had been the presence of Soviet troops over the last seven years that had forged an Afghan resistance from disparate and usually incompatible tribal and ethnic groupings, and that the removal of Soviet troops might permit a return to the traditional disunity. He must also have considered the fact that Soviet troops in Afghanistan were the first among the causes of Soviet external enmities. To the Chinese they were proof of a Soviet desire to

encircle China. In the United States they were the trigger of the movement to cancel the SALT II treaty and turn away from detente. They
had in effect cemented cooperation between Washington and Beijing against
the USSR. They were a rallying point for those in the Soviet Union who
called for massive increases in budget allocations for conventional arms.

They could even be seen as the challenge to Washington that had resulted in its doctrine of 'neo-globalism'. Soviet publications perceived a
new American strategy of 'neo-globalism' emerging in 1986. American
publicists had begun at that time to speak of a Reagan Doctrine,[29] a
phrase never used by Reagan himself. It referred to statements made by
Reagan in 1985 and 1986 that his administration would henceforth reserve
the right to provide aid to anti-Communist insurgencies in various places.
The Soviets took note of it in 1986 because of congressional votes to
supply assistance to such groups in Nicaragua, Cambodia, Angola, and
Afghanistan. Assistance of this kind was not a completely new departure. The newness of the Reagan Doctrine was the break with the idea
that such interventions must be covert. The Soviets were not slow to
note this about 'neo-globalism'.

In February 1986 the Reagan administration used its influence to ease
the departure of two dictators, Marcos in the Philippines and Duvalier
in Haiti. Reagan had done nothing to precipitate the events that overthrew these men but he decided to bring his foreign policy abreast of
what he perceived to be a general democratic world trend. On 14 March
he declared an intention 'to oppose tyranny of whatever form, whether
of the right or the left'. The Washington press quickly gave this idea the
title of Philippine Corollary (to the Reagan Doctrine). Reagan seemed
to have turned from his administration's early criticism of the Carter
human rights policy and from his earlier distinction between totalitarian
and merely authoritarian regimes. He was attempting, as he had in the
past, to align his policy with what is often narrowly and misleadingly
referred to as the Wilsonian tradition, called by its critics the idea that
foreign policy can be devised according to sentiment rather than national
interest, and by its friends the idea that foreign policy usually does best
when it is in tune with historically progressive trends.

The issue for Gorbachev was whether the various movements to which
the Soviets had given their support were part of a historic trend, or whether
Soviet policy was misreading the trend. The Soviet press had not looked
kindly upon the opposition of the 'bourgeois leader' Corey Acquino to
the Marcos regime. The influential Soviet commentator Aleksandr Bovin
referred acerbicly to the Reagan Doctrine as 'the doctrine of the revers-
ibility of history'.[30] Gorbachev himself called it an attempt to 'arrest the
course of history'. He told the Algerian weekly *Révolution Africaine* at
the end of March that nothing could keep the Soviets from helping nations

liberation movements. 'We have given and will continue to give them extensive assistance – political, moral and material.'[31] But he also said that if there were no American interference in the internal affairs of other states, regional conflicts would be on the wane. The question he must have pondered at this point was: could the Soviet Union put itself on the right side of history simply by putting a minus wherever the United States put a plus?

No less perplexing was a related question: could the Brezhnev era analysis of the relation between detente and the advance of class and liberation struggles be in error? For years it had been maintained that detente was no mere accident of history, but a reflection of the impact made by the achievement of strategic parity with the United States. Detente was therefore a way of recognizing the new correlation of forces. History had been slowed by the fact that the imperialists had been the first to get their hands on nuclear weapons. The achievement of missile parity by the Soviet Union, however, had removed the threat of 'linkage' by the Western imperialists of strategic blackmail in their attempts to resist the forces of history. The imperialists had not *chosen* detente but had it imposed on them by life. They should not be able to opt out of this historic bargain by a new arms buildup and aid to counter-revolutionaries. But how could one deny that they were doing precisely this? These questions were not at all abstract but were being decided every day, arms in hand.

Perhaps the analysis of the correlation of forces of the seventies had become obsolete and unsuited to present tasks, like the balance of cavalry forces before the Second World War. Many Soviet civilian analysts had been tending toward this conclusion for years. And western writers such as Elizabeth Valkenier, Jerry Hough, Francis Fukuyama, and Stephen Sestanovich had alerted their publics to the possibility of change in Soviet thinking. But it was not easy to determine what impact the analysts from the Soviet institutes had on the thinking of the leadership. Gorbachev had to weigh the possibility that the Third World might not be the avenue of advance it was thought to be in the seventies and, if he did not want to reason from this that the Communist cause was no longer historically progressive in the Marxist sense, he would have to conclude that perhaps the balance of forces could not really be turned on the basis of something that goes on in a relatively remote corner of Africa.[32] Events in places like the Ogaden desert had so disturbed US policymakers that the unravelling of detente was the result. No matter how much Gorbachev might insist that regional conflicts could not be a condition for disarmament agreements, the historic fact of the linkage was undeniable. The truth is always concrete, as Gorbachev said at the end of his address to the Twenty-seventh Congress.

Soviet analysis had been clear, at least since Andropov's period at the helm, that the American insistence on linkage of regional conflicts to arms control was the central source of current world tensions. It had been the United States, Andropov had argued, that had introduced ideology into interstate relations:

> The transfer of ideological antagonisms into the sphere of interstate (*mezhgosudarstvennykh*) relations has never brought gain to any who have resorted to it. Now, in the nuclear age, it is simply absurd and impermissible. The transformation of opposition of ideas to opposition of arms would be very costly for all humanity.[33]

The process of de-ideologizing interstate relations was, therefore, an imposition by political means of Soviet policy on the United States. Andropov claimed that this would be possible through steadfastness in the pursuit of the peace policy.

UNIVERSAL HUMAN VALUES

Gorbachev seemed to admit that the Soviets had been guilty of the same offence and that they needed to set an example of the correct course in action. The Twenty-Seventh Congress made a symbolic step in this direction. Anatoly Dobrynin was brought back from the embassy in Washington to head the Central Committee's international department, the seat of Soviet non-diplomatic international contacts. He replaced Boris Ponomarev, 80 years of age at the time, who had headed the department for three decades. The appointment was striking in Western eyes, as it seemed to mean the diplomatizing of 'Comintern' affairs as well as the removal of an ideological influence on diplomacy.[34]

Dobrynin gave an indication of the difference between his views and those of Ponomarev in a speech shortly after the Congress to a conference of Soviet scientists. Referring to the 'comprehensive plan' to abolish nuclear weapons in the next ten years, a plan advanced by Gorbachev in his speech of 16 January, Dobrynin emphasized the role of Soviet scientists in the struggle for the programme's success. This role he saw in essentially three aspects: first, an increased effort in people-to-people contacts with Western scientists. The Soviets normally used a broader conception of 'scientist' than is common in the West, one that might be closer to the German idea of science as *Wissenschaft*, that is, any body of disciplined knowledge, so that Soviet scientists would include, for example sociologists, psychologists, and historians. These were to take note, said Dobrynin, of the resonance of the Pugwash conferences on peace questions and other similar international meetings in order to use them to carry out Gorbachev's plan. Second, Soviet national security advice needed to be enriched by broader participation of scientists in media that were usually dominated by military opinion. He referred to the formation of specific arms control agencies in the foreign ministry and the international department. Third, Dobrynin implied that national security advice itself ought to be de-ideologized.

He insisted that complaints about excessive ideology in interstate relations had not emerged as a criticism of Soviet acts but those of the Reagan administration and its revival of the idea of a 'crusade against "godless" Communism'.[35] Once it was articulated and accepted in the context of criticizing the US, the de-ideologization idea could easily be broadened and applied to criticism of the Soviet experts as well. Dobrynin suggested that in an era of explosive technological developments in the military sphere, ideology became a particularly dangerous impediment to clear thinking. By contrast, new technology should stimulate new scientific thinking, which would be of necessity non-ideological.[36] The 'dialogues' of the past with Western representatives turned too often into ideological monologues, with lectures about the aggressiveness of Soviet foreign policy and the 'hand of Moscow'. Once again, the charge seems to cut equally well against Soviet ideologues.

Gorbachev had stressed in his 16 January speech that disarmament could not be linked to regional conflicts. We have already noted how Gorbachev's insistence on this point was weakening by increments. Dobrynin complained that American 'neo-globalism' only deepened regional conflicts, but he also admitted that these conflicts did not stem entirely from liberation movements. They involved, he said, undeniable tribal and religious elements. They desperately needed to be regulated and their material causes addressed by a 'comprehensive model of international economic security'.[37] Here, he said, was a pressing task for Soviet social scientists.

Dobrynin seemed to be calling for a new Peace of Westphalia. The treaty that ended the era of European religious wars in 1648 had opened an era in which national foreign policies would no longer be an instrument of religious Reformation or Counter-Reformation, but would be guided instead by the secular notion of national interest – a new Westphalia on a world scale. This in order to provide foundation for disarmament according to Gorbachev's 16 January plan. Soviet security interests must be governed, he said, by the idea of 'reasonable sufficiency (*razumnaia dostatochnost*)'.[38] They must recognize the mutual danger and the need for common security among the nuclear powers. The New Thinking, said Dobrynin, 'does not proceed from repudiation of the class analysis of the problems of war and peace', but represents a synthesis of the teachings of Marx and real humanism, in the defence of 'general-human' (*obshchechelovecheskii*) interests.[39]

Dobrynin's compromise formulae served only to underline the potential conflict with the ideological apostles of continuity as conceived pre-Gorbachev. To be sure, many of the components of the New Thinking had been percolating among Soviet writers in the early eighties. But it had always been possible to present them as part of the rhetoric of the traditional and periodic Soviet peace offensives. One could argue that Leninism had always taught that victory could not be achieved through war and that those who thought otherwise were guilty of Trotskyism.[40] One of the advisors closest to Gorbachev, G. Kh. Shakhnazarov, had no difficulty aligning many basic notions of the New Thinking with the claim that Marxism-Leninism never opposed the 'all-human' to the 'class' principle. In an article written in 1984 he had coupled these statements to a quotation of Chernenko's to the effect that war and peace were inseparable from the class struggle.[41] He maintained that the Western attempt to apply linkage had failed. But now the statements of Gorbachev and Dobrynin were using the same language to imply something different. And, if Gorbachev were to accept the idea that regional conflicts were spoiling the atmosphere for arms control, that suggested that the synthesis of the Brezhnev era, with detente and liberation struggles proceeding simultaneously, was now in ruins.

'ASYMMETRIC RESPONSE'

It followed that something more than 'one or two peace offensives', as Gorbachev put it, would be required to address the country's pressing security problems. Gorbachev left no doubt that the first of these was the American Strategic Defense Initiative. In this he continued in the view held by all the Soviet leaders since 1983 when the SDI was announced. But through the first three years he was in power, Gorbachev steadfastly denied that the Soviets were engaged in a comparable effort to create a ballistic missile defence. 'The Soviet Union is not developing space strike weapons or a large-scale ABM system', he declared in July 1985. He would declare the opposite to NBC's Tom Brokaw in the spring of 1988, admitting that the Soviets had an ongoing missile defence programme. Soviet scientists such as Evgeny Velikhov had participated in the American debate on SDI that followed Reagan's announcement in 1983 by suggesting that an American 'Star Wars' effort would force the Soviets into similar efforts. It was difficult even before 1988 to imagine that the Soviets could sit by and observe the American efforts, which even existed before 1983, without going full tilt on their own programme.

Yet in his first three years in office Gorbachev usually insisted only that the Soviets would 'find a proper response', and one that would be 'effective, less expensive, and coming within a shorter period of time' than the projected US defence.[42] Vadim Zagladin, in a television documentary shown at the time of the Geneva summit, said that 'there can be no doubt that the appropriate response will be found and, naturally, it will be equal to what the Americans present us with'.[43] Gorbachev had told a BBC reporter at the summit that 'some people hope the continued arms race will wear out the Soviet Union economically', a theme that he would sound repeatedly in subsequent years. Before the Twenty-seventh Congress he spoke of Western cynics who wanted to undermine the Soviet economy by drawing it into a space race. In April 1986, he warned the United States that 'the arms race will not wear us out', and a response would come, 'not necessarily in outer space'.[44] In August he spoke of a 'prompt response' which would not be what the US expected, but would make SDI worthless.[45] After the Reykjavik summit in October 1986, he spoke of the ascendancy of those with 'plans to wear out the Soviet Union'.[46] They want, he said, to 'obsolesce previous Soviet investment'.

The Reagan administration officials were indeed taking the view that the Soviet economy could not compete in defence spending in response to an abandonment of the SALT II treaty limits. Observers took note of the virtual disappearance in Soviet public pronouncements since 1985 of the formula 'everything necessary to defend the homeland', and its replacement by 'everything to prevent the strategic superiority of the United States'.[47] At the Twenty-seventh Congress, Gorbachev abandoned the formula 'acceleration of the perfection' of the economy for a more straightforward 'acceleration'. As 'perfection' was dropped, the Brezhnev era's economic efforts were described in terms of 'stagnation' (*zastoi*). Perhaps these changes were prompted by fear of an American 'exhaustion strategy'.

REYKJAVIK AND THE EXHAUSTION OF ANDROPOVISM

Gorbachev made repeated efforts to get the US to agree to a testing moratorium, including unilateral Soviet moratoria, from August 1985 to August 1986. When the US conducted tests at the end of the first six-month moratorium, Gorbachev made it clear that he considered American policy to be one of pure intimidation. The US, he maintained, was preparing a new generation of nuclear directed-energy weapons with which to gain by threats a free hand against Soviet interests.

But the Soviets, he said, would persist in striving for the victory of his January 1986 plan. Gorbachev continued to attempt to sell this strategy before an increasingly doubtful military and Politburo leadership. A kind of climax was reached at the Reykjavik summit in October, in which

Gorbachev attempted to tie a demand for US restriction of SDI with a comprehensive package of arms control measures, including a 50 per cent cut in strategic nuclear weapons and virtual elimination of Intermediate Range Nuclear Forces in Europe. In the most outlandish of all the Soviet–American summits, Gorbachev and Reagan pretended by turns to offer, on their own conditions, to eliminate all nuclear weapons. But in the end nothing at all came of the meeting. On his return to Moscow, Gorbachev tried to explain the collapse of the talks by charging that only the 'powerful forces in the ruling circles of the United States and Western Europe', had proved capable of 'saving' SDI from the clutches of his disarmament plan. He insisted, however, that Reykjavik had not been a failure, as apparently had been suggested in Politburo meetings assessing the summit. Progress had been made, although he lamented that it was hard to reconcile oneself to the loss of a unique chance to save mankind from the nuclear threat.

Gorbachev drew two lessons from the experience: first, that he was able to make a strong impression on the American leadership, not least Reagan himself. Indeed, very few leaders had met with Gorbachev personally without being affected by his personal qualities in face-to-face communication. Reagan was no exception. Gorbachev may have sensed himself closer to agreement with the United States than ever before. Reagan, in blaming the failure at Reykjavik on the other side, described the meeting afterwards in terms comparable to those used by Gorbachev. 'A moment of rare opportunity for arms control', said the American President, had been dashed because of Soviet opposition to the SDI.

The second and more ominous lesson appeared to be that the Soviet military and political leadership was bound to grow impatient with the campaign to woo Reagan for arms control, especially when it issued in a ludicrous charade like Reykjavik, with its wild promises of sweeping disarmament measures and its sudden letdown when the realization sunk in that Gorbachev's personal charm offensive could not carry the day. Gorbachev insisted against the sceptics that Reykjavik had not been a setback but an important step on a long journey, but he also seemed to realize that the format of the traditional Soviet peace offensive, in the name of the programme put forward by Andropov, was now exhausted.

At this point the Soviet position on the INF issue was unchanged from that of Andropov's in 1983 ('not one more missile' than those in the British and French arsenals). To make a breakthrough, genuine concessions would have to be made. Gorbachev may have thought that his bloc with the Andropov appointees was still solid, having survived the sweeping personnel changes of the last 18 months, and that it would support him in concessions to Reagan. However, at the end of December, signs accumulated that the period of *perestroika* unopposed was coming to an end. When Dinmukhamed Kunaev, Politburo member and Brezhnev loyalist, was removed from his post at the head of the party in Kazakhstan and replaced by a great Russian, Gennadi Kolbin, nationalist riots broke out in Alma-Ata, with fierce street fighting that lasted three days and three nights. When the Central Committee convened in January 1987, Gorbachev found that he could not get his agenda out of the Politburo to be placed before the plenum. For the first time Gorbachev was faced with substantial opposition. He could see that, if he accommodated it, he would be heading into the permanent position of first among equals, required to rule in concert with some new Suslov, most likely Ligachev. Moreover the experience of the various losers in Soviet succession fights was that being limited in some small way was the first stage on a slope that led to political oblivion. One could cite the cases of Trotsky, Zinoviev, Bukharin, and Malenkov.

If Gorbachev resolved to fight on, he could no longer continue to march under the banner of Andropov. If he meant to persevere in restructuring the leadership and finding his way to accommodation with the West, he would need to break with the Stalinist continuity of the last four years. Maintaining the campaign of changes would require a new rationale.

5 Gorbachev Bound: The Emergence of the Ligachev Opposition

Have you ever heard of the Catholic church apologizing for the Inquisition or the Crusades?

Bayardo Arce

The Central Committee plenum of January 1987 was the pivotal point in the rise of Gorbachev. Up to this point his reconstruction of the administrative and political leadership had pulled the whole Politburo along. There was a consensus that in order for progress to be made the superannuated Brezhnev coterie would have to have its ranks refreshed under a new leader. The Andropov appointees were swept along with the programme of 'Acceleration' because it was spiritually akin to the line of Andropov in 1982–83. So the sweeping changes of 1985–86 proceeded with an enthusiastic momentum. And Gorbachev clearly intended to go further still. He had inherited only the Brezhnev position as a mere first among equals; he strove, and *had* to strive, for something more.

The plenum that marked the first appearance of opposition to his plans had originally been planned for October 1986, in the normal six-month interval since the plenum of June, but it had been postponed twice, increasing speculation that the personnel changes that were designed to be its centrepiece had run into resistance in the Politburo. It was rumoured that Foreign Minister Shevardnadze was to be given control of the KGB, with Dobrynin, then in charge of the International Department of the Central Committee, replacing him. Chebrikov would be moved from the KGB to first secretary of the Ukraine, to replace the veteran Brezhnevist Shcherbitsky.

Chebrikov could hardly consider this a promotion, especially in light of the 'Berkhin affair' exposed in the press before the plenum. Viktor Berkhin, a correspondent for *Sovet shakhter* (Soviet Miner) in the Ukranian town of Donetsk, had been continuously harassed and even arrested by the local security organs, presumably for a colleague's slandering the organs in his reports to the paper. Berkhin's cause had been taken up by *Pravda* itself in a series of articles attacking the local public prosecutor, the militia, and officials of the KGB. In early January Chebrikov himself was forced to apologize for them.[1] The party's daily organ had defended a representative of labour from the 'arbitrariness and subjectivity' of the KGB! *Pravda*'s editor, Viktor Afanasyev, was later to be counted among the opponents of *perestroika*, but at this point he was on the Gorbachev bandwagon. It was also not insignificant that these things took place in the Ukraine, where both Shcherbitsky and Chebrikov had risen under Brezhnev. If Chebrikov had been moved to the Ukraine after the Berkhin Affair, this certainly would not have enhanced his prestige there. At the same time it was thought that Gorbachev would bring Gennady Kolbin into the Politburo as a full member along with his own man Boris Yeltsin. Kunaev, already removed from his post in Afghanistan, would now lose his Politburo membership.

LIGACHEV'S OPPOSITION

But Gorbachev ran into a wall of opposition to these measures, as he frankly admitted to a trade union meeting a month later. 'Suffice it to say', he told the gathering, 'that we postponed the meeting three times, for we could not hold it without having a clear idea of the issues.' Participants later told reporters of the hostile atmosphere of the plenum, where only about half of the conferees seemed to be with the General Secretary. Gromyko made a speech full of veiled criticisms of personnel and even foreign policy.

Why should there have been such conflict when all were agreed about the course of Acceleration? Were they quarrelling about the 'Rejkyavik package' of arms reductions linked to restriction of the SDI? This was to be a continuing problem, but it was more likely that the Andropov appointees had begun to put the 'regime question' ahead of these considerations, and had got the idea that Gorbachev's purges of the Brezhnev appointees would soon affect them as well. Perhaps it was time to slow things down.

Observers of the Soviet scene were unprepared for the idea of opposition to Gorbachev. On the contrary, up to this time most had been extraordinarily impressed with his control of affairs. Press and foreign office analysts in the Western capitals were accustomed to thinking long and hard about the strength of a Kremlin leader, out of a natural concern that agreements reached with a Soviet government would not be overturned. After it was determined that the leader was 'in control', this often resulted in indifference about leadership conflicts. The feeling was that these things would sort themselves out eventually. It was not always appreciated that leadership conflicts had their effect, often the crucial effect, on the shaping of policy.

Now for the first time it had to be recognized that Gorbachev was faced with rivals at the highest level. Up to this point, no inkling of this fact had been visible in the Western press. Who could the opponents of Gorbachev be? Months later, it was 'discovered', primarily on the basis of leaks by Gorbachev supporters, that Yegor Ligachev was fundamentally opposed on principle to Gorbachev's conception of *perestroika*, and that theirs was a clash of antithetical visions of the future of the country. Yet this was not visible prior to the January plenum, and not very clearly visible immediately afterward. One could safely surmise, however, that by virtue of his position as Second Secretary Ligachev was cast in the role of a Suslov, that is, as a defender of collective leadership against possible incursions by a too ambitious general secretary. It was difficult to imagine that an opposition bloc could survive without being aided in some way by the Second Secretary. Ligachev's memoirs, which began to appear in 1990, give confirmation of the accuracy of this picture.[2]

But who else comprised this bloc of opponents? The most likely hypothesis was that the former combination of Andropov appointees and Gorbachev appointees, which had moved so boldly against many of the old Brezhnev men, had run up against a frightening backlash in the form of the Alma Ata events, and had judged that a pause in the campaign was now in order. Gorbachev was cautioned and restrained by the members of Andropov's Class of 1983: Ligachev, Chebrikov, and perhaps Vorotnikov and Solomentsev. To these undoubtedly would have been added the voices of the Brezhnev appointee Shcherbitsky and of Aliev, who owed as much to Brezhnev as to Andropov. Gromyko, a veteran of the national security grouping (with Andropov and Ustinov) that had overseen the whole transition from Brezhnev to Gorbachev, must have counselled moderation. This profile of Gorbachev's opposition is suggested by the fact that these were the men most consistently targeted for attacks in the subsequent campaigns of *perestroika*. Among the candidate members, Yeltsin, Yakovlev, and Solovyov were consistent Gorbachev supporters, but General Sokolov was skittish about Gorbachev's initiatives toward the West, and others, like Dolgikh and Demichev, had been part of the Brezhnev team. In the Secretariat, Gorbachev had his old law school classmate, Anatoly Lukyanov, Yakovlev, Vadim Medvedev, Georgy Razumovsky and some others who would defend his measures, giving him perhaps the best bloc of support for the struggle ahead. Gorbachev subsequently showed that he was adept at inner-party combinations of every sort, but he had no need to rely on bureaucratic wire-pulling. He was capable of mounting a press campaign against opponents, using the force of his personality and his ability to rally the party for new political departures.

GORBACHEV AND LIGACHEV VERSUS ALIEV

Too great a gulf had been assumed to have opened between him and Ligachev, who was in fact at this time only seeking a pause. And Gorbachev's supporters, with the exception of Boris Yeltsin, did not immediately set out after Ligachev with fire in their eyes. There was in fact much evidence of continued cooperation between Gorbachev and Ligachev alongside the mounting tension.

The two men were united in their attitude toward Aliev. On Gorbachev's election in 1985, Aliev had appeared to many to be the new second secretary. He had delivered the Lenin Day speech, which indicates authority in ideological matters. Then with the emergence of Ligachev as second secretary, Aliev had to take a back seat. It would be anomalous for there to be no residue of antagonism and jealousy between these two men. Here Gorbachev had a card to play in encouraging Ligachev to press on with the personnel changes, which would most likely result in the eventual retirement of Ligachev's rival. Indeed, Aliev would be out of the leadership by autumn.

The campaign against him had other implications for Soviet policy on the Central Asian and Caucasian republics. Aliev was an Azeri, born in the Moslem enclave of Nakhichevan, who had risen to the rank of KGB general, thence to first secretary in Azerbaijan in 1969, a post that he held until 1982. Soviet Azerbaijan under Aliev made a point of settling Azeris in Nagorno-Karabakh, the mountainous area around which a situation of near civil war between Azeris and Armenians was to develop in 1988.[3] Appeals were issued to the ethnic Azeris in Persian Azerbaijan, calling for their 'national liberation' in a united Azerbaijan. This was in line with a policy of encouraging separatism in Mazanderan and other Iranian districts on the southern shore of the Caspian Sea. Aliev made himself a symbol of ambitions of the Brezhnev leadership in the Moslem world. He promoted the idea that the Soviet Central Asian republics, with their enlightenment and freedom from traditional customs, would be a magnet for the peoples of the Middle East and would give the Soviet Union extra prestige and influence there. He championed the notion, as once expressed by French Marxist Roger Garaudy, that the revolution would advance with a copy of *Capital* in one hand and the *Koran* in the other. Brezhnev and his associates had been in a revolutionary mood in 1979, closely supporting Khomeini with their Tudeh party comrades in Iran, at the moment when they decided to invade Afghanistan. Although the course of the Iranian revolution and the war in Afghanistan were to prove the undoing of this policy, the Soviets thought, at least for a time, that they could manage the situation and expand their influence. But they were engaged in what Alexandre Bennigsen has called 'the dangerous game of supporting Islam abroad while trying to destroy it at home'.[4]

As Aliev rose to Politburo status the fortunes of Soviet policy in the Moslem world were beginning to ebb. The war in Afghanistan went badly. Soviet troops from the Central Asian republics did not provide the needed edge. Afghan rebels who captured Soviet soldiers who were Moslems quickly realized that the latter had no stomach for the war. Their lack of money and their willingness to steal military equipment in order to sell it convinced their captors that Russia was actually a rather poor country. The Russian troops' seeming indifference about spreading Communism through the world made the *mujaheddin* think that perhaps it was they who should be coming to liberate the Russians. While their resources for that task were limited, those of revolutionary Iran were somewhat better. The Islamic regime of Khomeini, after a thorough purge of Communist *Tudeh* party members from its ranks, directed radio propaganda and Moslem literature into the Soviet Union. Most of it, ironically, was in Russian, the language common to all the Moslems in the Soviet Union. In Soviet press material on the threat of Islamic superstition and particularistic nationalism, great stress was laid on the importance of promoting the study of Russian, which was held to be the route to Soviet internationalism. However, Islam is also a supranational idea that can be spread through the same medium.

COMMUNISM AND ISLAMISM

The historic Russian victories in Central Asia and the Caucasus were under threat. These date from the expansion of Imperial Russia in the nineteenth century. Sovietization of the area in its contemporary form, however, really dates from the era of Stalin's great purges. The *Yezhovshchina* of 1937–38 (so-called for Commissar of Internal Affairs N. I. Yezhov) extended collectivization of agriculture to Central Asia and the North Caucasus. Prior to this time the Communists had tried to do their work without interfering too much in the cultural and religious life of the people. But the purges represented an attempt to get local customs under the influence of Soviet culture. Along with the arrest of many suspicious intellectuals went a wholesale decimation of the clergy and the destruction of a majority of the existing mosques. The authorities did not try to eliminate Islamic practice completely, but made cultural inroads against practices such as the veiling of women. They strove to weaken loyalties to tribe or family, and strengthen those related to the nations and to the Soviet centre, with its presumably federal relationship to the national republics. For them, nationalism was a less serious offence than religious obscurantism.

It was easier to persecute the reactionary aspects of the local culture than to substitute a modern secular culture for them. Soviet complaints about the difficulties of disseminating atheistic internationalist attitudes among the Central Asian Moslems leave the impression that the Soviet culture was inadaquate for the task assigned. Soviet writers regretted that they had been unable to stamp out local customs, such as the paying of bride-money, or visits to the faith-healer or to the fortuneteller. Nor could they prevent citizens from attending prayers or being married by mullahs instead of having a Komsomol wedding and a visit to lay a wreath at a tomb of war dead. Often people were steadfast through their youth, but in old age loneliness got the upper hand, causing them to seek companionship at the mosque. On some occasions the local Communist institutions failed to provide assistance in some personal or financial matter and it was only found in the religious community. Marx once referred ironically to religion as 'the sigh of the oppressed creature, the heart of a heartless world'. It is doubly ironic that this should apply as well in Soviet culture.

However, the Communist world view pretended to have found the answers to the spiritual problems of the human condition. Lenin put the struggle for a militant materialism as the first task of the cultural revolution that he thought essential to the preservation of Soviet power. But it is not easy to stamp out belief and instil secular humanism by revolutionary means. And there were deficiencies in the Soviet theory and practice of secular humanism. *Bezbozhniki* (militant atheists) may have felt that they represented progress in their confrontation with medieval religious obscurantism, but the *klub* (the local social centre where one can play chess and read the all-Union press), despite its undeniable attractions, could never replace the mosque as the font of spiritual refreshment. Nor could the Stalinist intellectual Sahara offer much to him who thirsts for free thought.

It would nevertheless be wrong to deny the attractions and benefits to the local intelligentsia of the Moslem areas of association with the centre. The emancipation of women, the promotion of literacy, and the prospect of education at institutions elsewhere in the Soviet Union all made their rival claims against a possible anti-Russian Islamic revival. Yet for the masses of workers, peasants, and herdsmen, who were and are much poorer than the ordinary citizens of Russia, the Ukraine, and the Baltic republics, an Islamism which lends them sympathy against their perceived oppressors also has its appeal.

This unpleasant fact become more evident as the war in Afghanistan went from bad to worse during the Chernenko period. Soviet leaders seemed to be considering for the first time that the masses of central Asia might sense as much kinship with the insurgents as with their Russian brothers. In 1986–87, articles in the press were complaining about Khomeini propaganda from Iran radio, Islamic literature smuggled in at ports such as Batumi, and even the outbreak of a small-scale insurgency in Tadzhikistan.

This was not at all propitious for Aliev and the Aliev line. He had been sent to Syria in 1984 and 1986 in an attempt to use his influence with Assad to lessen Syrian support for Iran in the first Gulf war, and to effect a reconciliation between Syria and Iraq. Despite its desire to tread softly with Iran, the USSR was unswervingly committed to Iraq's survival. Aliev, however, was unable to work miracles. And the Soviets were increasingly uneasy about the Gulf War in the beginning of 1987, as the Iranians repeatedly launched offensives to capture Basra and cut Iraq off from the Gulf.

Aliev was also indirectly affected by the Alma Ata events. The post-mortems in the press gave the impression that this was not so much a demonstration against the apparatus as a provocation fomented by the apparatus itself in order to protect its local nests of Brezhnev-era corruption. Far from being a model of secular modernism and a beacon unto the rest of the Islamic world, Kunaev's republic was a tangle of unseemly liaisons between party bureaucrats and local clans and mafias. 'Those accustomed to working in the old way, who only paid lip-service to *perestroika*', wrote one correspondent, had found it impossible to resist the winds of reform and resorted to stirring up Kazakh nationalism against the will of the party.[5] They had chosen an Islamic holiday for the demonstrations, prepared leaflets in advance and laid on buses to transport the demonstrators from schools and workplaces. In their ranks were many trade unionists and veterans of the Afghan war.

The reaction to the demonstrations was severe. Solomentsev, for the party Control Commission, was dispatched to the spot to oversee corrective action. He and Kolbin saw to it that the ringleaders were arrested and punished. A teacher in Alma Ata was sentenced to five years' imprisonment for distributing provocative literature. A former Komsomol official at the S. M. Kirov State University got seven years for inciting to riot. Presiding over various meetings, Solomentsev and Kolbin listened to complaints about food shortages, inadequate housing, and poor medical services. They had clearly been sent with instructions to uproot any nationalist seedlings that they might find. They concluded that there had not been enough attention paid to the Russian language, and that it was symptomatic that students in the dormitories were assigned rooms 'according to nation'. Yet despite this, they removed the Kazakh party Second Secretary, a Russian, to replace him with an ethnic Kazakh.

In the following months Kolbin took the local Communists to task for various petty corruptions. One *raikom* First Secretary was accused of keeping his own livestock in a *sovkhoz* herd for his personal use; another of illegally allocating a 30sq m apartment for his ill wife and son.[6] In several cases corruption charges resulted in death sentences. The Minister in charge of Higher Education for the Republic was dismissed in February; other Communist officials were expelled for trading illegally in apartments. By the summer over one thousand police officers had been discharged. And these punishments were matched in similar campaigns in other republics. The Moldavian First Secretary, Semyon Grossu, was denounced and expelled, as were officials in the entourage of former Uzbek First Secretary Sharaf Rashidov, and former Kirghiz First Secretary Turdakun Usubaliev. The Moslem republics were revealed to the Soviet public as being, not bastions of progressive internationalism, but nests of Brezhnev-era corruption.

NEGATIVE PHENOMENA

The struggle against these 'negative phenomena' served the purpose of a general housecleaning of Brezhnev officials in the republics. The remedy offered by the purgers, however, was merely a redoubling of efforts to find uncorrupted officials committed to the established goals in relations between the republics and the centre. In the course of the discussions some began to question that strategy. A leading authority on nationality questions, Yu. Bromlei, Director of the Academy of Sciences' Institute of Ethnography, argued that the party had taken the wrong tack in seeking to reduce the influence of national customs in the Union republics. Internationalism was growing inexorably in these republics, said Bromlei, with as much as a quarter of the students at the institutions of higher learning being natives of other republics. Many republics had a better rate of graduation from their universities than did the Russian ones. So the republic intelligentsias were a support for positive relations. Bromlei thought that national loyalties had a healthy role to play. 'Popular traditions', he wrote, 'are factors for overcoming negative phenomena in the moral life of society.'[7]

As examples of these negative phenomena Bromlei cited alcohol and drug use, economic parasitism, and superstition. Soviet propaganda had traditionally considered national identification as 'Uzbek', or 'Turkmen', to be more progressive than enthusiasm for international Islam. Bromlei's views exuded confidence in the party's abilities to manage relations with the nationalities. At a meeting in April he charged that the party had been weak and one-sided on these questions in the past, and had not done enough to promote understanding of 'our multinational culture'.[8] Comrades should be learning the local languages. Other speakers, including Gorbachev intimate Aleksandr Yakovlev, denounced the dogmatism and stagnation of the Brezhnev and Stalin eras, so that Bromlei's views on the nationalities seemed to mesh with the political campaign against the opponents of reform in the party leadership in the Union republics. No one in Moscow appeared to be frightened by the apparent explosion of nationalist sentiment in the Alma-Ata events. Their local apparatus organizers must have thought that the outpouring would send a chill up the spines of Gorbachev and the other purgers. But they were wrong. Neither Gorbachev nor his closest supporters gave evidence of being much bothered by nationalism.

In April, in the midst of the party discussion on nationalism, an Armenian academic by the name of Souren Aivazyan sent a letter to Gorbachev on the baneful condition of the Armenians in Nagorno-Karabakh, an Armenian autonomous *oblast* in Azerbaijan. They had been victims, he said, of the 'Pan-Turkism' of Aliev (Azerbaijan First Secretary from 1969–82) and his successors. This letter marked the opening of a local effort that was to result by August in a petition from Armenia on ecological and nationalist themes bearing 400 000 signatures. Gorbachev could not have believed the bizarre claim that Aliev had actually been promoting Pan-Turkism, but the allegation made by someone else was useful. It was now possible to blame (or to permit others to blame) the Aliev line for many things, including even the quagmire in Afghanistan.

So Gorbachev chose to encourage the Armenians in their dispute. The effect on Aliev's political career was immediate. He disappeared from view in May, missing conferences on consumer goods and transportation in which his participation would have been expected, and absenting himself from the Politburo grouping that saw Gorbachev off on a trip to Romania. He was nevertheless bitingly criticized in his areas of competence by Ligachev. By fall Aliev would be out of the Politburo. Encouraging the Armenian claims in Karabakh formed part of a broader pattern of response to the challenge made by Gorbachev's opponents, one that would be more generally applied in 1988. The idea was to cheer on 'national fronts for *perestroika*' against local officials, usually the first secretaries who had put obstacles in the path of the leader at the January 1987 plenum. Thus was Gorbachev playing the sorcerer's apprentice in nationality matters that would put the Soviet Union itself into mortal danger.

DE-STALINIZATION

One would not have known this from Gorbachev's fiery speech at the January plenum. At that time, he had spoken of the necessity to 'save the rising generation from the demoralizing effect of nationalism', especially from 'manifestations of parochialism, tendencies toward ethnic isolation, sentiments of national arrogance, and outbreaks similar to those which took place recently in Alma Ata'.[9] But Gorbachev had also spoken of the need to enhance Leninist conceptions of internationalism and to improve relations with other nations in the Union, referring specifically to problems in the composition of cadres. The inference was that part of the cause of current difficulties was the insufficient sensitivity of the previous central leadership.

Gorbachev had left no doubt that he was breaking with the assumptions of the last two years. Pressed by proponents of continuity with the past for a respite in the personnel changes, he replied with what the French like to call a *fuite en avant*, an 'escape forward'. Current resistance tells us, he was saying, that the problems were more deeply rooted than we had thought, that instead of talking about the further perfection of an already functioning system, we need to recognize the vast *failures* of that system that have been covered up by previous leaders for subjective reasons. These leaders failed because their understanding was conditioned by 'the state and progress of theory and on the atmosphere in theoretical science', which remained for the most part at the level of the 1930s and 1940s, 'when society was grappling with entirely different tasks'. 'Because of well-known circumstances, creative ideas disappeared from theory' and authoritarian views took on the status of holy writ.[10] Lenin's ideas were interpreted simplistically and were in fact violated at every turn. Degeneracy in theory could not but affect the solution of practical problems. Thus an 'incorrect attitude to co-operative property' and 'personal subsidiary holdings' had caused great difficulties in agriculture. Theoretical failings had promoted practical failings and weakened the top leadership in the party, which now had to be revitalized by an influx of new faces. Not a pause, but a redoubling of the tempos of efforts to change Soviet society from top to bottom was now in order in the year of the seventieth anniversary of the October Revolution. This must be pursued with 'Bolshevik audacity'.

Gorbachev did not pronounce the name of Stalin when he spoke of the 'well-known circumstances' in which theory had run aground. It definitely appeared that he was opening an offensive against at least some aspects of Stalinism, but could one be sure? Only a year earlier, in an interview given to the French Communist paper *L'Humanité*, when he was asked about vestiges of Stalinism in the Soviet Union, he had replied: 'Stalinism is a concept made up by opponents of Communism and used on a large scale to smear the Soviet Union and socialism as a whole.'[11] He went on to explain that 30 years previously, at the time of Khrushchev's denunciation of the Stalin terror in a secret speech to the Twentieth Party Congress, the problem of the 'cult of the personality' had been successfully resolved. It only remained to follow up on that, to prevent a return to the abuses of the past by insisting on modesty in public officials and defending socialist legality. In February 1986 Gorbachev had expressed satisfaction that this was being done. Now, in the first months of 1987, he was less satisfied.

He was in fact calling for another wave of de-Stalinization, a reprise of the themes that Khrushchev had attempted to sound between the secret speech of 1956 and his ouster in 1964. In this endeavour he could count on prominent intellectuals such as Aleksandr Bovin, Fyodor Burlatsky, and Georgy Arbatov, who were veterans of the 'first wave' of reform under Khrushchev. They had been recruited by Otto Kuusinen to work in the Secretariat, eventually as part of a coterie of consultants most closely associated with Kuusinen's protégé, Yuri Andropov. They had been disappointed by the halt in the de-Stalinization called by the Brezhnev leadership, and thought, as Burlatsky was to relate, that they would not see another attempt at reform in their lifetimes. None were specialists in the Stalin period, but they had all embraced the strategy of returning to detente with the West in the spirit of Andropov's initiatives of 1983.

It had not seemed necessary until 1987 that a foreign policy of detente be underpinned by an internal campaign of de-Stalinization. Even the ghoulish nuclear accident at Chernobyl in April 1986 had not prompted such a full-scale challenge to the historical underpinnings of the regime. In the first three days of the melt-down, no information from Soviet sources had appeared. Only when Swedish, Polish, and German scientists began to tell their publics about it had the Kremlin been moved to release information. Ukrainian chief Shcherbitsky had permitted a massive May Day celebration in Kiev at the height of the radiation danger, a celebration which, he said later, was authorized by Gorbachev himself. The General Secretary did not address the nation about the accident until two weeks later. To be sure, the event shook confidence in the radiant future of 'acceleration'. There was more determination about *glasnost*, with Vitali Korotich appointed soon after to be editor of *Ogonyok*. Yet the challenge to the ideological foundations of the previous regime did not come until almost a year later.

Gorbachev had given no indication of sensing such a necessity in his first two years in power. But now he was up against what he called a 'conservative' coalition bent on slowing the housecleaning of cadres. As in the Chernenko interregnum, a slowing of the pace of change served to bolster the authority of the most prestigious and influential Politburo personality, Andrei Gromyko, now no longer in control of the mechanics of foreign policy, but still an amply respected figure and a kind of visual personification of leadership continuity from the Stalin period. With real division in the Politburo ranks, Gromyko's voice was all the more powerful.

A comparison of the two waves of de-Stalinization provides further perspective. In 1955, Khrushchev was shaking up the cadres from above, replacing thousands of officials at the lower levels and numerous first secretaries as well. The First Secretary in Kazakhstan, Ponomarenko, was replaced by Leonid Brezhnev. Khrushchev was trying to find his way to a new foreign policy offensive based on reconciliation with a former enemy, Tito, a foreign policy that would, he thought, open up contacts with the neutral bloc in the Third World symbolized by the gathering at Bandung. In his opening to Tito, however, he was thwarted by Politburo opposition from a group whose most experienced spokesman was V. M. Molotov. Molotov, along with Suslov, had in 1948 supported Stalin's policy of expelling Tito from the Cominform and, morover, insisted on continuity with the main outlines of policy from the Stalin years. The debate was phrased in terms of the question whether the Soviets should observe the right to 'different roads to socialism'. Molotov did not at first think so. For Khrushchev he was not only a prestigious rival for pre-eminence in view of his many contributions over the years, but an obstacle to further progress in domestic and, especially, foreign policy. Khrushchev's response was to undermine Molotov and his other Politburo opponents by attacking their Stalinism.

Gorbachev seemed to be on the same track. His reference to the looming anniversary of the Revolution left the impression that he was preparing another edition of the secret speech, perhaps to be delivered on that occasion. The sophisticated legally trained Gorbachev could not, of course, consider himself comparable to the bumpkin Khrushchev. Gorbachev was generally regarded as the most intellectually capable General Secretary since the 1920s. But he was not really the same type of intellectual as were the old Bolsheviks. Lenin, Trotsky, Bukharin, Kamenev, Rakovsky, and many others were capable of writing pamphlet-sized theoretical tracts against their opponents. Lenin had written a book-length study outlining the true Marxist approach to epistemology, to get the upper hand over heretical opponents, knowing that his fine philosophical distinctions could never comprise the substance of a party stand. As the German translator remarked, tactical differences among Social Democrats could not be made to depend on 'whether Marxism is epistemologically in agreement with Spinoza and Holbach or with Mach and Avenarius'. Trotsky and Bukharin often made their points in speeches and articles, but occasionally in books. Even Zinoviev wrote a detailed history of the Bolshevik party. The pervasive sense of theoretical inferiority that Stalin felt in the presence of what he considered to be café types, helps explain why he took such pains to get rid of all of them (save Lenin, who died in 1924), burying all the many volumes of their works in order to substitute his own more meagre production. No leader since Stalin had bothered to advertise this sort of intellectual pretension, but it was known to be central to the atmosphere of Soviet Communism before Stalin. Gorbachev was now going to take the party back to Lenin; and, in order to stake his claim to the political leadership of the party, he had to present credentials as a contributor to theory.

This was all a hoax of course. Gorbachev got his 'writings' from a team of assistants led by Chernyayev, Shakhnazarov, and Yakovlev.[12] This may be why he was willing to let the discussion on Stalin develop without his close guidance. In 1989 Aleksandr Bovin answered a query of mine about Gorbachev's historical ideas by saying that, in his opinion, the general secretary had no business making theory. He was thinking of a General Secretary like Khrushchev. But Gorbachev had greater ambitions. He may not have been able to spark a discussion with his own ideas, but he was not shy about demanding that it be done by others. In February, he convened a gathering of editors, publishers, academics and electronic media people, telling them that it was time that they ceased to work 'in the old way'. There were too many 'white spots' (*belye piatna*) in the party's historical texts. The intelligentsia must provide a true rendering of the country's history, to show that, despite this or that unattractive episode, the party always moved forward and that in the end it did not allow itself to be ground under by fascism. The unmentionable names must be mentioned. Gorbachev did not tell them what to discover, only to find a new Soviet history to replace the tale told by his predecessors. But history, fortunately or unfortunately, does not work that way. One might have reasonably expected that opening the archives and encouraging those who could provide oral testimony or manuscript material would result in a number of lively monographs and other studies pointing toward a new outlook on history. But the archives were not opened; in fact, they were never opened until the Soviet regime fell in 1991. At any rate, the history that Gorbachev wanted at this time could not wait for patient study of sources. Something usable in the political struggle was needed immediately. So Gorbachev did not get a new history, but a new political journalism on historical themes.

The debate was opened by Yegor Yakovlev, editor of *Moscow News*, the weekly that would soon become a flagship of *glasnost*, with an article on the 65th anniversary of Lenin's Testament. This was the series of short dictations Lenin had made at the end of 1922, fearing his own permanent incapacitation or death, in order to help the party evaluate the qualities of its top leaders. In these notes Lenin speaks of the danger of a split in the party threatened in the main by the tense relations between its two most outstanding leaders, Stalin and Trotsky. Against this, the party must act to strengthen its collective leadership by doubling the size of the Central Committee. The most important six leaders, in Lenin's estimation, could be ranked in three tiers: Stalin and Trotsky, Zinoviev and Kamenev, and among the younger ones, Bukharin and Pyatakov. Stalin, as General Secretary, had concentrated 'unlimited authority' in his hands, 'and I am not sure whether he will always be capable of using that authority with sufficient caution'. Trotsky, 'probably the most capable man in the present Central Committee', nevertheless 'shows excessive enthusiasm for the purely administrative side of the work.' Lenin usually used this way of describing a fondness for throwing one's weight around. Zinoviev and Kamenev were at a distinctly lower level of importance, despite Zinoviev's pretensions to being, in the absence of Lenin, the most important leader. Bukharin, among the younger generation, was a 'a most valuable and major theorist of the party . . . but his theoretical views can be classified only with great reserve as fully Marxist, for there is something scholastic about him (he never learned and, I think, fully understood dialectics)'. Pyatakov, a man of outstanding ability, like Trotsky 'shows too much enthusiasm for the purely administrative side of the work'.

As to the matter of Stalin *or* Trotsky, the notes were inconclusive. However, ten days later, Lenin made the following addition:

Stalin is too rude and this defect, although quite tolerable in our midst and in dealings among us Communists, becomes intolerable in a General Secretary. That is why I suggest that the comrades find a way of removing Stalin from that post and appointing somebody else differing in all other respects from Comrade Stalin solely in the degree of being more polite and more considerate to the comrades, less capricious, etc. This circumstance may appear to be a negligible detail. But I think that from the standpoint of what I wrote above about mutual relations between Stalin and Trotsky, it is not a detail, or it is a detail which can assume decisive significance.[13]

From the standpoint of mutual relations between Stalin and Trotsky, Lenin had decided to oppose Stalin. The 'Testament' was not known to the Soviet public until 1956, and known in the West before that date only because of its publication in Max Eastman's *Since Lenin Died* in 1925 and Trotsky's *The Real Situation in Russia* in 1927.

Yakovlev, in raising the question of Lenin's appraisals of the leaders of his day, spoke of the similarity of that time and the present, when the Soviet system stood at a crossroads and pondered fateful choices. His own choice was clear from the analysis of the Testament. Trotsky, said Yakovlev, 'was never a Bolshevik'. In the Testament, Lenin had mentioned the fact that Trotsky had only joined the Bolshevik party in 1917, but he enjoined his readers not to hold that against Trotsky any more than the fact that Zinoviev and Kamenev had voted against the insurrection that brought the Bolsheviks to power in October. Yakovlev, however, found it to be significant, along with Zinoviev's and Kamenev's betrayal, which should not be forgotten. On Bukharin a certain selectivity was in evidence. Yakovlev quoted Lenin's remark about Bukharin's being 'scholastic', but he omitted its continuation to the effect that Bukharin had never studied dialectics. He quoted from the addendum on Stalin, citing the judgement that Stalin was too rude, and the warning that this was no detail. Yakovlev noted Lenin's harsh criticism of Stalin. He wrote admiringly of Lenin's being stern, implacable, relentless, adding mysteriously that 'others, in certain circumstances, can show sympathy and kindness, a momentary overcoming of their cruel nature'.[14]

The view that Trotsky was never a Bolshevik derives not from Lenin, but from Stalin's allies, Bukharin among them, in the post-Lenin succession struggle. Trotsky himself was fond of quoting Lenin's statement to the Petrograd Committee of the Bolshevik party on 1 November 1917:

> As for conciliation [with the Mensheviks and Social Revolutionaries] I cannot even speak about that seriously. Trotsky long ago said that unification was impossible. Trotsky understood this and from that time on there has been no better Bolshevik.[15]

On Yakovlev's reading Lenin feared Stalin and saw Bukharin as the alternative. The distortion was blatant. This was not a pure academic exercise designed to straighten out historical questions, but a tract in the service of *perestroika*, which had no need of Trotsky, but sought instead to invoke the name of Bukharin in its cause. Not only was the historical problem of Trotsky not of interest, it was an obstacle, except in that it might be useful in attacking Stalin. Indeed, Bukharin had called Stalin a Trotskyite in 1928.

Western observers, myself included, were astonished to see a discussion on history of such apparent frankness in the Soviet press. We were not moved to quibble about details, but preferred to see each new departure in the next two years as a stage in the direction of genuine investigation of the Stalinist past. Too many of us tended innocently to accept the oft-stated view that the Soviets were peering into their past in order to gain perspectives on how to move forward. This went along naturally with the proposition that Soviet foreign policy was an extension of Soviet domestic policy. But why should Western observers have been so impressed with the idea of a return to the de-Stalinization motifs of the Khrushchev era? At the time of the most ambitious wave of de-Stalinization the regime had yet known, at the 22nd Party Congress of 1961, Khrushchev was putting the world through the most frightful days of the nuclear era, with testing of a 60-megaton bomb, and the Berlin and Cuban Missile Crises.

At his worst, Stalin had never subjected the world to anything like that.

So a rerun of the Khrushchev era should not have seemed like such an inspiring prospect. On the other hand, Soviet intellectuals desperately wanted another thaw and another run at de-Stalinization. In March, this was stated in clearest terms by Georgy Smirnov, newly appointed director of the Institute of Marxism-Leninism, who was thought to be a highly influential advisor to Gorbachev on ideological matters. Smirnov had contributed for a quarter of a century to what might be called Stalinist devotional literature, in volumes with titles like *Shaping Communist Social Conditions* and *Soviet Man: The Formation of a Socialist Personality Type*. He deepened the characterization, only a few months old, of the Brezhnev period as one of stagnation. After paying tribute to the achievements of the centralized command economy, the industrialization of the country, the collectivization of agriculture, the victory over Hitler, he argued that the achievements of the past were under threat because of a 'braking mechanism' (*mekhanizm tormozheniia*) that Gorbachev was now struggling against. The beginnings of this phenomenon were to be sought in the decision of the October 1964 plenum that removed Khrushchev. The same forces were now reluctant to follow Gorbachev's campaign, still retaining persistent 'prejudices about the role of commodity-money (*tovarno-denezhnykh*) relations', and the problem of wage-levelling (*uravnilovka*) that reduced the incentives for better work, especially for engineers, doctors, and teachers.[16] Smirnov emphasized that he was not advocating a turn away from public ownership but a deepening of popular participation in the economy and society, citing as a model for emulation the enthusiasm that had greeted the collectization of agriculture in 1928.

Smirnov was prepared for a reprise of the last Thaw. In 1956, Khrushchev had taken pains not to question the collectivization of agriculture without which, he said many times, the Soviet Union would never have been able to develop its heavy industry, and would surely have been defeated by Hitler. Nevertheless Khrushchev persistently toyed with the idea of rehabilitating Bukharin, the great opponent of Stalin's collectivization. He was never able to resolve this contradiction, and never rehabilitated Bukharin. On this point, Smirnov was hewing to the tradition of the Twentieth Congress. Others were not so cautious. Academician V. A. Tikhonov, in an interview given in April to *Literaturnaya gazeta*, described the collectivization as a vast tragedy without a redeeming purpose, one that had destroyed the rational trade relations between city and country that had been produced by a combination of the peasant war of 1917 and the New Economic Policy. The destruction of this system was the work of 'people with utopian views' who were blinded by a simple-minded conception that 'commodity production and socialism were antipodes'.[17] Claiming that they were following Lenin's deathbed advice to strengthen cooperatives on the countryside, they [Stalin and his group] pretended that there was only one form of cooperative, the collective farm.

Inklings of nostalgia for the NEP were already noticeable from the early moments of Gorbachev's election, as we have seen. In 1987–88 they became a kind of orthodoxy with implications for Soviet collective farms, largely because of the example set by the de-collectivization of agriculture in China which, Soviet writers thought, demonstrated that market reforms held the key to all the dilemmas of food supply. But there was unfinished business in the argument: if the turn away from NEP had not been economically necessary, in the sense of preparing the country for the great trial of the world war, then it must have been done for no other reason than to provide an inflamed atmosphere in order to establish Stalin's Politburo case against Bukharin and the Right. The turn was defended at the time by citing the war danger posed by the British breaking relations with Soviet Russia in the Spring of 1927. Fear of a Kuomintang attack on the Far East in tandem with a Polish attack on the Ukraine caused the leadership to raise a frightful war scare, which it used as best it could against the Zinoviev–Trotsky opposition, accusing them of giving comfort to potential enemies by the fact of their opposition. Stalin was, moreover, faced with a deficit in grain deliveries (a 'grain strike') perhaps related to peasants' reaction to the war scare. All this could provide a rationale apart from Stalin's lust for power.

A debate of this kind never developed. Instead most writers simply took the Bukharinist position, before which the top leadership wavered. Despite all that was hoped for *glasnost*, it served mainly as a bludgeon against Gorbachev's opponents, while Gorbachev himself stood aside, and even defended Ligachev or others from the barbs of *Ogonyok* or *Moscow News*. In August, *Pravda* published an editorial citing Lenin to the effect that the *kulaks* (independent peasant proprietors) were the 'rabid enemies of Soviet power'. Their grain strike, said the editors, together with the threat of military intervention, had posed a mortal danger. Stalin's action saved the day and 'elevated the USSR to the ranks of the great powers'.[18] The *Pravda* board was trying, as it turned out, successfully, to hem in Gorbachev, at that time preparing his 70th anniversary historical speech, which would *endorse* the collectivization. In the same issue, V. P. Danilov, perhaps the leading Soviet academic authority on the subject, contributed an article, 'Sources and Lessons of Collectivization', arguing the reality of a *kulak* danger and maintaining that the government's hand had been forced by the grain strike. There had been excesses in a process that would have worked to the optimum only if it had been voluntary, but in the end, there had been no choice.[19] The argument against the resort to agricultural collectivization had a larger impact for having suggested that the whole campaign, and by implication any attempt to nterfere with the free market, had been a utopia. If this could be said about the turn of 1928–29, it was only a short step to saying it about the whole adventure of Soviet power. By 1989 this would constitute the core of the argument of a leading party theorist, Aleksandr Tsipko.

Another pointer was offered early in 1987 by the much-discussed film, *Repentance*. A murky, vague study of a Georgian tyrant, who may have been Beria, or perhaps almost any provincial boss, or perhaps Stalin himself (the director was not saying), it offered nothing to explain the reason for the emergence of the dictatorship, concentrating on its mendacity and its depredations against religion and the intelligentsia. Many striking scenes depicted the sufferings of the population at the height of the purges of the thirties. One of them showed a lumberyard full of logs sent from a prison camp, with women wandering through forlornly searching for carved messages from their inmate husbands. The film had a powerful, not to say mesmerizing, effect on reviewers and audiences alike. The message many came away with was that the crimes of the tyrant were the result of his utopia, which, in the name of an abstract idea, ground up the happiness of millions of real individuals. This theme was to overwhelm the consciousness of the Soviet intelligentsia during the next three years. The crimes of the Stalin era were so vast that no historical or political explication, or at least none that was offered, could reduce them. The only humane response was repentance. Eventually Soviet public opinion, far from embracing a new sense of its history and achievements, would sink into that paralysing mood.

NEW THINKING

However, in early 1987, Gorbachev was still moving forward with explosive optimism, especially on the cadre front. This extended to foreign policy with a vengeance. Since the replacement of Gromyko as Foreign Minister by Shevardnadze, perhaps half of the ambassadors had been recalled. Many senior officials in Moscow had also been changed. Gorbachev's January speech indicated that the renewal of cadres in foreign policy work would continue. There were signs of dissatisfaction with Gorbachev's performance at Reykjavik, but the General Secretary insisted that there had been no setback, but simply another moment on the journey to disarmament; at any rate, he said, there was no alternative to pressing on – 'This is something outside politics'. Instead of continuing to seek US acceptance of the linkage between an agreement on missiles in Europe and the American SDI programme, as Gromyko in his place would no doubt have done, Gorbachev made another sharp break with the methods of the past in February when he accepted essentially the

Reykjavik formula, elimination of all the INF missiles in Europe, and thereby called the Reagan administration on its offer of a 'zero option'. The initiative was eventually taken up by Reagan and the way was opened to the epoch-making agreement on INF that would be signed in December.

Gorbachev could argue that despite his retreat on the linkage to SDI and the accompanying concessions (destroying more warheads than the Americans; agreement to intrusive verification; exclusion of the British and French forces that had so concerned Andropov) he had only traded Soviet missiles capable of striking America's European allies in return for removing American ones posing a threat to the USSR itself. But he could not be certain for some months that Reagan would follow up his initiative so that the treaty could actually materialize. He got support from Andrei Sakharov, whom Gorbachev had released from his internal exile at Gorky. Sakharov argued essentially the same case as the civilian defence specialists who supported Gorbachev's initiative: there was no need to link Euromissiles with SDI because the latter was quite useless, not just against a surprise missile attack, but against a weakened retaliatory strike as well. Moreover, Sakharov noted prophetically that 'if disarmament begins, the SDI programme in the United States will lose its popularity'.[20]

The civilian specialists were brought to centre stage by changes in the Warsaw Pact military doctrine that were announced in May, changes specifying that in the future the Pact's security calculations would be based on the idea of 'reasonable sufficiency'. Like the 'zero option' for the INF negotiations, this was an idea generated among writers on the left in the West, who sympathized with the campaign against the deployment of Euromissiles. Along with related notions – 'defensive defence', 'defensive sufficiency', it was to be a central theme for explication of the thinking of the new strategists even after the INF treaty in December, when the discussion would shift to what many western analysts considered to be the 'heart of the matter', conventional forces.[21]

The civilians made use of the ideas and phrases used in the discussion of strategic questions in the West. This often meant taking advantage of the work of Western peace movement writers and thinkers, largely advanced as a response to NATO's forward defence concepts.[22] In numerous international meetings, something of a common language was developed to address security problems outside the context of Western security needs. General Jaruzelski, addressing a meeting of a Polish patriotic association in Early May, called for 'the evolution of the character of military doctrine so as to be accepted on all sides as strictly defensive'.[23] In the days of detente, the idea of 'strategic sufficiency' had been used by Nixon and Kissinger in a different context: to gauge minimum needs for nuclear forces in conditions calling for something short of all-out attack on the

Soviets.[24] When Nixon referred to a doctrine of strategic sufficiency in his first press conference, the inference was that the United States had renounced the search for strategic superiority and indicated its readiness to enter the projected SALT negotiations. Now, 'reasonable sufficiency' was employed to designate a response short of matching whatever the West presented as a challenge. The Soviet writers liked to use the term 'asymmetrical response', as in the case of a response to the SDI which would not match the programme but deploy 'countermeasures that doom that programme to strategic and economic failure'.[25]

The civilians were involved in the making of ideology, but it was ideology fashioned from American and other Western strategic ideas and phrases. They strove to defeat the devilish American 'exhaustion strategy', perceived to be the real reason for the SDI, by making use of the inner essence of its devilishness. If the Americans sought to build weapons systems that would force the Soviets to match them at even greater expense, the Soviets would answer by responding 'asymetrically' and matching the system (presumably by overwhelming it with more warheads) at lesser expense. Since the question was the budgetary challenge of military threats, an asymetrical response could be applied in the context of other military threats. Lev Semeiko, of the Institute on the USA and Canada, insisted that 'reasonable sufficiency does not replace parity, but on the contrary, presupposes the existence of strategic parity, that decisive factor for the prevention of war'.[26] On the other hand, as Evgenii Primakov argued, 'in this period sufficiency is realized at lower levels of parity'.[27] Nevertheless, said Primakov, Gorbachev's proposal of January 1986 had broken new ground. In the past, the Soviets had agreed to 'rules of the game' (*pravilami igry*), which meant that when the US made advances, the Soviets responded symmetrically. The US 'wanted to wear us out economically' with destabilizing systems such as the MX, Trident 2, and cruise missiles, and with aid to the contras and *dushmany* (Afghan fighters). Now, he argued, the Soviets respond, not militarily but politically, by changing Western public opinion about the Soviet threat.

Yet this business of mental appropriations is not so easy to control and predict, as we have seen with Dobrynin's ideas. The Soviet civilian writers managed to change Western perceptions greatly, but were also themselves changed by Western opinion. In many cases, ideas originating from the European left were introduced by the Italian Communists. The 'zero option', which originated in German Social Democratic circles, was accepted by the PCI in essentially the same form as put by Reagan. The Italians advocated intrusive verification and 'mutual security'. They argued that NATO was basically 'a defensive and geographically limited alliance'. They endorsed the Harmel Report of 1967 with its call for the twin objectives of defence and detente. They expressed the hope that the

Eureka Project for scientific and technical research, widely ballyhooed as the European SDI, could take on a real all-European dimension.[28] The PCI leaders thought that Gorbachev's New Thinking had had a precursor in Enrico Berlinguer's theory of New Internationalism, put forward in 1982.[29] This idea had been condemned by the Soviet ideologue Suslov, but Soviet writers would, by 1989, be apologizing to the Italians and admitting the filiation of Berlinguer's and Gorbachev's ideas, expressed in the formula that 'the New Internationalism has prepared the New Thinking'.

The introduction of new phrases ('concepts') into public discussion of national security matters opened a debate on their implications (to 'infuse content' into them). One could with good logic argue that the USSR had provided reasonable sufficiency since Lenin's day. But any new phrase permitted the civilian analysts to gain some ground in their tug-of-war with the military for authority in strategic matters. Andrei Kokoshin, of the Institute of United States and Canada, and Valentin Larionov, of the General Staff Academy, tried to provide an argument in support of the Jaruzelski Plan and the Warsaw Pact May decision by citing an analogy with the battle of Kursk in 1943, an engagement in which large amounts of German armour were destroyed as a result of being lured into positions where they were blocked by obstacles and subjected to fire from anti-tank weapons. For Kokoshin and Larionov, Kursk showed the importance of a deep-echeloned defence with well-prepared positions, and the capacity to counterattack the enemy forces after yielding ground to an initial breakthrough.[30]

They were not able to argue for the superiority of the contemporary anti-tank weapons. Nor were they able to join ranks with those who, like Zbigniew Brzezinski, advocated tank-free zones in Europe. This beautifully simple solution removed the essential implements without which offensive operations were inconceivable. In their 'Kursk' scheme, tanks were necessary to contain the breakthroughs and undertake counteroffensives. And the military could agree on the necessity of counteroffensives, which require virtually the same weapons and forces as offensive operations. (It was outside the rules of the debate to remind the authors of the crucial role in the battle of Kursk played by intelligence. The Soviets had enjoyed detailed information as to the time of the Nazi attack and the composition of forces.) So the military people could endorse the new concepts by stressing the importance of 'sufficiency for defence', which ironically could be interpreted in almost the same spirit as the pre-Gorbachev formula 'everything necessary for the defence of the Soviet Union'.[31]

Nevertheless, the discussion permitted a degree of pressure on the military by the civilians that would not have been possible otherwise.[32] As the debate was beginning to heat up, the military were further embarrassed by the unauthorized flight of a West German amateur pilot, Mathias Rust, from Helsinki into Red Square. As if this were not humiliating enough, Mr Rust's flight was made on 28 May, celebrated in the USSR as Border Guards' Day. Gorbachev quickly seized the opening to remove a number of military men. Marshal Koldunov was relieved of his duties as commander of the Air Defence forces and as Deputy Minister of Defence. A few days later Marshal Sokolov retired as Minister of Defence and candidate member of the Politburo. For his replacement, Gorbachev reached far down into the promotion list for a relatively minor figure, Dmitry Yazov, a Deputy Defence Minister in charge of personnel. Despite his seeming lack of stature, Yazov would nevertheless remain a critical voice in defence of the military in coming years, and a key conspirator in August 1991.

A TURN

The most vociferous critic of the military at the time of Rust's flight was Moscow party chief Boris Yeltsin, who in June followed up the sackings of the responsible military men by wholesale attacks on the Moscow Air Defence district. The commanders, he charged, were intolerably arrogant, and in the ranks there was excessive hazing of the recruits by the more experienced men. The army was lagging behind the reform spirit in the party and commanders acted as if the decisions of the party congresses and plenums did not apply to them. In the commissar apparatus, the political officers showed 'zero *perestroika*'. Yeltsin at this time was thought to be close to Gorbachev, so the criticisms had a special bite, in view of the fact that Gorbachev had told a Komsomol congress in April that the 'braking mechanism' had strong support in the Central Committee, the Government, in the Union republics, even in the Komsomol itself. Yeltsin was a point man in the campaign against these forces.

At the end of June, Gorbachev seemed to have won a sweeping victory over his opponents when a Central Committee plenum promoted his close confidant Aleksandr Yakovlev from candidate to full membership in the Politburo, along with Nikolai Slyunkov and Viktor Nikonov. General Yazov was made a candidate member. Gorbachev seemed to have broken the logjam of opposition and reduced Ligachev to impotence. But after the plenum, the intensity of the campaign eased noticeably. Yeltsin was apparently reduced to passivity. Ligachev told the plenum of October 1987 that was to oust Yeltsin from the Politburo that the Moscow party chief was the 'only alternate member of the Politburo who takes no part in the proceedings'. Yeltsin himself testified that the upsurge of enthusiasm for *perestroika* that had followed the January plenum had dissipated after June, when 'people began to lose faith'.[33]

Rather than a ringing defeat of the forces arrayed behind Ligachev, as was then supposed in the West, the plenum's personnel changes were more likely part of an arrangement between Gorbachev and Ligachev, bringing in Yakovlev as a Gorbachev supporter alongside two others who were not so clearly committed to the campaign for sweeping personnel changes, and including the promotion of Yazov. After the plenum, Ligachev's prominence rose. By the end of summer he was chairing meetings of the Secretariat. Yeltsin's subsequent complaints about the nefarious role of the Secretariat reinforce the picture. Yakovlev was, as another specialist in ideological matters, a potential rival to Ligachev, but he was advertised as being mainly responsible for international matters. Gorbachev and Ligachev were coming more and more to give the appearance of representing opposed principles, but the relationship between them was, and continued to be in 1988–90, more nuanced than was thought in the West. Gorbachev's turns would be incomprehensible if one assumed nothing but enmity between him and Ligachev.

There might be a better explanation for the evident turn by Gorbachev. He was confronted with an anti-*perestroika* rejection front in the Soviet bloc. On his trip through east central Europe in the spring he had found a mixed reaction to his reform ideas. In Czechoslovakia there was general resistance to reform and criticism of measures taken up in the Soviet Union. Vasil Bilak, a central committee secretary complained that 'false reforms will weaken socialism, the unity of the party, and ties with the USSR'. Gorbachev's visit with Nicolae Ceauşescu in Romania was inordinately tense, with Ceauşescu leaving no doubt that he considered the Soviet Union to be going through a process that did not promise much. Gorbachev, he suggested, had a long way to go if he wanted to bring the Soviet regime to the polished condition of the Romanian system. Things were not much better in East Germany, or in Hungary, where President Karoly Nemeth and Prime Minister Karoly Grosz were not enthusiastic (in 1984–85 Grosz had been thought to be a supporter of Romanov against Gorbachev). There was not much real encouragement from the normally pliant Bulgarians. Aside from Poland, where General Jaruzelski was unreservedly enthusiastic about the 'Soviet springtime', the rest of the bloc was arrayed against the Soviet reforms in a solid front.

The other bloc leaders were trying, unsuccessfully, to gauge Gorbachev's reform intentions from the discussion on Stalin going on in the *glasnos* press. But no clear picture was emerging. Not many advances for theory Gorbachev was planning his speech on Soviet history to mark the 70th anniversary of the Revolution in November, along with a book on *perestroika*. He may have thought of the example of Lenin holed up in his Finnish redoubt during the summer of 1917, studying texts of Engels and preparing *The State and Revolution*. Lenin was trying to show how the Marxism of the Socialist International was inadequate to the task of leading a revolution in Russia. Gorbachev was trying something similar for his revolution. He was not, however, getting much help from the historical discussion going on in the Soviet press.

STALIN DEFENDED

In fact, a rather considerable confusion prevailed about the Stalin legacy. In January the newly appointed rector of the Moscow State Institute of Historian-Archivists, Yuri Afanasyev, had published excerpts from his inaugural address, 'The Energy of Historical Knowledge', in *Moscow News*. They contained a ringing denunciation of the state of affairs in Soviet historical writing, where, Afanasyev charged, stagnation had prevailed since the 1960s when promising new trends had been suppressed by social science ideologues. In March Afanasyev had given an interview to *Sovetskaia kultura*, suggesting that historical investigation should concentrate on two periods, 1917–29 ('under Lenin and after Lenin') and 1956–65, the period of the Twentieth Congress and Khrushchev's reforms. The idea seemed to be to find alternate paths to the one chosen by Stalin and to determine why the critique of Stalin had run aground. Revival of the critique might point the way to a different future for society. The goal of the inquiry was that of a cleansed and reshaped Communism. Afanasyev granted that many other topics were worth pursuing by historians, but he rejected attempts to appreciate, for example, the problems of the tsarist autocracy as treason to the 'entire social democratic tradition of the Russian intelligentsia and in more general terms the whole progressive democratic tradition'.[34] He granted the wisdom of the observation of Plekhanov and Martov: 'They believed that Russia was not ready for socialism, and that capitalism in Russia had yet to reach maturity. And Lenin knew that this was not a stupid argument.' These and other opponents of Lenin were worth publishing, along with the works of 'major figures who were both brilliant and useful to the revolution'.

On the appearance of the January interview, *Moscow News* got an immediate phone call from Fyodor M. Vaganov, head of the Main Archive Administration, telling the editors that its publication had been a mistake. Professor Vaganov was invited to give his own views, but nothing appeared until May when a letter from four historians, P. Soboleva, A. Nosov, L. Shirikov, and S. Murashov, was delivered to the editorial offices, apparently with the imprimatur of Vaganov.[35] The authors presented a passionate defence of the achievements of the historians of the Brezhnev years. They centred their criticism of Afanasyev on the question of the historical role of Trotsky. They said that Trotsky had conspired to ruin the Revolution in October 1917.[36] Zinoviev and Kamenev had warned against the insurrection. 'Their anti-Leninist stand during the October Revolution', said the authors, 'was leading straight to the formation of a Trotsky–Zinoviev bloc aimed at frustrating Lenin's [that is, Stalin's] plan of socialist construction.'[37] Reference was made to the debate on the Peace of Brest-Litovsk in 1918, the discussion on the trade unions in 1920, and the dispute over inner-party democracy in 1923–24. Underlying all these discussions, the authors argued, was the question of Socialism in One Country, the possibility of which was denied by Trotsky and the Trotskyites. But the story told by Vaganov and his co-signers was a remarkably faithful rendering of the version given in the Stalinist *Short Course* of 1938. In fact, there was no theory of Socialism in One Country until 1924 and no objection to it until 1925; moreover, the objection was raised by Zinoviev rather than Trotsky.[38] For emphasis the authors cited the English historian E. H. Carr to the effect that the line followed in the 1920s by Stalin, the 'Leninist' line, was the only possible one. This gave an indication where the ultimate authority might lie in case of any of their assertions being disputed. Their position was militant and 'zero-*glasnost*'.

Afanasyev did not reply with a citation of historical facts or a clarification of the issues raised by the four historians. He charged them with indifference and even hostility to Gorbachev's call for openness and the elimination of 'white spots' from Soviet history. He correctly argued that their views were essentially those of the *Short Course*. About the issue of Trotsky, the centrepiece of the letter against him, Afanasiev could only plead agnosticism: 'there is not a single word on either Trotsky or Trotskyism in my statement'.[39] He could not straighten out the historical issues. His speciality was not Soviet but French history. Even an historian of Soviet Russia, to speak with confidence about the matters raised by the anti-*glasnost* historians, would have to be able to cite documents, to which the Soviet historians had no real access. The issues are, moreover, rather complex and not easily sorted out for polemical purposes, if, that is, the polemics are to be reasonably honest. Exposure to the ample works of Trotsky, Bukharin, Zinoviev and the others, available in many editions and languages in the West, would have created an experience among Soviet historians comparable to coming out of Socrates's cave. But they do not settle all the questions, not even the most important ones. Moreover, clearing up and de-mystifying the issues bearing on the figure of Trotsky involves emancipation from a pall on the mind associated with his infamous name. One could see this in conversations with Soviets who assumed that any fair-mindedness or rudimentary historical sympathy demonstrated toward him, such for example as might be afforded a Gerard Winstanley or a Danton, made one a Trotskyist. One must recall the pervasive fear in George Orwell's *1984* of the human devil, Immanuel Goldstein, a figure meant to refer to Trotsky. The whole question also confronted Russians with a primal anti-semitism in their political culture, with a personification of the demon Jew, as in an old White poster, a horned imp atop a pyramid of skulls.

Small wonder then that not many wanted to pursue the matter. In fact, the next two years were to see, if anything, an increase in fearful warnings about the devil Trotsky, in order to accommodate the rise of enthusiasm for Bukharin. Stalinist diehards thought it a mistake for the *glasnost* writers to have opened the historical question of the rise of Stalin and the alternatives to Stalin in the first place. These were arcane and explosive matters. However, once the Stalin legacy was revealed to the public for the first time, at the same time that the public was at last permitted to express an opinion about Communism, the clock was ticking. The party had in effect only a last chance to make sense of the mountain of atrocities that would soon fill the pages of the magazines under *glasnost*. And in the end the party would not find an answer to the question: To what end have these monstrous sacrifices been offered?

Yet Gorbachev did not sense the danger as he entered on the task of justifying *perestroika* in literary form. He had to worry about the struggle in the Politburo, and he no doubt judged that there had been enough accomplished on the second phase of his rise to the top, the phase begun in 1987. The June plenum had approved his call for an extraordinary party conference to be held in June 1988, with delegates to be elected by secret ballot in 11 of 15 republics (excepted were Ukraine, Belorussia, Uzbekistan, and Kazakhstan). This was an impressive gain for *perestroika*. Perhaps the dosage administered was about right for the time being. The campaign could be renewed in the future.

6 The Thought of Mikhail Gorbachev: A Treatise and a Speech

The man who deals only with cold calculations becomes a sterile egotist. It is on poetry that the security of the throne depends.

Gerhard von Scharnhorst

In short, the party's leading nucleus, headed by Joseph Stalin, safeguarded Leninism in an ideological struggle.

Mikhail S. Gorbachev, 1987

Gorbachev used the months of August and September to finish supervising a book on *perestroika* and to prepare his speech on the 70th anniversary of the Revolution. This would permit him to put a certain theoretical foundation under his sweeping campaign of personnel changes with a reinterpretation of Soviet history in general and the Stalin period in particular. He had set this task for the policy of *glasnost* at the January 1987 plenum when he argued that the problems of the present 'go back to when creative ideas disappeared from theory', and would have to be made good by a searching investigation of history. Vitali Korotich, editor of *Ogonyok*, underlined the point in February when he said that 'until we resolve the question of Stalin we'll never move forward'. Yet Gorbachev appeared to be limited in this quest for a bold new analysis of the party's past by two facts: first, the discussion of spring and summer 1987 among the proponents of *glasnost* had not provided many insights, and had in fact even withered in the heat of counterattack by the Stalinist academic ideologues; second, the defence of orthodoxy tended to encourage his opponents who had brought about a tense standoff in the factional struggle in the Politburo. Ligachev's later account directly links his own Politburo opposition against Gorbachev to 'the problem of the slander of history', and names historian Yuri Afanasyev as the principal culprit.[1]

However, there is actually no point in assuming, as many Western observers did, that Gorbachev was really fighting the defenders of Stalinism tooth and nail and not simply trying to place himself in a position to benefit from the attacks of the *glasnost* writers against them. In fact during this period he continually and often indignantly cautioned both sides against excess, even while he gained from their antagonism. Gorbachev was an astute political man in the context of the Stalinist system even if he was not a student of Marxist theory in the sense of a Lenin or a Bukharin. Where Lenin might write a document and argue his opponents into the ground in party meetings by defending its fine points, Gorbachev was more likely to water down a document in pre-publication manoeuvres. As much as he wanted to be known as having a flare for theory, his views on the problems of party history could not escape relation to his rather straitened political circumstances. Perhaps this helps to understand why neither the speech nor the book made a real break with the intellectual traditions of Stalinism.

HOMAGE TO HEROES

The speech was first given on 2 November, under the title *October and Perestroika: The Revolution Continues*, to a joint session of the Central Committee and the Supreme Soviets of the Union and the Russian Republic. In its first few paragraphs, the words 'heroic' and 'heroism' occur often enough to convey the purpose of Gorbachev's quest. 'Seventy years is not such a long time in the history of world civilization, but in its scale of achievements, history did not know such a period as unfolded in our country after the victory of the October Revolution.'[2] Today, Gorbachev continued, we return to the days that shook the world to find a 'spiritual support and instructive lessons', and we conclude that 'the socialist choice of the October Revolution has been correct'.

Gorbachev praised Lenin's 'art of political guidance' as demonstrated in the July days of 1917, when the Mensheviks and Socialist Revolutionaries used the forces of the Petrograd Soviet against the Bolsheviks. Lenin had first to consider with great anguish the abandonment of the slogan 'All Power to the Soviets', and then to retrace his steps when the Soviets elected Bolshevik majorities ('acquired a truly popular essence'). Gorbachev passed up a chance to pronounce on the change of slogans in April 1917, the 'rearming of the party' as Trotsky called it. Up to then the party had marched under the slogan of 'The Democratic Dictatorship of the Proletariat and the Peasantry' which envisioned no inroads on private property, and of course no socialism in Russia. Lenin's adoption of the slogan of power to the Soviets caused his Menshevik opponents to say he had been reconciled with Trotsky, as Trotsky's theory of Permanent Revolution had long held that a democratic revolution would be driven by strikes to the point of a decision for a socialist regime. Indeed Trotsky joined the Bolshevik party several months after Lenin dropped the Democratic Dictatorship slogan. This 'rearming' was the subject of one of the first disputes of the struggle to succeed Lenin, when in 1923 Zinoviev suggested that the Revolution had been made according to the slogan of the 'Democratic Dictatorship' and that to say otherwise was to fall into Trotskyist Menshevism. In many ways, the campaign against Trotsky was the central motif for the rise of Stalin, so here was an issue on which Gorbachev could create some new history. Yet he chose to pass the matter by.

The period of the civil war was treated by Gorbachev in traditional fashion as one of great heroism. There was no suggestion of the idea of War Communism as the regime's first utopia, such as would appear among the champions of economic reform in the following years. Gorbachev called the transition to the New Economic Policy in 1921 an outstanding example of Lenin's dialectical analysis. The tax in kind (*prodnalog*) of that period had continuing interest for today's problems, said Gorbachev, and released the creative energy of the masses, enhancing the initiative of the individual. Lenin's views on the tax in kind and his recommendations on encouraging the cooperatives served as an argument for the introduction of more market mechanisms in the present Soviet economy. This could be described as enhancing socialism, as in Lenin's famous slogan 'through NEPist Russia to socialist Russia'.

NEP had been introduced into Soviet Russia at a time when some manner of retreat before the peasant had seemed inevitable. When the black market in Petrograd had been suppressed in 1920, food riots had resulted, and after them a city-wide general strike. When strikers' representatives went to the Kronstadt fortress to seek support, the whole garrison rose up in revolt against the Soviet regime. The terrible Kronstadt revolt, which the Bolsheviks only succeeded in suppressing with great bloodshed, was the stimulus for the NEP which eventually normalized food deliveries to the city. Permitting the peasant to trade surpluses on a free market and individuals to work in trade had established the relative stability of the Soviet economy of the twenties. In one of his last writings, Lenin envisioned taking the idea even further through the institution of cooperatives, of which he imagined millions over the whole land. 'Given the social ownership of the means of production', Lenin had written, 'given the class victory of the proletariat over the bourgeoisie, the system of civilized cooperators is the system of socialism.'[3]

NEP had begun with the idea of a retreat, but at the time of Lenin's death no leading Bolshevik raised opposition to it, including those who would later be critics on the left.[4] To suggest its continuing relevance in a Soviet society that was industrialized and urbanized, as Gorbachev was doing, was a break with established thinking. Stalin had railed against those who considered the law of value, that is, the market, as a regulator of the command economy. Gorbachev was now reversing this and advocating a market that filled the role of provider of light industry and consumer goods to an economy hitherto geared primarily to heavy industry production. This was to be the most striking innovation in a speech that would on other topics observe most of the traditional limits.

After several paragraphs on the importance of discerning the full and unvarnished truth which, as Soviet audiences have come to know, does not necessarily mean that truth is to follow, Gorbachev approached the extremely sensitive subject of the succession struggle after the death of Lenin. That was a difficult and complex time in the life of the party, he told the delegates. And things were further complicated by the fact that there were personal rivalries in the leadership. These had been referred to by Lenin in his Testament, said Gorbachev, when the founder of Bolshevism had stressed that 'this is no trifle, or it is a trifle that can assume decisive importance'. Gorbachev was quoting the last line of Lenin's remarks dictated to his secretary Lidia Fotieva on 4 January 1923; the potentially decisive trifle to which Lenin referred was his own suggestion 'that the comrades think about a way to remove Stalin' from the position of General Secretary. But Gorbachev had twisted the meaning, changing the admonition to remove Stalin into a mere observation on the existence of differences among the leaders. With this legerdemain, he saved Stalin from the most politically damaging sentences ever spoken against him.[5]

Gorbachev repeated the Stalinist position on Trotsky: the latter had showed overweening pretensions (probably true enough), fought the Leninist line (that is, Stalin's line), and excluded the possibility of building Socialism in One Country. These charges showed Gorbachev's reliance on the *Short Course*, Stalin's official history of 1938.[6] The publication of one pamphlet, Trotsky's *Towards Socialism or Capitalism?* of 1925, in which Trotsky offers suggestions for the construction of Socialism in One Country, would be sufficient to refute the last one. Both Stalinists and Trotskyists have, for very different reasons, maintained that Trotsky denied the possibility of building socialism in the absence of help from a revolution in the West, so neither has been eager to produce the document.[7] Gorbachev's judgements then lead him to the conclusion that 'the Party's leading nucleus, led by Stalin, upheld Leninism in an ideological (*ideinaya*) struggle'.[8]

Gorbachev also made a slight political rehabilitation of Bukharin, but only to credit his role in helping Stalin to defeat Trotsky. As to Bukharin's conflict with Stalin, erupting as the collectivization of agriculture began in 1928–29, once again Stalin was right. Bukharin's actions, said Gorbachev, proved Lenin's judgement of his inability to think dialectically. Lenin's Testament was quoted to this effect. So according to Gorbachev, Bukharin had been right against Trotsky but wrong against Stalin. No criticism was made against the collectivization of agriculture as a policy, but unavoidable 'excesses' were admitted in its execution, for the Administrative-Command system was taking hold and it was capable of treating the peasant without proper Leninist consideration.

However, Gorbachev continued, in the course of pursuing these necessary policies, the methods appropriate to the period of struggle with hostile exploiting classes (War Communism?) were mechanically transferred to peaceful (!) socialist construction, along with the 'theory' of the intensification of the class struggle with the progress of socialist construction.[9] This was traceable in many respects to the lack of democracy in the country. It led to the rise of a personality cult around the figure of Stalin, causing 'serious damage to the cause of socialism and the authority of the party'. Wholesale repressions followed, to be covered up by the fiction that Stalin was unaware of them. But there remained, said Gorbachev, an historical responsibility. 'The guilt of Stalin and his immediate circle before the party and the people for mass repressions and lawlessness is enormous and unforgivable. This is a lesson for all generations.'[10]

Gorbachev paid homage to the Twentieth Party Congress of 1956 for exposing the repressions imposed by Stalin, commending in particular the courage of Nikita Khrushchev. Thus Gorbachev finally included himself among the 'children of the Twentieth Congress', along with the whole generation that had pinned their hopes on the reforms of the fifties and sixties. And his analysis of Stalin was remarkably similar to that of Khrushchev. In 1956, Khrushchev had argued that Stalin had been essentially correct against all political opponents – 'against the Trotskyists, Zinovievists, Bukharinists, and the bourgeois nationalists' – and justified to disarm them ideologically. 'Here', Khrushchev had said, 'Stalin played a positive role.'[11] However, years later, when Stalin's opponents were long since routed ideologically, 'then the repressions against them began'. These repressions were carried out not only against the enemies of Leninism but against 'honest Communists' who had helped Stalin and the party in all its previous struggles. So Stalin's worst sin had been the liquidation of the Stalinists. Khrushchev, who had originally been one of these Stalinists, reluctantly granted that even the anti-Stalin Communists had been at one time comrades of Lenin. Where Gorbachev referred to Trotsky and his circle as 'petty-bourgeois', Khrushchev had remarked that 'after all, people around Trotsky were people whose origin cannot by any means be traced to bourgeois society'.

Khrushchev had repeated the political formula of S. M. Kirov, leader of the Leningrad Communist organization, whose assassination had served as a prelude to the bloody purges of 1936–38. Kirov, at the plenum of October 1932, had objected to Stalin's demand for the execution of Mikhail Ryutin, author of a lengthy 'platform' document that had demanded the removal of Stalin according to Lenin's last wishes. Stalin did not get his way with Ryutin because a Politburo group of 'liberals', looking to Kirov for a lead, had restrained the dictator. Kirov made his case in *Pravda* a few days later: there was a time when we fought various oppositions, he said, but now they were all defeated and harmless. 'The question of *kto-kogo* (who will defeat whom) has been answered decisively for socialism not only in the cities but in the villages.'[12] Against him Kaganovich and Molotov had argued that, as the proletariat gets closer to the goal of socialism, the class enemy steps up his efforts, resorting to terror and assassination, so that the struggle takes on an increasingly intense and criminal character. This would be Stalin's formula in the purges, which got under way only a short while after Kirov himself had been mysteriously assassinated at the end of 1934. Khrushchev, in facing down the opposition of the same Molotov in 1955, forced him to repeat Kirov's formula at the time of the Ryutin case, that 'the question of *kto-kogo* had been answered against capitalism' in 1932.[13] Khrushchev thus used Molotov's historical record against him in the struggle of 1955–56, in which the secret speech on Stalin played a prominent part.

Gorbachev followed Khrushchev's analysis in large part, without contributing any new information. Khrushchev had revealed numerous communications of Lenin's bearing on Stalin, and was much more honest about the Lenin Testament's passages on Stalin than was Gorbachev, quoting the whole of the addendum of 4 January 1923 about removing Stalin. He also provided documentary material of great interest to historians about the period of the purges. Gorbachev's statement on Stalin was basically Kirovist: Stalin had been correct on political questions against all, only incorrect to have shot his opponents and 'liberal' supporters.

And Gorbachev endorsed most of Khrushchev's efforts at reform of the 'administrative-command' system, noting only that their failure must be attributed in part to the lack of democracy. He even spoke well of those who removed Khrushchev from power in 1964, giving them credit for overcoming the 'voluntaristic tendencies and distortions' in Khrushchev's domestic and foreign policies. Commenting favourably on the efforts at economic reform in 1965, he only noted a loss of momentum during the Brezhnev years, despite the very valuable achievement of missile parity with the United States.

He ended by decribing the development of *perestroika* which, he said, had just passed through its first stage. But there was no Kirovism here, the *kto-kogo* had not been completely decided. In fact, as *perestroika* proceeded, the struggle *intensified*: 'it would be a mistake not to notice a certain increase in the resistance of the conservative forces'.[14] The struggle against them at the grass-roots level of party organizations would finally decide the question of whether the current attempt to reinvigorate Soviet society would triumph. Gorbachev was probably taking note of the removal of Boris Yeltsin from the leadership, about which we will have more in the next chapter. His tone revealed a certain sense of being embattled by critics, and after the speech he would complain, as Stalin did in 1932, of being beset on two sides, on the right and on the left.

The speech played to mixed notices, no doubt because it was expected to improve on Khrushchev's analysis, and perhaps to be accompanied by revelations of interest to historians about important facts of Soviet history. Andrei Sakharov lamented that 'In his speech, Gorbachev did not tell the truth that should have been told about the 1930s and 1940s.' French Sovietologue Michel Tatu considered it a cautious review of topics for future historians to study and an equally cautious statement for the General Secretary at a delicate moment in Kremlin politics.[15] In the United States, historian Richard Pipes, who had been in the Reagan National Security Council, remarked: 'I think it is a step backward in some ways.' Stalin biographer Robert C. Tucker by contrast considered Gorbachev's analysis 'far more radical' than that of Khrushchev, in that it appeared to serve as a basis for more liberal policies. He was thinking about the remarks on the tax in kind that indicated a willingness to experiment with forms of private ownership, as in the days of NEP. Bukharin biographer Stephen Cohen warned that 'to dismiss it because it was not a full exposé of the past is to miss the point . . . historical truth isn't the issue. The issue is the great political struggle under way in the Soviet Union.'[16]

The first stage of *perestroika* was the second stage of Gorbachev's ascent, perhaps comparable to the stage at which Khrushchev found himself in 1955–56. Where Molotov had stood, blocking the path to power of the ambitious leader, there now stood Molotov's former protégé Andrei Gromyko. As Khrushchev made his critique of Stalin in the process of seeking an opening to Tito and the nonaligned movement, Gorbachev was retracing Khrushchev's steps in the process of pursuing his radical programme of arms reductions, according to the outline he had given in January 1986. This was a continuing goal as well as a source of justification for the Gorbachev revolution. On the other side, his contribution to de-Stalinization was not particularly impressive. It was certainly not on the basis of his pronouncements in the historical speech that his supporters would later, and not without a certain temerity to be sure, accuse his opponents of being 'Stalinists'. But the speech was only a part, and apparently to him the less important part, of his intellectual labours in 1987. The real intellectual foundation for *perestroika* would be in a literary form.

THE CULTURAL SLOPE

In the same month as his speech on the Soviet past, Gorbachev's writers produced *Perestroika: New Thinking for Our Country and the World*. In this volume Gorbachev sought not only to address Soviets on their movement for reform, but also to speak to the world, especially to the United

States. It was not at all like his historical speech, containing practically nothing on the Soviet past and neglecting to mention Stalin's name even once. Chapter headings announced a discussion of the Lenin heritage and 'Lessons of History' but virtually none of the topics covered in the speech was taken up in the book. Its organization was designed to demonstrate that Soviet foreign policy was an outgrowth of domestic policy, with a discussion of the aims of *perestroika* providing a foundation for an exposition of the New Thinking and its potential impact on the world. Its central message was that the United States and the Soviet Union must not permit the arms race to lead them by the nose into situations that could bring universal ruin: 'We are all passengers aboard one ship, the Earth, and we must not allow it to be wrecked. There will be no second Noah's Ark.'[17]

The backbone of the New Thinking, said Gorbachev, was the recognition of the priority of Universal Human Values. 'It may seem strange to some people', he remarked, 'that the Communists should place such a strong emphasis on human interests and values.' Among those to whom it would have seemed strange were those who had heard his Lenin Day speech of 1983 when Gorbachev was being groomed as future ideological secretary. On that occasion too he had argued that the great question was war and peace, but he had quoted Lenin to the effect that it could not be decided 'except in view of the class antagonisms of modern society'.[18] Now he maintained that weapons of mass destruction marked an 'objective limit' to class struggle on the international plane. 'Peaceful coexistence of states with different social systems should no longer be thought of, as in the past, as a "specific form of class struggle".' The Twenty-seventh Congress had broken with that idea, as with the idea that the source of the two world wars of our century lay in the contradictions of the two social systems.

Gorbachev was distorting the views of his predecessors, even including Stalin, who had in fact always maintained that contradictions among the capitalist countries were the source of war. Gorbachev proceeded to 'correct' his predecessors by stating that view. But he also recognized the intimate relation between war and revolution. The Russo-Japanese war had triggered the Russian revolution of 1905 and the First World War the Revolution of 1917. The Second World War had unleashed a wave of revolutions and a world-wide anti-colonial revolt. The forecast of capitalism's final destruction as a result of a third world war, was now a fantasy that must be buried. Did this mean that Communists had given up class analysis? Not so. It was still useful in judging events in the capitalist world. But in the countries that had established socialism, classes no longer existed, so the outlook according to Universal Human Values was the most appropriate one. Gorbachev was reiterating the theory of Dobrynin, promoted since 1986, which would be the ideological centrepiece of the struggle between him and his opponents until the issue was resolved in Gorbachev's favour in September 1988.

Gorbachev's brief exposition of the 'Lessons of History' repeated in sharper form his endorsement of the Stalinist regime's historic achievements. There was no mood of repentance here. The collectivization of agriculture had been necessary to modernize food production, to provide recruits from the peasantry for industry, to force the countryside to invest in the city. All this was necessary to defeat Hitler, and it was not the West but the Soviet Union that defeated Hitler, 'due to our better steel, better tanks, and better planes'.[19] There were excesses and 'people suffered', to be sure, but there was also a keen feeling of pride that the modernization of the country had been done without outside help. And the achievements continued after the war, including Khrushchev's unmasking of the personality cult at the Twentieth Congress. Gorbachev even praised the course taken after the removal of Khrushchev in 1964: 'a line toward stabilization was taken. And it was a well-justified line.'[20] Having produced a substantial and beneficial effect, however, these efforts simply exhausted themselves. An atmosphere of complacency gave rise to stagnation and retardation. *Perestroika*, standing on the shoulders of those who had made the gains of the past, now faced the task of resuming the advance of the socialist cause by way of 'democratization and again democratization'.

The January 1987 plenum had marked a key turn, said Gorbachev, because it had shaken up those who had thought that *perestroika* was 'just another campaign'. A signal was given to them by an increase of *glasnost*, which was not just another campaign but a new norm for Soviet life. This involved the liberation of the intelligentsia from their old constraints. The intelligentsia was, for Gorbachev, 'our great and perhaps unique achievement, our inestimable spiritual capital'.[21] It was important to adopt 'a new style of working with the intelligentsia. It is time to stop ordering it about, since this is harmful and inadmissible.' To those who claimed that society could not benefit from an atmosphere in which every person was his own priest, it must be demonstrated that a passive intelligentsia was a far worse thing. Soviets must learn the habits of civilized discourse, including the patience 'to hear out our friends'. This would be difficult, said Gorbachev, because, in the final analysis, 'we lack political culture'.

An extraordinary admission! Gorbachev's adviser Fyodor Burlatsky had sketched this idea in an article in July, 'Learn Democracy', intended to prepare the way for new conceptions of the history and role of the intelligentsia. A new idea of the Marxist reconciliation between consciousness and spontaneity was to combine the norms of a planned society with those of a society based on contract. Specifically Burlatsky was referring to the necessity of the Five-Year Plans to make room for 'commodity–money relations' (*tovarno–denezhnykh otnoshenii*), but he did not argue as an economist – rather as a pained critic of Russia's traditional culture. He began by quoting Pushkin: 'Tradition is the soul of power', and noting that 'our political culture has a long and, to put it bluntly, rather contradictory tradition.'[22] On the one side a few glimmers of direct democracy, as in the *veche* (popular assembly) of medieval Novgorod, and on the other, three centuries of Mongol domination, and a line of Tsars from Ivan the Terrible to Nicholas II who preserved almost intact the tradition of a grinding authoritarianism. No surprise that Rousseau had considered Russia a 'land of slaves', and the nineteenth-century Russian philosopher Petr Chaadayev had called it 'a country without a future'.

Even the cult of the personality, said Burlatsky, was not only imposed from above, but also reflected the level of the masses' political culture. Since 1985, he said, we see a new political culture emerging before our eyes, but many of the older uncivilized attitudes leave their residues. Some want the working people to absorb entirely the costs of transition to a more efficient economy, in the form of unemployment in the tens of millions. So it is callously suggested that these people be thrown into the street, since the effort to retrain them would entail an enormous national programme. Even the atmosphere of increased democracy in the workplace still produces barbarous distortions; situations arise in which people take advantage of *glasnost* to settle old scores with otherwise irreproachable managers. Burlatsky was pointing to a problem which, economists would be claiming by 1989, was destroying all authority in industry. If their observations were correct, it would not be too much to say that the reforms were making industry less efficient than the regime of stagnation.

Burlatsky quoted Chaadayev on the uncivilized nature of Russian civic life. He did not give the rest of Chaadayev's argument. Russia had failed to participate in the rise of civilization, Chaadayev had argued, because she had not had a Renaissance, a Reformation, a tradition of free Universities, town charters, and other corporate liberties developed in the West from the Middle Ages, nor a real Enlightenment comparable to the French movement of the eighteenth century, nor in fact any real constitutional tradition. She had missed these developments, Chaadayev contended, because of her choice of Byzantine Christianity instead of Roman Catholicism. The latter would have given her a cultural heritage respecting the division between church and state – rather than Caesaro-Papism – and a respect, by way of Roman law, for individual rights and private property.

Chaadayev was also cited by the leftist intellectual Boris Kagarlitsky in his prize-winning history of the Soviet intelligentsia, *The Thinking Reed* of 1988, with essentially the same intention as Burlatsky. Kagarlitsky called him 'Russia's first original thinker'.[23] He joined Chaadayev's arguments to those of Russia's first Marxist, G. V. Plekhanov, on the backwardness of Russia's economic and political life under a tsarist regime best understood by reference to Marx's theory of the Asiatic mode of production. Russia was an Asiatic system, less advanced in the Marxist scheme than the states of antiquity. Lacking a nobility capable of defending itself by imposing a constitution on the central authority, she had only a servitor class bound to the Tsar as vassals. Stalinism, in Kagarlitsky's argument, developed a *nomenklatura* as a privileged estate, more like the old nobility than a 'new bourgeoisie', as some leftists of the thirties had thought. The Stalinist 'partocracy' was the lineal descendant of the Tsarist 'statocracy.'

Where Burlatsky was a direct influence on Gorbachev and a sounding board for his thoughts, Kagarlitsky spoke to a potential mass anti-Communist constituency. He was stating an idea that would be a prominent feature of the anti-Communist position by 1989, by which time calls for the end of the 'partocracy' would be displayed prominently on placards in the anti-Gorbachev demonstrations. What is common to these two views, and apparently to Gorbachev's as well, is the characterization of the Soviet Union as a kind of *ancien régime*, requiring a 1789. The French Revolution was a kind of Reformation and a revolution at the same time, an introduction of modern political culture and a revolt against the privileges of the established church. Complaints about Soviet political culture seemed to suggest that the system needed not merely a new and more democratic regime to build on the gains of past history by means of reform. If the USSR was a species of *ancien régime*, how could a revolt against stagnation fail to develop into a material and spiritual revolt against Communism itself? Or, another question on political culture introduced by the invocation of Chaadayev: would a Russian road to a western political culture also be a road to a more Latin kind of culture? Is Russia, in view of its never having had a Reformation, a country with a Latin political spectrum? Should Russia look to Latin Europe for the best understanding of its moral and spiritual plight?

INTERDEPENDENCE

Gorbachev left no doubt in *Perestroika* that the goal of his domestic efforts, especially in the economy, was the attainment of 'world technological standards'. As he saw it, this was connected to the problem of the Cold War. With Detente, Soviet policy-makers had trusted too naively in the permanence of good relations with the West, which they thought would permit them to concentrate efforts on certain technologies with the confidence that others could be got in trade with the West – a kind of international technological division of labour. But in reality, said Gorbachev, 'we were punished for our *naivete*'.[24] The failure of Detente meant that the Soviets would continue to suffer restrictions on the sale of advanced technology. There was even talk about the exhaustion and collapse of the Soviet system. This prompted the Soviet government to set the goal of reaching world standards in machine tools, instrument-making, electronics and electrical engineering. Gorbachev did not feel foolish in stating the response to the challenge of the outside world in this way, while outside observers could only smile at the idea of the Soviet Union reaching world standards in non-military electronics, for example, from which even the United States is quite far.

But Gorbachev was much more confident than he would be only two years later. 'Our society is now economically and politically strong', he asserted. The same was true, he argued, for the socialist bloc as a whole. Some of the countries had gone through 'serious crises in their development', as with Hungary in 1956, Czechoslovakia in 1968, and Poland both in 1956 and in the early eighties. There had also been 'serious falterings' in cases such as that of Yugoslavia in 1948, and later with China and Albania. Gorbachev made no attempt to assess the increasing fragmentation under way in the group of Communist countries and the near-disappearance of anything that could be described as an international Communist movement, an idea that had become so ephemeral that it was out of the question to call an international conference for fear of the humiliation when prominent parties refused to attend. He called the Council for Mutual Economic Assistance (CMEA) alive and vital, while in reality it was in a state of disintegration and decline that would accelerate in the following two years. He looked to China as a possible model for future reforms in the USSR, speaking of 'very interesting and in many respects fruitful ideas' being realized there.[25] He was referring to the village agricultural reforms which were of great interest to economists writing in the Soviet press, and which to some extent fuelled his historical interest in the Soviet experiment with private agriculture in the twenties.

Gorbachev did not exhibit very much new thinking about the Third World. There was only scant evidence that he shared the views that had been expressed for some time in Soviet specialized journals, views amounting to a reversal of Brezhnev-era projections. The suspicion was voiced that much of the 'non-capitalist development' of countries with an anti-Western and revolutionary leadership was not leading to a recognizable socialism, but only a permanent dependency on the socialist bloc. It was even said that Marxism was not particularly applicable to situations in which the revolutionary forces were not based on class, but on religious or tribal loyalties. This literature developed prior to Gorbachev's election, and some utterances of his seemed to show that he took it seriously. The agenda of the Geneva summit of 1985 put revolutionary developments in the Third World into a category of regional conflicts dangerous to both the Soviet Union and the United States. In *Perestroika*, Gorbachev placed the emphasis in more traditional terms. The Western states were ignoring changes in the Third World, he charged, in order to collect 'neo-colonialist tribute'.[26] Capitalism had doomed many countries to economic stagnation by saddling them with a huge burden of debt, representing a 'time bomb' whose detonation could have desperate results.

The United States in particular had to realize that regional conflicts were not the work of Moscow's intrigue, but a reflection of local conditions. This was a familiar opinion frequently expressed in the days of old thinking. Gorbachev's way of stating it could win him a degree of the confidence of hard-line critics. He defended the Sandinistas, the Palestinians and the anti-apartheid movement in South Africa. The United States 'needs regional conflicts', said Gorbachev, in order to step up its anti-Soviet propaganda and a policy of force.

At the same time he called regional conflicts a 'seedbed for international terrorism' for the eradication of which he pledged Soviet support and readiness to engage in bilateral anti-terrorist agreements. And he hinted at an acceptance of the idea of global interdependence:

It's my conviction that the human race has entered a stage where we are all dependent on each other. No country or nation should be regarded in total separation from another, let alone pitted against another. That's what our Communist vocabulary calls internationalism and it means promoting universal values.[27]

It would not do to take this statement of the matter as a firm commitment to a specific new policy departure. Rather Gorbachev was advancing a controversial idea in the manner of *glasnost*. The 'new concept' is introduced and said to be completely compatible with the old thinking, in fact a reflection of its original spirit. Then, the new concept is filled with content in debate, in the course of which there develops a factional lineup of supporters and critics. Gorbachev is thus offered a menu of definitions, from which he is not necessarily required to choose, that is, not until the political situation should demand it.

In this case, the use of the term 'interdependence' so widely employed in the West to offer a contrast with the ruinous 'beggar-thy-neighbour' economic policies of the thirties which, at their worst, lowered the level of world trade, broke up a naturally harmonious world market into trade blocs, promoted national economy based on racial community, and dangerously increased the atmosphere of conflict leading to war. 'Interdependence' was usually cited to make a case for freer trade. Did Soviet writers who toyed with the term really mean it with all its possible implications? At its broadest, these would include acceptance of a world division of labour according to the law of comparative advantage, and denial of any Soviet intention to organize a separate division of labour based on the CMEA, as had been the declared purpose of that organization since its inception. Or: was the term merely synonymous with Dobrynin's Universal Human Values (*obshchechelovechestvo*) as described in his 1986 speech? In that case, it was merely the expression of a global community of fate in the face of problems such as nuclear annihilation, ecological disaster, AIDS, international terror, and the like. Or: were we talking about proletarian internationalism as conceived since the early days of the First International of the 1860s? And there were still more possibilities.

Western observers before 1989 were likely to debate what exactly Gorbachev meant by a given phrase and what this revealed about his true 'intentions'. And this too contributed to the shaping of the meaning. For his part Gorbachev was free to choose as the occasion demanded. Western discourse on the Gorbachev phenomenon is dominated by the image of a leader who boldly leads into new intellectual territory. Yet it would have been more accurate to recognize Gorbachev as one who flirted with ideas, permitted his thoughts to be shaped by others in the course of debate and against the backdrop of real political pressures, disavowed supporters or merely abstained from their defence, acted at a distance and chose more precise positions only as the occasion warranted. The opposition of Universal Human Values and class struggle was the centrepiece of the jousting between Gorbachev and the opponents who surfaced at the time of the January 1987 plenum, and a victory on its behalf was never announced until the September 1988 plenum. Yet most of the pronouncements issued in speeches and articles proclaimed that the two ideas were by no means mutually exclusive. This prompts the conclusion that thinking about these new concepts was not done in a laboratory, but under the harsh constraints of real political life.

A COMMON HOUSE OF EUROPE

Gorbachev framed his discussion of European affairs in terms of a rather Gaullist-sounding remark of François Mitterrand made in 1985, to the effect that Europe must once again become 'the protagonist of its own history'. Gorbachev insisted that this did not mean a Europe defined as Western Europe, but one including Russia, presumably a Europe from the Atlantic to the Urals, as de Gaulle once put it. 'We are Europeans', he asserted, 'The history of Russia is an organic part of the great European history.' The Kievan state of the Middle Ages was united with Europe by Christianity, he said, reminding his readers of the celebration of the millennium of the baptism of Rus' to be held in Moscow in 1988. A union of east and west Europe had defeated fascism. It was wrong, he charged, for Margaret Thatcher to have spoken about the USSR getting into the war 'only' in 1941, whereas before that England was fighting Hitler single-handed. Gorbachev claimed credit for the political fight against fascism in the Popular Front era and the Soviet aid to the Spanish republic in its civil war. The Hitler–Stalin pact of 1939 had not been inevitable, but only resulted from the failure of the British and French to cooperate with Soviet Russia. 'And who handed over Czechoslovakia to the Nazis?' Gorbachev called the Munich agreement of 1938 a result of British intentions to turn Hitler to the east to crush Communism.

And who divided Europe after the Second World War? There was a Western version of the origin of the Cold War, said Gorbachev, that claimed it was exclusively the fault of the Communists. 'But what about the Fulton Speech of Churchill? Or the Truman Doctrine?' He reminded his readers that the Soviet Union had always opposed the division of Europe into military blocs, and that the formation of NATO had preceded the Warsaw Pact. The superpower arms competition had been a painful burden for the human race, and there had been reason to rejoice when its tensions in the nuclear sphere were brought under some control by Detente, culminating in the Helsinki process of 1975. But 'we were naive', said Gorbachev, to think that the Detente process was irreversible. A second Cold War developed in the recent period, partly, he admitted 'by way of self-criticism', because of the 'weakening of the economic positions of socialism in the late seventies and early eighties'.[28] This had aroused an American 'desire to exhaust the Soviet Union economically'.[29] The principal instruments for this, he said, were destabilizing weapons programmes such as Trident 2, and most of all the SDI.

The liquidation of the Cold War, he argued, presupposed the integration of the Soviet Union into the rest of Europe. The House of Europe from the Atlantic to the Urals, united by the common heritage of the Renaissance and the Enlightenment, could be the centre of gravity of world civilization. This new Europe must be reconciled to the existence of two German states:

We, naturally, are bound to be alerted by statements to the effect that the 'German issue' remains open, that not everything is yet clear with the 'lands in the east', and that Yalta and Potsdam are 'illegitimate'. Such statements are not infrequent in the Federal Republic of Germany. And let me say quite plainly that all these statements about the revival of 'German unity' are far from being 'Realpolitik', to use the German expression. They have given the FRG nothing in the past forty years. Fuelling illusions about a return to the 'Germany of the borders of 1937' means undermining the trust in the FRG among its neighbours and other nations.[30]

In July 1990 when, as a result of the decisions of the previous year, Gorbachev decided to clear the way for the absorption of the East German regime into the Federal Republic, and to permit the continuation of this state as a member of NATO, he described the bilateral agreement with Helmut Kohl to this effect as having been achieved 'in the spirit of that well known German word, *Realpolitik*'. It can hardly be said that Gorbachev knew where he was going, or more accurately, where he was being taken, when he theorized airily on these topics in his book in 1987.[31]

He was thinking in terms of a revival of *detente* and arms control, rather than remaking the map and the power balance of Europe. The House of Europe slogan was another pass at the Europeanization of the Soviet Union, as the idea was conceived, mostly in the West, during the sixties, when it appeared that American policy was preoccupied with Vietnam and de Gaulle was capable of creating fresh contacts with the Soviet Union and new possibilities for continental cooperation.

Gorbachev stressed that he had no wish to try to weaken the historic ties that bound the United States to Europe, but he did not refrain from mentioning ironically the Greek myth of the abduction of Europa in reference to the 'onslaught of mass culture from across the Atlantic'. The 'profoundly intelligent and deeply humane' European culture was thus threatened by the 'primitive revelry of violence and pornography and the flood of cheap feelings and low thoughts'. This is not the place to frame a judgement of these disturbing notions about the Atlantic culture, except to note that travellers to the major cities on both sides of the Atlantic would probably not have noticed so severe a contrast as Gorbachev described. The point, however, is that Gorbachev seemed to realize that the division of Europe was served by high tension between the superpowers, and that, if that tension could be greatly reduced, American influence in Europe would inevitably decline in proportion. He seemed not to have realized how much Soviet influence would also be bound to experience a parallel decline.

THE REFORM OF COMMUNISM

So, what can be concluded about the thought of Mikhail Gorbachev at this juncture in his rise to power? It seems clear enough that, despite his encouragement of reinvestigation of the whole problem of Stalin and Stalinism, by his remark of January 1987 about the problems in Soviet life going back to the time 'when creative ideas disappeared from theory', he did not feel prepared either to break with Stalinism or to offer a new perspective for Communists on the Stalin legacy. He was less honest about the Lenin Testament than Khrushchev. The foreign policy themes covered in the book were obviously Gorbachev's real interest and competence. The speech, by contrast, took Gorbachev through historical territory with which he had only passing familiarity. He would have needed help of the kind that Pospelov and other party ideologists gave to Khrushchev on preparation of the secret speech of 1956, in order to thread his way through this thicket of historical problems, and he would have needed a capacity for study comparable to the leaders of Lenin's time to arrive at some new approach. In a certain sense Gorbachev's failings were those of the entire intelligentsia lifted to prominence by the purges of the thirties. It should be remembered that Stalin made a point of shooting anyone capable of even the appearance of originality in writing about Marxism. Stalin had seen to it that a new history was written and that writings in conflict with it were made to disappear. The mental outlook of succeeding generations of Soviet Communists was shaped by this natural selection.

Thoughtful articles published since the January plenum on the problem of collectivization clearly had their impact on the position that Gorbachev adopted in the speech. But Gorbachev did not get much help on the crucial question of how it was that Stalin had got Soviet Russia in his grip in the years after Lenin's death. Instead of a real historical debate from which Gorbachev could draw ideas, he got only intimations from writers such as Yegor Yakovlev, Yury Afanasyev and Dmitry Kazutin. These were powerfully refuted by F. M. Vaganov and the historians who insisted on the Stalinist version presented in the *Short Course* of 1938. Even had Gorbachev wanted to present a more truthful version, he would have had to argue it against opponents drawing on these historians' experience in arguing the Stalinist case. Gorbachev the political man probably did not want to take the matter any further, in view of his turn to a more conservative position after the June plenum. We now know that extensive amendments were made to the preliminary draft of his speech before he was able to clear it in the Politburo. In the end the speech was not what it claimed to be, that is, not an attempt to use the weapon of science to lead the political struggle in an enlightened way, but only a mendacious theoretical articulation of an intermediate position in a tense political struggle.

Gorbachev's rehabilitation of Khrushchev and the Twentieth Party Congress was of great importance to much of the intelligentsia, and to many *feuilleton* writers who had in their youth pinned hopes on the thaw of those years, only to be disappointed when Brezhnev and Suslov closed down de-Stalinization after 1964, afterward even coming close to a restoration of Stalin's legacy to its pre-1956 condition. This was a reversal of signs for Gorbachev from his time as Chernenko's ideology secretary, when Molotov was rehabilitated. It will be recalled that Molotov was a target of the secret speech in 1956. And reviving Khrushchev was a way of undermining the intellectual authority of all the Brezhnev men, but especially Molotov's protégé of 1939, Andrei Gromyko.

Yet in reviving Khrushchev's critique of Stalin, Gorbachev did not carry it even a fraction of a step further than its status of 1956. Khrushchev had argued essentially for Kirovism, Stalinism without its resort to terror, as it reposed in 1932–34. He had not cleared up the mystery of Kirov's death (which most Western historians assume to be the work of Stalin) indicating only that there were many things about it that needed to be investigated. Gorbachev did not even do that much. He mentioned Kirov's name only once, to give him credit for having opposed Trotsky.

The really novel element was the reference to the tax in kind that had introduced NEP in the twenties. But Gorbachev refrained from offering any encouragement to those who wanted a full political rehabilitation of Bukharin, as a symbolic way of championing NEP's mixed economy of which Bukharin had approved. Nor did he indicate any enthusiasm for the Communism of Bukharin as a kind of historical alternative of a more moderate kind than that of Stalin.

In fact, one can see some elements in common between Stalin in 1932 and Gorbachev in 1987. Stalin had been stopped at the fall plenum from disposing of Ryutin by a combination of his own supporters and local party officials. These 'liberals', while in general satisfied with him at the helm, nevertheless feared making any new departures (such as shooting party members) that would upset their situation. As it developed, they would have been happy to reduce the leader to the status of a revered figurehead. Stalin fought them by reaching into the ranks of the party at the local level and calling for a radical campaign against them, which he was able to create at the height of the purges in 1936–38. Without comparing Gorbachev himself with Stalin, one can see a similarity with the factional situation of Gorbachev at the January plenum. His response was not to use the police and vigilance campaigns against his party opponents as Stalin did, but to try to root them out from below by democratization. The following years were to show that democratization was a dagger aimed at the heart of the Communist party's leading role as traditionally conceived. Yet Gorbachev gave not a hint of any intention to destroy the hegemony of the party at this time.

His main focus was foreign policy, his main intention to demonstrate that the new Soviet initiatives on arms control were not just part of another peace offensive, but a result of his revolutionary efforts to transform Soviet society root and branch. Yet the foreign policy outlook he described was a mix of new and old thinking. The new ideas – a foreign policy based on universal human values rather than class analysis on an international plane, an embrace of the word 'interdependence', a recognition that Soviet society lacked political culture – these had yet to be filled with substance. As was the case with his calls for democracy, no one could say for certain what they meant. Alongside these ideas, there was a stern defence of the record of the eastern bloc countries: no apologies for the Soviet invasions of Hungary in 1956 or Czechoslovakia in 1968. The Hitler–Stalin pact was defended, as was the status quo in the two Germanies. We are not contrasting his views with some ideal case, but with positions he ended up with only a few years later. In 1987 Gorbachev wanted only a reprise of the opening to western Europe that had been available to the USSR in the days of de Gaulle. He imagined a new unity of action with Western left parties, Social Democrats and Greens. He seemed to have no premonition of the fate that was to befall the Communist bloc in 1989.

Would bolder action to come to grips with the sins of Stalinism have helped Gorbachev's drive to revitalize Communism? Perhaps it would have been entirely irrelevant. Yet Gorbachev did pass up a chance that many observers east and west thought he would seize, to rehabilitate all the anti-Stalin Communists shot in the thirties, and to cause the nation to face squarely the Stalin legacy for the first time. There was no real consensus among the Western observers as to what the truth about Stalinism really was. Some would have said that there had been alternatives to Stalin in the twenties and thirties, and that Leninism did not have to result inevitably, through its inner logic, in Stalinism. Others would have insisted on the contrary that 'soft Communism' is a myth and that no other result than Stalinism at its worst could have issued from the utopia of a centrally planned socialist society imposed by force.

Gorbachev's own mind did not appear to be settled on the troublesome questions. Perhaps we can appreciate his dilemma by imagining him to be torn between two contradictory impulses, like the two souls of Faust. The first, Gorbachev the Heir of Lenin, might speak to him in the following voice:

> The time has come to square accounts with old Koba (Stalin) at last. Time to say to the comrades that everything they have been reading and saying about him and his era for the last half-century has been a tissue of lies. Let's start at the beginning. Stalin was not Lenin's choice but his enemy; Lenin had resolved to crush him politically in order to clear the way for a collective leadership that could work with (and also

restrain) Trotsky. And almost everything written about Trotsky is non-sense. True he was a Menshevik before 1917 and he had many squab-bles with Lenin as they elbowed for room in the exile movement. These are interesting for historians but not crucial for interpreting anything later. Least of all are they some sort of original sin, as the ambitious Zinoviev made them out to be in 1923–4. And not everything Trotsky said against Lenin was wrong. He issued some prophetic warnings about the preoccupation with centralism in our movement. Trotsky was a key to the events of 1917: Lenin's April Theses looked very much, as Plekhanov and others remarked, like Trotsky's idea of the Permanent Revolution. Trotsky led the October rising, rather than betraying it by delay, as we have been saying since 1924. It was only a tactical ques-tion. These and other matters from the pre-1917 period were only dredged up and distorted in 1923 by those who feared Trotsky (primarily Zinoviev, but also Stalin and Bukharin). While he was well and active, Lenin never took these things seriously. So the whole crusade against 'Trotskyism' was just a device of the struggle to succeed Lenin. The absurdity of the whole thing can be appreciated when one recalls that Zinoviev called Stalin a Trotskyite in 1923, Bukharin accused Stalin of Trotskyism in 1928, and Stalin called them all (and millions of others) Trotskyites when he shot them in 1936–8.

Trotsky was a bother, to be sure, since he thought himself a univer-sal genius, and he would undoubtedly have thrown his weight around had he remained in the leadership. But it is ridiculous and even de-praved to say that he would have been worse than Stalin. He and the others would have solved the problems at least as well as Stalin did, and probably much better. It can hardly be doubted that he, Kirov, or any one else could have devised a more sensible strategy for the Ger-man Communists than cooperating with Hitler against the social demo-crats. No one but Stalin was capable of shooting millions, and the cream of our officer corps among them. It's impossible to say whether Sta-lin's dragnet didn't fish up some real fifth-columnists who would have hurt the country in some way, but thinking like this is a dead end at any rate. Trotsky said things in exile that can be debated, but they must be judged alongside the truly monstrous things that were going on in the USSR at the time.

So let us drop this whole pathological preoccupation with Trotskyism for good, along with the 'learned' critique of Bukharin, Radek, Zinoviev, Kamenev, and the others, including the 'bourgeois nationalists', that we have been taught. Politically this will free our hands. We will still want to experiment with market mechanisms in the economy, which will put us under the sign of Bukharin spiritually, but even Trotsky agreed with Bukharin on such matters in 1925 and again in 1930–1. As

Bukharin admitted in 1928, the differences among them were as nothing against their differences with Stalin on the regime. Renouncing
Stalinism may not save us in the face of a real democracy that will no
doubt refuse to forgive the impositions of the last fifty years, but it
will give us a chance to make Communism dynamic and interesting
again. It may absorb some of the attention of the intelligentsia and
Komsomols that might otherwise be preoccupied by their consuming
hatred for us. The people who will want to tear down our monuments
to Lenin will not be thinking of him as they do it but of his 'best pupil'
Stalin. Perhaps we should insist on a distinction.

Gorbachev's other soul, Gorbachev the Political Man, would receive
these pleadings with a smirk, and respond to the arguments of his brother
spirit as follows:

Very nice oratory! To make the party young again, and all by means
of a wholesome balm, surely would be best. Yet where can this substance be found? How quaint that you think so much can be accomplished just by owning up! Have you thought of the opprobrium we
would bring down on our heads by coming to terms with all of Stalin's
victims, especially the political ones, and very especially Trotsky? Don't
you understand that a thousand mysteries of the history of our country
are tied up with that name? I'm not saying that I know all the ins and
outs of it, but I do know that political investigation of the history of
Stalin's rise to power puts us in a position of apologizing to Trotsky.
Not only does this make us Trotskyists (I admit I don't exactly know
what I mean by that – but haven't we all used the term at one time or
another to describe some species of political folly?), but it exposes us
to the jagged edge of Russian nationalism, with which we have co-
existed by a policy of careful concessions since Stalin's time. We don't
want to rub raw the sore of anti-semitism on the skin of people who
have labored with us in great struggles.

And a great many Russians recall us marching to Berlin shouting
Stalin's name. Don't be so foolish as to underestimate the strength of
traditional Stalinism in the USSR of *perestroika*. You are a man who
has been in politics for many years so you must know that myth has in
our history been a material force in itself. Besides, if we are pursuing
a line of neo-NEP and creating a market, our patron saint is Bukharin
and more useful to us is a Bukharinist argument like that of 1928–9
that Stalin and Trotsky are both wild men, their War Communism vs
our NEP, etc. In fact, in a debate between 'reformers' (Bukharinists)
and 'conservatives' (Stalinists), our Gorbachev can maneuver in the
centre and win support from both sides.

We have to remember that the crucial problems we face are not historical ones. If we centralize our thoughts, we realize that the pressure of the U.S. exhaustion strategy is the central challenge. We are trying to remove the threat of U.S. weapons, the SDI and the like, by removing the 'Soviet threat' from the minds of the west and the Chinese. Our domestic policy, including our policy toward the intelligentsia, is all supposed to make our conciliatory moves credible, assuming of course that the best way to convince the world that we are changing is for the changes to be genuine. Everything else, including ideological policy, needs to be subordinate to that. Anything that smacks of Communist ideological revival works at cross purposes.

Remember that the past cannot dictate our present course. That would be dogmatism, and we need to be creative. As old Hegel said, the truth is always concrete.

Somehow the second voice seems to speak with the greater resonance. Not that Gorbachev could be reduced to a political man of such narrow calculation. Yet no one who has nuclear weapons can avoid thinking about them before thinking about anything else. Even matters of 'principle' are forced through this prism. Communists who in the past have spoken about the will of the people being greater than the imperialists' technology have generally been regarded, especially by other Communists, as madmen. While a course of action that promised to reduce the nuclear threat did not promise to solve other pressing problems, in particular the problem of the survival of Communism in a regime of democracy, it still compelled support. Moreover, the other voice, that of the heir of Lenin, was not one that Gorbachev or his generation had ever really trusted.

To be sure, it spoke to their sense of what the sociologists call 'charter myth', but not to their experience. There was something false, even pernicious about it. It was the voice of the Lenin Day speeches, full of rectitude and urgings to continued sacrifice, backed up by unanimous decisions and stormy prolonged applause offered by a congregation in which there might not be one believer. The comrades called themselves Communists, yet they all clung hard to their cramped ideas of reason of state. They were realistic enough to tremble at the thought of standing up to the Russian nationalists with a frank defence of the old heretics in their movement who happened to be of Jewish background. And they worshipped at the font of the sober realism of the nuclear era. In its name they were prepared to sacrifice anything.

7 Between Yeltsin and Ligachev

> No nation acquires the power of judgment unless it can pass judgment on itself. But this great privilege it can only attain at a very late stage.
>
> Goethe

Gorbachev's historical address and his book on *perestroika* do not verify the view then accepted, in which Gorbachev was fighting resolutely and continuously against the forces of stagnation and Stalinism. A bitter confrontation was erupting, but the issues of principle were not so clear. The advocates of reform, not at all sure of the specifics of what they wanted, were opening up a complex struggle in periodicals such as *Moscow News*, *Ogonyok*, and *Argumenty i fakty*, suggesting a many-sided critique of Soviet reality and Stalinist history carried out in Gorbachev's name. On the other side, the coalition that had prevented Gorbachev's personnel changes at the Politburo and Central Committee meetings of January 1987 seemed only to be sure that they did *not* want to defend the old order. But they could not decide to what degree they would be willing to endure press criticism under the rubric of *glasnost*. Or, another way of putting it: they did not mind changing things, even drastically, as long as they were permitted to keep their positions of influence; they did not want reform to be used against them. Gorbachev stood between these two phalanxes in the fall of 1987, manoeuvring toward the more traditional positions that seemed strong since the July plenum, even at the risk of disappointing his more ardent followers. But he had no interest in real reconciliation. For the delightful part of it was that the intensity of the conflict seemed only to make both sides more reliant on him.

His centrism raised difficult questions. In the bad old Stalinist days, a manoeuvre away from a previous course was never considered a betrayal of principle, rather evidence of the use of dialectic to navigate stormy seas. Those who could not make the turn could be called dogmatic or mechanical, and cast off as unnecessary ballast. Disposal was never a problem, since there was no political life, and sometimes no life at all, outside the consensus of the top leadership. Under the new conditions, however, with their promise of democratization and increased freedom of expression, anyone who refused to manoeuvre and stood on principle was in effect offered a chance to fight on. This new situation fundamentally altered the historical patterns of Soviet politics. One can only imagine the result, if one of Stalin's or Brezhnev's discarded associates had been free to appeal to the public against him. Gorbachev himself showed no sign of understanding the new situation. The anger and frustration with which he responded to the appearance of the 'Yeltsin phenomenon' demonstrate that he was not at all prepared to accept a rival for his position as champion of reform.

Boris Yeltsin could be said to be the first Western-style politician to come forward under Gorbachev's regime. He displayed the ideal qualities that American consultants look for in candidates: the stance of an outsider coupled with impeccably orthodox views. He thought of himself as a radical populist and a product of *perestroika*. But his opposition to the Communist apparatus was not shared by any major party figure. The top leaders viewed his case with curiosity, not to say stupefaction. On the old rules he was a man without a future, but on the new, he was a figure capable of transforming the very nature of the Gorbachev revolution.

No sooner did the Yeltsin Affair burst into the foreground at the plenum of October 1987, held just prior to Gorbachev's historical speech, than a spirit of confidence returned to the opponents of Gorbachev. Five months later they would have a platform of opposition to the excesses of *perestroika* and a personification of the values of unreconstructed Stalinism in the figure of Mme Andreyeva from Leningrad. Her views were not echoed by the Politburo oppositionists. They wanted to be free to disavow her, but also to use her as a club with which to beat Gorbachev, as they thought he used the *glasnost* press to beat them. It was disconcerting to Gorbachev that the alternatives to his leadership were becoming so much more sharply outlined, but there was a positive side to the emergence of rivals on the right and left. Gorbachev's repeated rejection of Yeltsin won him a following among Communists who may have doubted his commitment to the cause of Lenin. Thus Gorbachev was able to split the traditionalists arrayed against him. And the attacks made against Mme Andreeva by the *feuilleton* writers in the *glasnost* periodicals weakened the cause of the traditionalists still further. These developments seemed to work in Gorbachev's favour, on the assumption, that is, that Yeltsin would obligingly fade into obscurity. Over the next three years Gorbachev and the whole of the Communist party were to learn to their chagrin that the Yeltsin problem would not go away.

YELTSIN VERSUS LIGACHEV

From 1985 to 1987 Yeltsin was known in the party and the public as a Gorbachev man, but Yeltsin himself would later divulge that he and Gorbachev had never been intimates. Yeltsin hailed from Sverdlovsk in the Urals, where whole factories had been shipped in 1941 to escape the advancing Nazi armies. He grew up in the atmosphere of heroic feats of shock-work, accomplished by wretched armies of starving workers who toiled into the night despite lacking even the basic necessities of civilized life. This produced in his character a deep respect for administrative prodigies by hard-driving managers, alongside a resentment about the joyless life of the average worker in this terrible period. 'Winter was worst of all', he wrote in his memoirs describing this time, 'There was nowhere to hide from the cold. Since we had no warm clothes, we would huddle up to the nanny goat to keep warm. We children survived on her milk. She was our salvation throughout the war.'[1]

Young Yeltsin was a model pupil who earned the best grades, but he did go through the experience of being expelled from school on one occasion. At his primary school graduation, during a carefully planned and rather solemn ceremony that was moving along without a hitch, he suddenly requested permission to speak. He was allowed on the stage to say what everyone thought would be a few gracious words from an exemplary student. Instead he delivered a detailed critique of one of his teachers who, he complained, intimidated the students with demeaning punishments and crippled them psychologically with mockery that destroyed their confidence. The event was ruined. A few days later Yeltsin was issued a citation that would have prevented him from acquiring his secondary education, but he answered by further protest, finally succeeding in getting the matter before a local party committee, where he was able to make his case at length. There he was vindicated and the teacher in question was dismissed from the school. No doubt somewhere in Yeltsin's adult consciousness lay the conviction that he could prevail in a righteous cause against severe odds in the appropriate public forum. One could view his struggles of 1987–88 in this light.

Yeltsin graduated as a construction engineer from Ural Polytechnic Institute in Sverdlovsk in 1955. He joined the party in 1961 and worked his way up through the ranks, becoming first secretary of Sverdlovsk *obkom* in 1976. He met Gorbachev, then First Secretary at Stavropol at this time, with whom he worked out exchanges of materials, metals, and timber from the Urals for agricultural products from Stavropol. He continued to visit Gorbachev after the latter's appointment to the Central Committee. It is likely that he was one of the group of people who advanced with Gorbachev because of a connection to Kirilenko. At any rate, he arrived in Moscow to take a position in the secretariat in April 1985. He was part of a caucus of first secretaries united in supporting Gorbachev and opposing Grishin and Romanov as candidates for the post of general secretary, concerting his efforts with Ligachev to that end.[2] By the end of the year he was rewarded by being moved in to replace Grishin as head of the Moscow party committee.

Men who have held this post in the past have usually figured prominently in Kremlin politics. In the struggle between Zinoviev and Trotsky in 1923–25, Lev Kamenev held the Moscow bastion for Zinoviev. In 1925, when Stalin parted with Zinoviev to form a bloc with Bukharin, Kamenev came under pressure and was eventually removed for a Bukharin man, N. A. Uglanov. When Stalin broke with Bukharin in 1928–29, Uglanov was attacked as part of the Right Danger and removed in 1930 for a Stalin man, Lazar Kaganovich. During the dispute at the 1932 plenum over whether to shoot Ryutin for his criticism of Stalin, Kaganovich spoke for Moscow and advanced the bizarre theory that the class struggle intensified as the country approached socialism, the idea that would serve to underpin the purges of 1936–38. Khrushchev got his biggest break when he was brought in from the Ukraine by Stalin in 1949 to head the Moscow organization, in order, as Khrushchev saw it, to help undermine Malenkov and Beria.[3] In one of the few inner party crises of the Brezhnev period, the Moscow chief Nikolai Yegorychev, thought to be closely allied with Aleksandr Shelepin (who had helped in the *coup* that brought down Khrushchev and installed Brezhnev), criticized Brezhnev's policies during the Middle East war of summer 1967. His defeat, which proved to be a stage in the removal of Shelepin, caused Moscow to be put into the presumably more trustworthy hands of Viktor Grishin, who held it subsequently for Brezhnev, Andropov, and Chernenko.

Succeeding Grishin in 1985, Yeltsin came into office wielding an iron broom, fully in accord with the expectations of Gorbachev. By his own estimate, in his term of office he replaced some 60 per cent of the *raion* secretaries, acting in concert with Gorbachev, who removed perhaps 66 per cent of the local secretaries in the Union.[4] Yeltsin railed against nests of corruption where the Moscow 'mafia' held sway. He crept around anonymously, appearing in queues in front of food stores, riding with workers in buses and subways to hear their complaints, making impromptu speeches against the special stores and other privileges of the *nomenklatura*. He worked long hours, a practice which forced the other officials to stay late on the job with him. He presented to those who had to keep up with him the visage of a manager of Stalin-era shock-work, alongside that of an impossibly idealistic bumpkin and demagogue. To the ordinary inhabitants of Moscow he seemed to be an opponent of the bureaucrats and the mafias of the second economy, a breath of fresh air and a harbinger of new hopes.

His demagogic appeal was in the manner of traditional party populism. He denounced nepotism in the form of special schools for children of the leaders of the apparat, an act which could not have endeared him to people like Gromyko whose children were prominent intellectuals. He pared down the fleet of ZIL limousines that the big men used to race around Moscow through a succession of green lights. At the same time his economic measures, quite unlike his policies after taking power in 1991, sought to use the state power *against* individual initiative, as in his opening of state stalls at the markets to compete with those selling food from private plots. He tried to act against the practice of employing large numbers of *limitchiki*, temporary residents of Moscow who took the hardest and least pleasant jobs that permanent residents normally declined. But his measures only caused the various enterprises to break down for lack of people to perform key jobs. Yeltsin won great popularity for opposing the least attractive attributes of the system, but he also broke the many complex and informal threads that held the system together. When matters got worse, he could blame it on the resistance of mafias of one sort or another, but in fact he was also resisted by many commercial employees who formerly greased the wheels in different small ways and were forced by his threats of arrest to work to rule – and therefore to make the service worse. Yeltsin's economic measures were in this sense a microcosm of Gorbachev's effect on the economy as a whole.

Yeltsin must have expected that, in view of his vigorous support for the leader and his measures in the name of *perestroika*, he would be made a full member of the Politburo at the plenum of June 1987. It was thought that the Gorbachev group would be strengthened by the promotion of both Yeltsin and Aleksandr Yakovlev. But only Yakovlev made the list, along with Slyunkov and Nikonov, who proved in the next two years to be closer to Ligachev. It is likely that there was a compromise with Ligachev, who had come to be highly critical of many of Yeltsin's actions. Gorbachev got his way on a Law of State Enterprise envisioning in a three-year period a transition to a system where the major firms made their own negotiations for raw materials (something that complicated the lives of Soviet managers no end, and which they hated) and after a kind of tax in kind, sell the remainder of their products on the market. Many firms were to be on a footing of *khozrashchet*, that is, to have to pay their own way without subsidies. Holding Yeltsin out of the Politburo for the moment may have been the price Gorbachev had to pay to win Ligachev's compliance for this reform.

The armed forces were reeling from the effects of the Rust Affair, with Gorbachev's drastic reshuffling of the top military posts. In the spirit of this campaign, Yeltsin continued the attacks on the military well after Gorbachev had got his personnel changes, an action that must have solidified the opposition to him. Yeltsin now felt Ligachev using his position as Second Secretary bearing down on him. Yeltsin complained to the June plenum that the Secretariat 'sends people out to cull negative material in party organizations'.[5]

At the end of August, Yeltsin provided facilities for a conference of 52 grass-roots groups with the title 'Social Initiatives in *Perestroika*'. These unofficial organizations (*neformaly*) ranged from left clubs in big cities with names like 'Perestroika', 'Obshchina', and 'Straight Talk', to smaller groups: 'The Farabundo Marti Brigade', 'The Che Guevara Brigade', and 'The Forest People'. The declaration adopted by the meeting asserted their unswerving commitment to the idea of socialism as proclaimed in October 1917. Moreover, it recognized the 'constitutional role of the CPSU'. The Informals claimed, however, that the party was divided between 'those who bear direct responsibility for the abuses and errors of the past', and 'healthy and progressive forces', and described the activity of the informals to be a support for the latter.[6] The burgeoning Soviet New Left thus saw itself as an opponent of the forces gathered around Ligachev. It also declared itself the enemy of the *Pamyat* organization of Russian nationalists, and suggested the construction of a monument to Stalin's victims.

Yeltsin could not escape being identified with this popular movement, which seemed to be an extra-parliamentary wing of his own campaign against the apparat. Viktor Afanasyev's *Pravda*, now much more in line with the resistance to *perestroika* than at the time of the Berkhin Affair, denounced the conference as a provocation. Ligachev cannot have enjoyed witnessing the rise of a spontaneous force on the left, after having been able to block Yeltsin's appointment to full membership on the Politburo. But in the run-up to Gorbachev's ideological speech, he and Chebrikov were more occupied with speaking on historical topics, defending the achievements of the Stalin period against criticisms made in the *glasnost* press. Ligachev singled out *Moscow News* and *Ogonyok* for promoting 'one-sided articles on the generation that built socialism'. Chebrikov renewed his warnings about the activities of the foreign intelligence agencies who sought to flood the USSR with copying machines, VCRs, and other equipment suited to make propaganda against the Soviet power. He saw 'subversive centres from abroad' trying to take advantage of the political naiveté of the socialist democracy in order to lead it into hostile acts.[7]

In Politburo meetings Yeltsin and Ligachev clashed over the question of privileges and perks for the apparat. They also disagreed about the anti-alcohol campaign, for which Ligachev had developed genuine enthusiasm, but which was now proving to be a mistake. Ligachev enthusiastically reported alcohol sales plummeting in this or that region, largely because of its unavailability, while Yeltsin cited reports of people drinking moonshine or, in some cases, hair tonic and rubbing alcohol.

The bickering reached a peak in September, when, as Gorbachev related to the Nineteenth Party Conference a year later, Yeltsin sent him a letter asking his intervention in the dispute and threatening to resign his posts. The letter was largely an attack on Ligachev, citing the harmful effect of his influence on the primary party organizations and his 'systematic persecution' of Yeltsin dating from the June plenum. Ligachev was not the only harmful influence in the Politburo, said Yeltsin; there were others who were hostile to *perestroika*, but were slyly professing to support it. 'But is their conversion wholly to be trusted? This suits them and, if you will forgive my saying so, Mikhail Sergeyevich, I believe it also suits you'.[8] Yeltsin concluded by asking to be relieved of his posts without having to submit the matter to a plenum of the Central Committee.

Gorbachev put him off, hoping to keep matters from coming to a head prior to the 70th anniversary celebration. It was a crucial time for another reason: preparations were being made for the Washington summit at which Gorbachev was expected to sign an agreement eliminating intermediate-range missiles in Europe, *without* any linkage to the SDI, as he had offered to do in February.

Instead of holding his silence in line with the General Secretary's wishes, Yeltsin changed his mind and decided to make his strike. The attack opened at a Politburo meeting where Gorbachev presented a draft of his historical speech to the assembled members. Everyone made at least perfunctory and generally positive remarks. There were some corrections. Ligachev and Chebrikov had some input: the remarks in the final version on collectivization of agriculture, the heroic aspect of the industrialization drive, and other matters were identical in substance to those in speeches made by the two in August. Yeltsin, however, went further. He made some twenty separate comments on historical questions, styles of work in the apparatus, and other matters. According to Yeltsin, Gorbachev was dumbfounded and responded by breaking off the session and storming out of the room. The other comrades sat in their places in stunned silence. When Gorbachev returned, he delivered a scathing attack on Yeltsin for his work in the Moscow organization, his negative attitude, and his deficiencies of character. No one else raised a voice against him. Sheepishly Yeltsin promised to take note of the criticisms, but in the following days, he realized that he was being ignored. He concluded that Gorbachev had resolved to wash his hands of him.[9]

However, Yeltsin was not finished. He decided to present his case to the plenum on 21 October. This time a chorus of censurers had been forewarned and had time to prepare a critique. Yeltsin was given the floor immediately after Gorbachev's introductory remarks. He spoke of the need to restructure party work in the primary organizations and most particularly, in the secretariat, where the 'style of work' of Ligachev was inadequate to the task. He sensed an exhaustion of *perestroika*:

> At first there was enormous enthusiasm, an upsurge, high and strong including the plenary meeting of the party's central committee in January Then, after the June plenum, people began to lose faith, and this wor ries us very much indeed.[10]

The comrades should remember that the central problems in the history of the party stemmed from the fact that in the past party authority had been given to 'one man' who was fenced off from criticism.

What worries me for one is that, while we don't as yet have such an atmosphere in the Politburo, recently there have been signs of a definite growth, I would say, of eulogizing the General Secretary on the part of some members, some permanent members of the Politburo ... We cannot permit this. We must not permit it.[11]

Historians will debate this plenum in the future, attempting to establish the degree to which Yeltsin had thought out the implications of taking on Gorbachev and Ligachev at the same time, not to mention comparing Gorbachev's regime to that of Stalin. They will no doubt see a far-seeing calculation in Yeltsin's mind, drawing their conclusions from his later activities in the electoral arena where he could be said to have profited from the public perception of having powerful enemies in the apparat. But when he made these extraordinary remarks he had no assurance that he would ever be able to bring his case out before the public in a free election. He must have been hoping for vindication before a strictly party forum, perhaps as he once hoped when he put his case against the abusive primary school teacher before a forum of about the same size.

The response from the members of the plenum was a lengthy recital of denunciations, 25 speakers in all, each of them passionately defending Ligachev and castigating in his own way the political naiveté of Yeltsin. Of the 25, Georgy Arbatov sounded the only positive note when he said something in passing about Yeltsin's courage, but he was quickly corrected by the subsequent speakers. Ligachev led off, ominously reminding the delegates that it had been Gorbachev who had proposed Yeltsin for his present post.[12] Speaking for the trade unions, S. A. Shalaev spoke glowingly of the secretariat's work under Ligachev, comparing it with reverence to that of previous chairmen Suslov and Chernenko. On the other hand, Ryzhkov reminded the delegates that under Brezhnev and Chernenko meetings were often wound up in 15 minutes, whereas now they sometimes go on for seven hours. 'One sometimes leaves dripping wet.' The Ligachev secretariat erred, he said, in supporting Yeltsin too much! This was a way of blaming Yeltsin on Ligachev. Vorotnikov spoke of Yeltsin's having 'too much confidence, excessive ambition, recourse to leftist phrases'. Gromyko simply said that Yeltsin held 'wrong views'. Instead of detailed remarks about these, he mused vaguely about history and attributed the mistakes of the Stalin period to a lack of party unity! Shcherbitsky wondered if Yeltsin's errors did not stem from getting his information from *Moscow News*.

Yeltsin sat through it all and then rose for his rejoinder. But this was milder than his initial statement. Even so, he was interrupted sharply by Gorbachev who especially resented the reference to a personality cult around him. Didn't Yeltsin know what the personality cult was? 'Are you so illiterate politically that we must organize elementary school studies for you here?' As to the historical report, didn't Yeltsin realize the effort that had gone into it? It had been 'born in pain' in countless meetings to discuss various points. Shevardnadze had used the same phrase, adding emphasis, 'born in pain, in the literal sense'. In the literal sense! That must really have been something!

Gorbachev finished up by ominously dropping a few names. Politburo work was not easy, he said, and the comrades must remember what a large step forward they were making by comparison even with the immediate past. They must remember the nadir of the Brezhnev Politburo, as some members present could describe it for them, specifically Gromyko, Shcherbitsky, and Solomentsev. Gorbachev slipped in an unpleasant reference to three future targets. At the Nineteenth Party Conference in June there would be a call from the floor for the dismissal of former Brezhnev men Gromyko and Solomentsev. Gorbachev nevertheless stressed party unity, calling reports of conflict in the leadership 'empty talk, foreign radio gossip. They want to provoke quarrels among us, set Gorbachev against Ligachev, or Yakovlev against Ligachev, or whomever.'[13]

The plenum ended by resolving that Yeltsin's speech had been 'politically erroneous'. It was a defeat for Yeltsin but also a serious defeat for Gorbachev, who was not only on the verge of losing his point man in Moscow, but now also in the position of having to defend Ligachev against him. Ligachev himself emerged more powerful than ever. His old rival Aliev was finally removed from the Politburo, a development that was generally overlooked in the press speculations about the affair.

In fact it would be almost two years until the Soviet public and the rest of the outside world would learn the details of the discussion. Lukyanov made a vague report to TASS ten days later letting the world know that Yeltsin had indeed submitted his resignation. He permitted the impression that Yeltsin had made a real attack on Gorbachev,[14] rather than a rather mild criticism of the false adulation of 'two or three Politburo members', as Yeltsin had put it in his rejoinder. Yeltsin himself denied attacking Gorbachev, telling CBS that he had done nothing of the sort: 'No, no! I categorically deny it. I never said it. A categorical no!' When questioned by reporters, Aleksandr Yakovlev was elusive, saying only that 'if internal matters were discussed by one and all, there would be no sense in having a party'. The *Obshchina* group circulated a petition complaining about an 'information apartheid'.

A number of spurious versions of Yeltsin's speech were sent around.

One was published in the British *Observer* and broadcast in Russian on the BBC. *Le Monde* got a version with a very confident Yeltsin explicitly attacking Ligachev, Chebrikov, Raisa Gorbachev, and even Shevardnadze for his failure to end the war in Afghanistan.[15] This no doubt originated in Chebrikov's KGB shop, a production in the same spirit as the videotapes of Raisa that were circulated earlier in the year, with the wife of the leader looking very much the stylish bourgeoise out shopping in the West.

THE WASHINGTON SUMMIT

There was also an apparent spillover of the Yeltsin affair into the area of foreign policy. Only a few days after the plenum, while preparations were going forward for the December summit in Washington and an INF treaty, Gorbachev almost threw the process into reverse by changing his position of February and once again insisting that the INF question be linked to an agreement on SDI. The Chief of the Soviet General Staff, Marshal Sergei Akhromeyev, gave an interview in the office of Valentin Falin, then head of the Novosti press agency, on a closely related topic. He expressed deep concern that the American SDI programme, combined with an agreement to cut long-range missiles by half would 'radically step up the military threat to the Soviet Union'.[16] Some internal pressure, undoubtedly from the Ligachev forces, must have caused Gorbachev's about-face. Those who sought to slow down Gorbachev's reforms seemed also to be committed to the Reykjavik position linking SDI to any substantial arms reduction agreement. This bolsters the view that the quarrel with Ligachev began right after Reykjavik.[17]

Then Gorbachev changed position again, sending Shevardnadze flying to Washington to agree to a summit on 7 December. American officials and observers of the arms control process were amazed by the changes of course in one week. William Hyland called Gorbachev's behaviour 'reckless almost to the point of irresponsibility', wondering on second thought 'whether he is reckless, incompetent, or both'. The fear was that Gorbachev would come to Washington and repeat his previous attempts to get curbs on the SDI, in which case it was likely that the summit would simply explode.

We have only vague testimony about the Politburo deliberations of that extraordinary week. After their victory over Yeltsin, Ligachev and his supporters pressed for a reinstatement of the Reykjavik package, with the endorsement of the most respected military opinion. Whereupon Gorbachev apparently gave in for a few days, perhaps wondering if the New York stock market crash of hundreds of points which was under way that week would affect the American position. Gorbachev may then have gone along, arguing that it would not, and won the day shortly after. This would apply a pattern of challenge and response that fits what we know of the crises at the January plenum and the Andreyeva Affair a few months later. Gorbachev, confronted with fierce opposition, has to yield temporarily, but fights on and ultimately wins the day by demonstrating that the course of his opponents only repeats the intransigence of the past and will not yield results. Gorbachev was tearing out of their hands the old diplomacy of detente and its lessons of linkage. His opponents were clinging to the idea that SDI was the central problem, toward which all the other desiderata must be subordinated. They thought that this was the Gorbachev position. But Gorbachev was actually going away from the position of Gromyko toward the analysis of Andrei Sakharov, according to which the SDI could not be stopped by negotiation, but only by so changing the climate between the superpowers that SDI would lose its domestic support in the United States. Gorbachev was attracted to this argument, perhaps out of fear of another Reykjavik fiasco.

So he saw to it that the Washington summit would produce real results. The agreement on intermediate-range missiles that Gorbachev signed in December 1987 was a landmark success in the history of arms control. To get it, Gorbachev had to make numerous concessions on intrusive verification that permitted what was previously feared as an intelligence tour of the western USSR. But Gorbachev had succeeded through his concessions in gaining the initiative. He succeeded in weakening the resolve of Western governments by eliminating weapons that they had only recently deployed amid fierce popular protest.[18] He could claim that he had only traded missiles that could strike American allies for missiles that could strike the Soviet Union itself. He could also claim to have started the ball rolling for a possible de-nuclearization of Europe.

Gorbachev and his advisors had correctly seen the opening for their initiatives in NATO's dual track decision of 1979, to deploy the missiles but to seek arms control negotiations for their elimination. The West German 'zero option' position, adopted by the Reagan administration in 1981, further committed it to arms control, oddly, in the midst of its unprecedented arms buildup. Henry Kissinger and a number of other authoritative Western figures worried that the 'zero principle' threatened to unhinge NATO's deterrence of the USSR's superior conventional forces. This deterrence had been conceived for more than two decades in terms of the idea of flexible response to any future Soviet conventional attack, with relatively inexpensive nuclear weapons such as those affected by the INF treaty. These weapons were, moreover, an instrument of the various NATO conceptions of deep-echelon interdiction of a massing Soviet attack force. Negotiating them away without reducing the conventional superiority of the Soviet forces could only serve Soviet interests. Gennadi

Gerasimov was also correct to point out that those who feared the zero principle were in a position of being at odds with the dual track decision of 1979.[19]

In achieving the INF treaty Gorbachev also gained a march on Soviet authorities like General Ogarkov, who saw western defence strategies based on deep-echelon strikes as a mortal threat to the USSR. Instead of responding to the perceived threat by military preparations, Gorbachev seemed to have made a big step toward removing it 'by political means'.

Yet the victory in the arms control field was Gorbachev's only one. The Central Committee recital of 21 October against Yeltsin was certainly no advertisement for the advance of political culture under the regime of *perestroika*. The Stalinist dinosaurs had strutted their stuff in the traditional manner. Their trouncing of Yeltsin raised the question of who owned *perestroika*, and one could easily conclude that in the eyes of the Central Committee, Ligachev, not Gorbachev, was now its most reliable interpreter. Instead of citations of phrases from the Twenty-seventh Congress on Universal Human Values, one heard exhortations like those of Gromyko to maintain the traditional unity of the ranks against 'our class and political opponents', and to march together 'as preceding generations of Communists willed to us'.[20] It seemed that successes in improving relations with the United States were Gorbachev's only asset in this struggle, promising as they did a way out of the hard road of military preparedness travelled by the country for so long. Even the most unimaginative party official who deeply feared the ongoing changes in the internal regime nevertheless felt the attraction of lessening this burden.

A CENTRIST SOLUTION

The rally to Gorbachev's foreign policy was not, however, a rally to Gorbachev. It was a rally to party unity as defined by Gromyko to form a synthesis of the views of Gorbachev and Ligachev on the development, pace, and scope of *perestroika*. Not acceleration of the pace of reform, but consensus was the watchword. After a few months nothing more would be heard of the slogan of *uskorenie*. Yeltsin was induced to resign his positions in November, prompting the further expectation that Moscow would become a preserve for Gorbachev's opponents. Lev Zaikov was appointed to replace him. Zaikov was at that time in both the Politburo and the Secretariat. He had come up through the ranks in Leningrad, starting out as a worker in a factory and, after having risen to the position of director, attaining a night school degree in engineering. He had come into the Central Committee in 1981. When Romanov was brought to Moscow to work in the Secretariat in 1983, Zaikov was made first secretary of the Leningrad *obkom*, a promotion that led some to think that he had been Romanov's protégé. One French observer concluded on the contrary that, since he had been involved in defence industry work throughout his professional life, he must have been a co-thinker of Ustinov, who helped defeat Romanov in 1984.[21] Mikhail Poltoranin, the point man for point man Yeltsin as editor of *Moskovskaya pravda*, complained that Zaikov and Ligachev were 'like Siamese twins'. On the other hand, Gorbachev developed the habit in the first months of 1988 of referring to 'Comrade Zaikov and I' being of this or that opinion. So there were reasons to suppose almost anything about Zaikov's commitments in the present struggle, but the most sensible supposition was that Zaikov was in an enviable position at the fulcrum in a balance between Gorbachev and Ligachev. He would have been unwise to jeopardize this position by tilting too far toward one or the other, especially when they themselves were filling the air with expressions of their unshakeable unity.

Gorbachev's new line was that *perestroika* was besieged on two fronts: on the one side by the braking mechanism and on the other by 'political avant-gardism'. *Pravda* complained at the end of the year that

> there have emerged leftist, avant-garde sentiments and aspirations to do everything at one stroke, and with this, panic in the event of failing to do so ... Conservatism and skipping stages are two sides of the same coin.[22]

Many, however, sensed that the balance had swung to the side of Ligachev. Yakovlev disappeared for several weeks at the end of the year, making no statements at a time when he would have been thought the most prominent spokesman for the cause of accelerating reform. The economist Nikolai Shmelyov feared the ascendancy of a 'torpid and irresponsible bureaucracy' and pleaded: 'The issue of vanguardism is now in the background. Is it now possible to shift the issue of conservatism there as well?'[23]

Ligachev went to France in December to attend the Twenty-sixth Congress of the French Communist party, receiving *Le Monde*'s Michel Tatu, one of the keenest Western students of Soviet politics, at the Soviet embassy. He spoke at length of the aims of *perestroika*, defining it as 'democracy with economic reform'. He described himself as a supporter of socialist pluralism, especially pluralism of opinion in the spirit of *glasnost*. And yet: had he not criticized the new spirit of freedom in the Soviet press? 'It's true, I've criticized *Ogonyok* and *Moscow News*, but we've also criticized *Pravda*, *Izvestiia*, *Sovetskaya Rossiia*, and others ... if the press can criticize everyone, why can't one criticize the press?' Ligachev pronounced himself in favour of the fullest efforts to fill in the gaps in the present knowledge of party history. 'To recover the truth in its totality – I repeat, in its totality – is indispensable for us if we are to draw lessons from it for the future.'[24]

One would conclude that Ligachev was by no means a hard-bitten Stalinist hoping to shut down the campaign for *glasnost*. Stalin's purges, he said, 'had nothing in common with the principles of socialism, nor with the norms established by the Constitution of 1936' (the constitution written by Bukharin and Radek before they were arrested). Yet his call for the full truth meant *balance* between revelations about the crimes of Stalinism appearing in the *glasnost* press and historical perspective about the correctness of the general line of Stalin's political fight for socialism, as regards 'the heroism, the drama' of those years, as he had put it to a meeting of cultural workers in 1986. One must realize, he had argued, that the reforms 'were travelling and would always travel along the path to socialism . . . more socialism, the maximum socialism'.[25] Gorbachev would repeat these phrases over the next two years, 'more socialism and again more socialism'. One could say that he only did it for tactical reasons, but to prove this one would have to be able to read his heart. In public utterances there were only very subtle differences between the two men.

In the *Le Monde* interview, Ligachev presumed to speak for himself and for Gorbachev. The only fundamental difference between the two seemed to lie in the Regime Question, that is, Ligachev's view that reform, however desirable, could not be achieved by 'shaking up the cadres'. One might recall that Gorbachev's speech at the preceding January plenum had held the opposite message. Things had definitely taken a turn from those days. Ligachev pronounced himself the guarantor of the continuity of the cadres. As a footnote, he casually remarked to his interviewer that he, not Gorbachev, chaired sessions of the Secretariat, something that was not known at that time.

The opinion pieces on the editorial pages in the American newspapers were filled by then with speculation about the real differences between Gorbachev and Ligachev. Some said that Ligachev was a Stalinist, others that Ligachev was for *perestroika* but not *glasnost*, or that he did not like the New Thinking in foreign policy. One could find differences in emphasis in speeches by the two men, but they were committed in their public utterances to the same line, down to the details and the specific phrases. Ligachev was trying to be a good second secretary in the mould of Suslov and was making his contributions to ideology, usually now emphasizing the need for frankness about the past alongside respect for continuity. Gorbachev was arguing essentially the same thing. Now it was even possible to put the innovative *feuilletons* of the *glasnost* press under the heading of 'political avant-gardism', with Gorbachev and Ligachev standing squarely in the centre. Yet the reason that their intellectual consensus could not hold was that they could not agree on the regime question, in this case on the question of whether or not the cadres should continue, post-Yeltsin, to be shaken up. A letter to *Moscow News* from Yuri Volkov, a Philosophy professor and supporter of *perestroika*, put the issue into sharp relief by stressing the need to continue the renewal of cadres. It was necessary to have a second wave of changes and even a third, he complained. The new leaders of the apparat decry the sins of the past at first, but after a few weeks on the job, they behave exactly like the people they replaced.

While Ligachev was in Paris, Gorbachev was in Washington signing the INF treaty. Yet, after this triumph, he did not return from the summit in a triumphant mood. Having reached a historic accord with many advantages for the USSR, Gorbachev had nevertheless abandoned the attempt to tie it to American restraint on the SDI. Reagan spoke glowingly of the INF agreement and the hopes for a START treaty involving 50 per cent reductions in strategic arsenals, but he also said: 'This could be another historic achievement, provided the Soviets do not try to hold it hostage to restrictions on SDI ... Building a defense against nuclear weapons is a moral as well as strategic imperative and we will never give it up.'[26] The Soviet press hailed the new agreement on INF, but cautioned about a revived spirit of anti-communism in the United States. Speaking to Soviet television on 14 December, Gorbachev himself warned that 'certain circles in the U.S. and other western countries' were demanding 'compensation' for the INF in another round of nuclear force modernization in Europe. These people were under the impression, he said, that differences on SDI had been resolved and were actually calling for speeding up work on that programme.[27]

CONVERTS AND INTRANSIGENTS

On Reagan's side there was general approval of the agreement, but anger on the part of some conservatives, who continued to be dissatisfied with the Soviet human rights record and the apparent lack of Soviet movement on Afghanistan. Gorbachev would soon move to satisfy them by his announcement of 9 February that the troops would be out by 15 May. Gorbachev had admitted in an interview with Tom Brokaw that the Soviets were doing research work on their own programme of ballistic missile defence, but that they would scrupulously observe the provisions of the ABM treaty, and would not deploy their system. When the moment for that came, said Gorbachev, the Soviets would have an 'asymmetrical' and cheaper response.

The hope expressed by the two sides in Washington was that START might begin by June 1988, with reductions to 4900 ICBM and SLBM warheads per side. But this did not dampen the feeling of urgency in the West for modernization of the other nuclear systems, in the spirit of NATO's 1983 Montebello accord. Margaret Thatcher was convinced of its necessity, German Chancellor Kohl expressed reservations, with French President Mitterrand warning against 'over-arming' at a time of good prospects for disarmament. Serious and influential opinion in Western countries regarded the modernization of short-range missiles to be an imperative to maintain the complex structure of deterrence according to flexible response in the face of the continued existence of a Soviet conventional superiority.

Soviet press opinion considered Thatcher to be the leading voice for this view. But Gorbachev enjoyed a reputation for having convinced Thatcher that business could be done between the two. That prompted confidence that the New Thinking could conquer even the most sceptical. Accordingly, special efforts would be made in the following months to open up a dialogue with those considered by Moscow to represent the political Right in the US. *Moscow News* would carry interviews with Zbigniew Brzezinski and Richard Pipes that dealt, not only with problems of military security, but with the progress of Soviet internal reform. There was a desire to impress even the hardest critics of the reality of the reconstruction of Soviet life.

One important conversion was already accomplished. Ronald Reagan told the *Washington Post* at the end of February that Gorbachev was very different from the Soviet leaders to which the West has historically been accustomed.

> having met most of them ... I think that one difference is that he is the first leader that has come along who has gone back before Stalin and that he is trying to do what Lenin was teaching ... with Lenin's death, Stalin actually reversed many of the things. Lenin had programs that he called the new economics and things of that kind. And I've known a little bit about Lenin and what he was advocating, and I think that this, in *glasnost* and *perestroika* and all that, this is much more smacking of Lenin than of Stalin. And I think that this is what he is trying to do.[28]

In the first news conference of his presidency, on 29 January 1981, Reagan had said that Lenin and the Soviets aimed at world domination since the founding of their state and had reserved 'the right to commit any crime, to lie, to cheat', in the name of this cause. But the White House staff had been following Soviet developments of the last few years and discussed some of the historical questions being debated in the USSR. Apparently they were also familiar with the interpretation of Lenin provided in Stephen Cohen's biography of Bukharin, which so interested Soviet advocates of a 'Bukharin alternative',[29] We will have more about this below. At a reception during the Washington summit, Gorbachev was introduced to Cohen; the Soviet leader told him that he had read his book. Probably no Western scholar has ever had a similar experience with a Soviet leader. It is not every day that an historian can have an impact on history by virtue of his ideas.

Nevertheless Gorbachev had not made enough converts in the Soviet Central Committee. His summit performance was graded as a balance of pluses and minuses, with the failure of SDI linkage a sore toe on which Ligachev or other rivals could step at will. Gorbachev's assessment of the 'Revolution' was that *perestroika* had completed its first, essentially theoretical, stage. The point now was to carry this message to the masses in the process of deepening the restructuring of Soviet society. Did that mean that the discussion on the origins of Stalinism was to go no further than the 1987 historical speech? It was hard to say. 'We are criticized from the right and the left', he told a meeting of mass media executives and ideological workers. The phrases of *'ultra-perestroika'* had proved to be useless. There was no point, he said, in stepping up the reshuffling of personnel. At the same time he took note of a critique 'from the Right' erroneously arguing that the foundations of socialism were being undermined. It would be unrealistic to think that the last battles had been fought against this Braking Mechanism.

Gorbachev claimed that some economic successes had been registered: rates of growth for the productivity of labour had increased, even while the work force itself was decreased. Gains in remuneration of labour were now effectively tied to productivity gains. In engineering and technology a start had been made toward the complete independence of those sectors. Gone were the days when the Soviets would sell oil willy-nilly to foreign buyers in order to buy equipment and spare parts, rather than investing in primary engineering to make the equipment and cure the nation of its 'import plague'. On the other hand, in order to find new ways of working it was necessary 'to improve the morale of the people', by telling them the truth about present problems and about the history of the country. The 70th anniversary speech had been a 'principled stance' on history but it was 'not dogmatic'. There would be further research by a special commission, hopefully to result in a series of historical essays and 'a good, truthful book that could become a textbook of CPSU history'.[30]

Only in Gorbachev's remarks about further investigations into history was there evidence of a will to free himself from the embrace of Ligachev. In his speech to the February plenum he stressed that reforms could not go forward 'without reliance on theory and without ideological substantiation'. He also gave emphasis to a 'new role for universally shared values', which must be understood to be compatible with materialist dialectics.[31] These two things were for the moment as compatible as Gorbachev and Ligachev. From this point on, however, speaking about 'Universal Human Values' would identify one as a critic of Ligachev, and defence of the perspective afforded by class struggle would make one a supporter of the Second Secretary. As for Ligachev, he was enjoying to the full his status as ideological watchdog. The February Central Committee plenum was devoted to education, a field on which Ligachev sought to make an imprint. He spoke expansively on the need to reduce the power of the Education Ministry, and for enterprises to make more investments in schools. Teachers, he said, ought to have their own union (teachers were at the time virtually the only profession without one). As to history, however, the main lesson for him was that it was from the older generation that the youth learned devotion to socialism.

Ligachev complained of excesses in the treatment of the past in the press. Describing the work of the February plenum, *Pravda* gave Gorbachev and Ligachev roughly equal billing, but its editorial, probably written by Viktor Afanasyev, spoke Ligachev's language:

> Haste in making assessments is not acceptable, nor is throwing before the public substandard works that merely follow the prevailing winds and obscure the truth. Unfortunately, such writings can be encountered in certain of our newspapers and journals. These publications are making up various fabrications without having on hand facts or scientific evidence.[32]

This was the new tack of those who sought to slow down the discussion of the Stalin period. The debunkers did not have documents to back up their assertions! Where, one might have asked, were they to get such documents without access to the party archives and the publication of the works of Stalin's opponents? This complaint was apparently a bone of contention at the plenum. Irkutsk First Secretary V. Sitnikov described a debate that was 'enthusiastic, heated, I would say fast and furious'. Dzumber Patiashvili, Georgian First Secretary, spoke of the 'shameless distortions' of historical facts used by some writers in order to 'blacken the names of certain revolutionary and war heroes'.[33] The two sides were able to agree about avant-gardism, ultra-*perestroika*, and other epithets associated with Yeltsin, but they remained deeply divided about the *glasnost* press itself.

STALINISM AGAIN

As had been the case one year before, the logjam was to be broken by renewed controversy on the Stalin question. Not that this debate had ever really let up. The discussion was, for all its astonishing aspects when viewed from the perspective of what had gone before, a stringently controlled thing. Even the editors of the flagship periodicals of *glasnost* that so shocked the 'conservatives' were acutely aware that they were doing something for which others had been shot in the past, and were careful to administer the right dose of debate on the right topics, to the degree that this was possible. Even so, it was hard to control. Thinking about Stalin was explosive for the intelligentsia and for the broad Soviet reading public that dwarfed the readership of the best known and prestigious periodicals in the West. Even a thick journal like *Novy mir* had a print run of over a million. In 1986–87 the press had been filled with revelations on many topics, but the most explosive topic was Stalin and the ideological foundations of Soviet Communism. In 1988 the Ministry of Communications, citing a paper shortage, cut back subscriptions to many of the leading *glasnost* periodicals. These gentlemen knew that any discoveries about the period could only further inflame public opinion against the authorities. It was only the unusual Soviet family who did not have some member who had been affected by the repressions of the Stalin era.

In 1987 the biggest sensation had been the publication of Anatoli Rybakov's sprawling social novel *Children of the Arbat* (*Deti Arbata*), written some twenty years before and debated in cafés by people who had perhaps only read reviews, but who judged the political climate by what they heard about official attitudes toward the book. It was always on the verge of being published, but it had remained 'on the shelf' with other potential classics until the advent of *glasnost*. When it was finally published, it failed to impress Western critics who gave it disappointing reviews. It was received much differently in the Soviet Union. The extensive 'non-fiction' sections detailing the situation in the Stalin regime in 1934 were a special treat for Soviet readers who were curious about the eve of the great purges and what had always been thought to be the triggering event, the assassination of Sergei Kirov, head of the Leningrad party and potential rival of Stalin.

The material in Rybakov's chapters dealing with Stalin amounted to an education in the political background of the purges. Rybakov acquainted the reader with Stalin's relations to Zinoviev, Kamenev, Bukharin, Yagoda, and others. He traced the rise of Stalin to the struggle between Zinoviev and Trotsky in 1922–23. He showed familiarity with the attempt of Bukharin and Kamenev to make a bloc against Stalin in 1928 (while he put the date mistakenly in 1926). He gave information on the 1932 crisis over the question of whether Stalin should be allowed to shoot Ryutin. He described Kirov as the rallying point for the Politburo group that restrained Stalin on that occasion, and gave a many-faceted analysis of Kirov's power in Leningrad, an 'undestroyed bastion of opposition', as Rybakov's Stalin calls the second Soviet capital. Rybakov left no doubt, in a plot that leads to the assassination and to which the assassination gives meaning, that Stalin had Kirov murdered.[34] He also speculated acutely on the views of Kirov about Stalin. Rybakov has Kirov saying:

> His line was correct, but his methods were unacceptable . . . the party made a great error in not taking Lenin's advice and removing Stalin as general secretary. They should have done it. Trotsky would still not have got the upper hand, for he was regarded as an outsider by the party. Nor would Zinoviev and Kamenev have become the leaders, for the party did not trust them. The party would have been led by its true Bolshevik core, the present Politburo, plus Bukharin and Rykov, and even Stalin, but as an equal member. Well, it was impossible to correct that error. Stalin was now immovable.[35]

In this thought attributed to Kirov lay the ideal that underpinned much of the historical analysis brought out by *glasnost*: Bukharin's line and general interpretation of genuine Leninism, without the terrible Stalin regime. It amounted to a kind of nostalgia for the days of the Stalin–Bukharin bloc of 1925–27, with Stalin only a staff sergeant in the armies arrayed against Zinoviev and Trotsky. The question that was not explored was: why, if this balance had been so perfect, did it not last? Or, if it had not lasted, could it have been so perfect?

Rybakov's Stalin is very much a product of the Russian *ancien régime*, inspired as much by Ivan the Terrible, whose reputation he wants to save from Marxist historians, as by Lenin. The characterization was not subtle enough for some reviewers, while others found insight in the idea of Stalin as as an admirer of the despots of the past and a student of Machiavelli.[36]

In 1987 there had been no massive manifestation of historians who wanted a review of party history. We have already noted the failure of Yuri Afanasyev's efforts before the counterattack of Vaganov and others. The watchword among professional historians of the Soviet period was the struggle against stagnation and dogmatism in historical science, but they did not prove able to present new material to underpin the search for a new basis for the reforms. A meeting had been held in April 1987 with many of the most famous historians. P. V. Volobuev, who had suffered from the reign of the party ideologists of the Suslov era, understandably decried the 'anti-Leninist methods' of the time, praising the prodigious research that had gone into Khrushchev's secret speech, but lamenting the failure to follow it up and dispense with the dogmas of the *Short Course*. The respected historian of the collectivization, V. P. Danilov, found inadequate the critique of Tikhonov (see Chapter 5) and the novelist Mozhaev, according to which Stalin's excesses were attributable to his having adopted the earlier views of Zinoviev, Preobrazhensky, and Trotsky. Danilov insisted that it would not do to ignore the perspectives of people like Bukharin and Trotsky, while dismissing them by referring to their positions in terms of 'the Left and Right danger'. V. S. Lelchuk saw the debate on the NEP as the central question, with a considerable body of opinion in the party and government, as he saw it, prepared to dispense with it and undertake a vast industrialization drive. When Trotsky criticized NEP in 1923, he had been condemned, only to see others raise the same critique at the end of the twenties. 'Was this accidental or not?'[37]

There was dissatisfaction with the old official versions of Soviet history, but no new version. The General Secretary had been expected to give a lead, but the speech of November 1987 had been a disappointment. Then a second round of *glasnost* debate on the Stalin question was reopened in January 1988 when the Journal *Znamya* published Mikhail Shatrov's play *Dalshe, Dalshe, Dalshe* (Onward, Onward, Onward). Lenin, Stalin, Trotsky, and many other historical figures appeared on stage to debate their actions, with a condemnation of Stalin by Lenin as the centrepiece. Some of the myths about Trotsky were dispelled, but as with Shatrov's earlier play, *The Brest Peace*, the case against Trotsky was prominent. In *Onward, Onward, Onward,* Shatrov attacks Stalin for his Trotskyism. The whole question of Trotsky's 'betrayal' of Lenin in October 1917 is reviewed, but the version presented in the *Short Course* of 1938 is not really disputed, except in the charge that Stalin too was trying to obstruct Lenin in carrying out the October insurrection. It is denied that Trotsky's disagreements with Lenin after 1917 stem from his being in the pay of foreign intelligence agencies, as was at one time asserted, but it is suggested instead that these are attributable to Trotsky's pre-1917 non-Bolshevism. There is frank treatment of many important matters. Shatrov makes it reasonably clear that Stalin was responsible for the murders of Kirov and Trotsky, and for the Ukrainian famine of 1932–33. As with most of the previous explorations of the *glasnost* literature, the revelations were powerful. However, instead of Stalinist propaganda the reader was now offered Bukharinist propaganda.

Nevertheless, that was enough to provoke determined resistance. V. Glagolev, in *Pravda*, professed not to be impressed with Shatrov's play, in which he found 'only a semblance of historical objectivity'. While it was true that 'even time will not soften the tragedy of 1937', there was no reason to abandon the struggle for real knowledge of the laws of historical development.[38] Whatever Viktor Afanasyev's propensities had been at the time of the 'Berkhin Affair', *Pravda* in his hands was now a determined supporter of the Ligachev line.[39] At the February plenum, Gorbachev himself referred disparagingly to 'recent lopsided appraisals of history'.

These appraisals, lopsided or not, were building an historical case against the existing apparat. *Kommunist* published a critique of the Zhdanov cultural policy of the forties. *Moscow News* presented interviews with the daughters of a number of Old Bolsheviks shot in 1936–38, including the daughters of Smilga, Lomov, and Krestinsky. These could be added to the rehabilitations of N. I. Muralov and V. I. Mezhlauk that were made the previous year. V. M. Kuritsyn, a professor at the Moscow MVD militia school, published an article on the purges that traced their horrors to the decision to collectivize agriculture and the onset of the 'administrative-command system'.[40] The 'decisive role' in the industrialization was played, not by Stalin, but by the labour of millions of Soviet people, said Kuritsyn. The excesses against the kulaks, he maintained, were needless in view of the fact (and here he followed Tikhonov) that they were already essentially liquidated by the civil war. He spoke of Stalin being resisted in his repressive course in 1932 by Kirov, Ordzhonikidze, Kalinin, Kosior, and Postyshev and encouraged by Molotov and Kaganovich.[41]

Fyodor Burlatsky argued for a renewed appreciation of Khrushchev as the political heir of Bukharin and Kirov.[42] David Gai described the affair of the Doctors' Plot in 1953, and Stalin's order for the arrest of his doctor, Professor Vinogradov, after the latter had proscribed any political activity for him. Gai remarked that 'Stalin remembered his role in isolating the leader of the revolution [Lenin] from party work – under the pretext of concern for the latter's health.'[43] Gai was hinting about the idea of Stalin' having had something to do with Lenin's death, a speculation of Trotsky's just prior to his own assassination.

THE BUKHARIN ALTERNATIVE

More and more juicy morsels about Stalin's crimes were being exposed with their shocking impact on a public opinion that not only had not seen things like this in the Soviet press, but had been schooled to believe only in the positive contributions of 'the generation that built socialism'. Where was this leading? There had to be some political lesson to give meaning to the Grand Guignol of Stalinism. And *glasnost* was arriving at the view that the lesson was a Bukharin path as an alternative to Stalinism. This had not been argued by Gorbachev in November. But in the wake of Gorbachev's speech, attempts were made to interpret its meaning in a way favourable to Bukharin. Lev Voskresensky noted the service of Bukharin in routing the Trotskyites (a service to the rising power of Stalin). He acknowledged Bukharin's mistake in opposing the turn to collectization in 1928, but called attention hopefully to Bukharin's change of heart when he joined (with Stalin) in fighting the 'Right Danger'. Voskresensky had visited Bukharin's widow and now gave an account to Soviet readers of Bukharin's 'Testament', a few paragraphs that he had asked her to memorize on the eve of his arrest in 1937. 'I knew nothing about Ryutin's and Uglanov's secret organizations', Bukharin had said. 'I was fond of Kirov and I haven't schemed in any way against Stalin.'[44] Bukharin had been tried in 1938 as an organizer of a Bloc of Rights and Trotskyites that had existed since 1932, in fact as being part of the Ryutin group. According to what is known from sources available in the West, Bukharin was indeed aloof from the Ryutin conspiracy in 1932, and did not scheme in any way against Stalin. On the occasion of Kirov's murder, he even wrote an article blaming it on Zinoviev and Kamenev, and called for stern action against these 'Charlotte Cordays' of the Russian Revolution.[45] This would be supplied by the first Moscow trial of August 1936, after which Zinoviev and Kamenev were shot. Bukharin was by no means the sentimental gentleman that some of the Soviet intellectuals had been describing.

Nevertheless, he was at this moment emerging as the patron saint of *glasnost*. In February, a plenary meeting of the Soviet Supreme Court repealed the 1938 verdict against Bukharin and his co-defendants. 'Why', asked Yevgeni Ambartsumov, 'didn't this happen before?'[46] He answered by citing the resistance of 'the so-called champions of Marxist purity'. These men had kept the truth from the Soviet people, and even now were denying them access to the writings of Bukharin, which might serve as 'an ideological foundation for *perestroika*'.

Glasnost had arrived at its own myth of Bukharin: the consistent opponent of Stalin and defender of democracy and market forces in the Soviet economy. Gorbachev's opinions on Bukharin's ideas, however, were not on record, but those of people assumed in the West to be supporters of Gorbachev definitely were. The question was: did Gorbachev support them? There can be no doubt that the rehabilitation of Bukharin enhanced the image of Soviet soul-searching that had made an impact in the United States. It was largely the Bukharinist interpretation of the relation between Lenin and Stalin that moved Reagan to his remarks about the profound changes in the Soviet leadership. The effort to convince the American public that the Soviet reforms were genuine was now bolstered by citing the Bukharin revival. An American columnist informed his readers that there had once been a great struggle for power between Stalin and Bukharin over the permanence of NEP. The Moscow trials of the thirties had been staged to persecute those who still supported NEP. He concluded: 'To rehabilitate those whom Stalin killed in the spy trials and their aftermath is to rehabilitate also the policies that Bukharin and his friends died in defending.'[47]

For purposes of myth-making, the succession struggle after Lenin's death was being reduced to the economic issue: for or against NEP, while in fact it was a complex, many-phased battle involving changing positions on economic policy, historical debates, foreign policy and Comintern matters, alongside the crucial question of what kind of regime they were to live under. It was a debate in which Bukharin, except for a short phase in 1928–29, supported Stalin and rendered him indispensable help in building his power. Far from struggling to make NEP permanent, Bukharin actually regarded NEP as a regime of transition to a system without a private sector. None of the other contestants, neither Trotsky, nor Zinoviev, nor even his primary opponent in debate, Preobrazhensky, whose views represented the extreme position in favour of heavy industry and planning, wanted to end NEP. Stalin ended up going far beyond all of them. In 1926–27 Bukharin supported a series of measures, including restrictions on the hiring of wage labour and a superprofits tax, in order to hem in the private sector. When Bukharin engaged in economic debate with Preobrazhensky, the latter was a member of the defeated opposition of 1923 that had made a stand for inner party democracy, against whom Bukharin could present himself as the main theorist of the Stalin group. The fight against Preobrazhensky's economic ideas was also a mopping-up action against his old ideas about the abuses of the party regime. By branding everyone who disagreed with Stalin as one or another species of Trotskyite, Bukharin helped to establish 'Marxism-Leninism' with much of the falsification of history that was to be found in the Stalinist *Short Course* of 1938.

The fact was that Bukharin was with Stalin against Trotsky in 1923–25, with Stalin against Zinoviev and Kamenev in 1925–26, with Stalin against Zinoviev, Kamenev, and Trotsky in 1926–27. After a brief Politburo fight when Stalin abandoned NEP in 1928, Bukharin embraced the collectivization and Five-Year Plan, and refused contact with his old supporters. As he protested in his Testament, he permitted no succour to Ryutin or anyone else who gave even a hint of being against Stalin. In fact he wrote effusively in support of every measure of Stalin's, including those which led to the Moscow Trials. Recent interviews with his widow make clear that he did not protest the shootings of the defendants in the first Moscow Trial, because he thought they deserved what they got. More could be said about Bukharin, and moral judgements of the Old Bolsheviks are by no means easy for the historians who must remember the straitened circumstances in which they had to operate. Yet Bukharin found himself in the impossible position of all those who hitched their wagon to Stalin's star and told their lies for him, calling for the sternest measures against his opponents, and hoping that they could survive by asserting their loyalty to him.

In view of Buhkarin's indispensable work in building the ideological and political foundations of Stalinism it would be less accurate to consider him an alternative to Stalinism than a pole of Stalinism, a Right-Stalinism. The Stalinist police state was robust well before the 'administrative-command' system made its appearance with the five-year plans. Stalin was a Bukharinist in 1925–28, and turned again to the right in 1933, when Bukharin returned to favour with the internal return of the neo-NEP policy of the second five-year plan. Shooting Bukharin in 1938 marked another turn toward a new and harder policy internally and toward the pact with Nazi Germany. Years after Bukharin was gone, Stalin again turned to the right in 1948. This process of turning right and left was a distinguishing characteristic of Stalinism throughout the life of the dictator.

While the historical Bukharin personified a moderate pole of Stalinism, the mythical Bukharin of *glasnost* represented a softer Communism that was compatible with democracy and a mixed economy, Stalinism being reduced in this characterization to the leftism of post-1928, with no attempt to take account of Stalin's turns. This soft Communism was held up as a reformed Communism capable of taking up the tasks of the historical social democracy. This at any rate was the hope raised by neo-Bukharinism. In fact, Communism reforming according to this myth was a collapsing Communism, as the revolutions of 1989 were to show. One could experiment with minor market elements as Stalin had in the past, but a serious transition to a market economy with the one-party dictatorship in place would have made the Soviet Union a kind of fascist state. If this transition were accompanied by the development of real democracy, this would mean the end of Communism. A case of either/or.

In some respects an accurate historical picture of the various positions essayed by Bukharin, however important for posterity, would miss the point. Bukharin became known as the representative of a line of solicitude for the Soviet consumer and the enterprise factor in the Soviet economy, as the main opponent of Stalinism from a stance more recognizable to democrats in the West. The West, insofar as it had wanted to see Soviet behaviour change, has wanted it to change in this direction. The emergence of the Bukharin myth thus represented something about which the West could be hopeful.

American foreign policy toward the Soviet Union had never been based on detailed understanding of Soviet intellectual processes, despite the fact that the Soviet Union had been studied as no other country has been. The really influential discussions were not carried on with those who studied the problems but with generalists who distilled the studies of others. And this distillation was unavoidably adjusted to suit definite constituencies and their predispositions. No one wanted to wrestle with the nuances of Soviet politics except to enlighten a specific policy purpose. That is, the tendency was for knowledge about Soviet politics to become transformed into myth. Even while some might have sensed this, they nevertheless had to rejoice that the myth was producing good things. At the same time, one had to hope that one was not being led into another in the series of disillusionments fed by exaggerated hopes that the Soviet Union had changed for the better. Some observers expressed unease at the memory of previous disappointments. This memory could not be completely banished by the bracing words: 'That was then; this is now!'

Soviet writers, on the other hand, were fond of claiming that there was a way to approach politics and history scientifically, but in practice *glasnost* did not produce science but only more ideology. The myth of Bukharin was now building up a crescendo of excitement preparing the way for a massive assault on the 'administrative-command system' of the Soviet economy, the system that had been brought into being when the Bukharin alternative was defeated in 1928. Whatever one might say about this economy, and one could say much against it, it was the system under which the country had been modernized and urbanized, under which the society had lived for 60 years. It would be a departure to dismantle it before the peoples' eyes. But that was now in prospect. No more 'acceleration of the perfection' of the Soviet system. And, the advocates of *perestroika* had reason to hope: no more Ligachev!

8 Another Escape Forward, 1988

> Gorbachev is a peculiar pendulum. He must swing to stop, not to start, the clock: If the clock starts to move decisively, backward or forward, his days of indispensability are numbered.
>
> Melor Sturua

The struggle of 1987, fought out between the Rejkyavik and Washington summits, had ended for Gorbachev in a limited but ominous setback. His most enthusiastic bulldog, Yeltsin, had been reduced, and as far as could be determined in view of past practice, virtually eliminated from the top leadership. Gorbachev himself had been forced to join in the campaign against Yeltsin, in what amounted to an acknowledgement of the limitation of his own powers and especially of his ideological ambitions, a limitation imposed by the Second Secretary Ligachev and the forces defending collective leadership. A future of carrying the bags of this collective leadership yawned before the General Secretary like a chasm. Would Gorbachev be willing to accept this reduced role without a fight? If not, the only way out would be through another 'escape forward'. At the beginning of 1988, therefore, he launched another campaign against the ideological foundations of his party opponents. Wriggling out of his compromise with Ligachev on party history, as expressed in his November 1987 speech, he encouraged the *glasnost* periodicals to press on for a second round of their examination of Stalinism. Railing against the diehards who had clung to the 'Reykjavik package', that is, linkage of the American SDI to any future arms control agreements, he urged broader consideration of the need to put foreign policy in terms of Universal Human Values and New Thinking.

The second round of his fight against collective leadership would produce his greatest victory, at the September–October Central Committee plenum, a victory in much different circumstances from those of January 1987. By 1988, the Soviet economy appeared to be responding to the reforms. After an initial spurt in the growth of national income to four per cent in the first year of the Gorbachev leadership, 1986–87 had returned matters to the condition of mid-1985, but between the promulgation of the Enterprise Law of 1987 and the September–October plenum of 1988, growth shot back up.[1] The most encouraging reports were from the European republics, especially Lithuania, Belorussia, and Moldavia, with the Central Asian republics showing less growth than the average. This performance seemed to verify the correctness of the party's course of 'acceleration' originally agreed in 1985 and augmented in 1987 by laws on enterprise autonomy and cooperative initiative.

The general goal of Gorbachev's 'acceleration' was to reinforce 'intensive factors of production' in order to provide by 1990 a platform for the next step of the campaign to reach Western technical levels. This was conceived by Gorbachev's economic advisors as no less than a struggle to reorient Soviet society itself. But each measure of Gorbachev's seemed to produce the opposite of its intended effect. Hand in hand with acceleration went a pious kind of social engineering to perfect the sculpting of 'Soviet Man', a passion with which both Gorbachev and Ligachev were taken. The campaign against alcohol was typical. To be sure, Gorbachev's predecessors had fought this fight themselves, usually with advertising campaigns and various limited restrictions on the sale of alcohol. But Gorbachev went much further. Beginning in 1985, sales were limited to five hours a day, with the number of shops that were permitted to make sales reduced by half. Drunkenness was prohibited at work, and often punished by firing. Subsidies were removed in order to make alcohol more expensive. The effect, however, was to make moonshine more profitable, to fill the coffers of a rapidly expanding organized crime network, and most important, to deny the state the tens of billions of rubles it would have collected in revenues from the state shops. So much sugar was used to make the moonshine that it had to be rationed. And alcoholism itself actually increased.

This was one programme that worked too well, and not the only one. The campaign for quality control in industry through a system of state inspection (*Gospriemka*) resulted in the rejection of 15 per cent or more of the monitored production. There was a serious decline in gross output. Factories were sometimes shut down for days to suit the inspectors, their managers complaining that 80 per cent of the defects were caused by the inputs rather than by workmanship.[2] Angry workers struck at the inspected factories.

The new problems of the Soviet economy were cheerfully explained by Gorbachev and his supporters by saying that the legacy of the period of the Stagnation under Brezhnev was more onerous than they had first thought. Real figures on budget deficits and inflation did show a more dire situation than had been anticipated in 1985, when 'acceleration' was first announced. Yet the graceful decline of the Stagnation years does not fit the profile of the total collapse that eventually set in, nor can arguments designed to explain the long-run deficiencies of the Soviet command economy serve to explain the catastrophe that overcame the economy in the winter of 1990–91. For better perspective on these problems it is necessary to look to the reforms themselves, the pressures prompting them and the philosophy underpinning them.

The quaint but powerful opinion of many of the intellectuals and experts with the most influence on Gorbachev was that the economy suffered from the fact that the Soviet worker did not really consider himself to be the owner and beneficiary of the assets of the first workers' state. If Soviet agricultural productivity was only one-fifth of that of the Western countries, this could not be explained, they held, as had been done in the past, by charging it up to insufficient investment in machinery or blaming unfavourable weather. As one statistical survey put it, the gap between Russia and the West was the result of 'the attitude toward the land and work, and employees' low stake in improving performance'. Investments in agriculture were in most cases to replace farm equipment that had gone out of service, 'not only because of the poor quality of the machinery', as one writer put it, 'but because of peoples' careless attitude toward it'. Abel Aganbegyan, at that time the economist reputed to be closest to Gorbachev, was convinced that the way forward must be found by initiative from below:

> The most important lesson to draw from an analysis of the past concerns the need for the democratization of society as the indispensable element in a successful *perestroika*. Only through democratization can the majority of working people be brought into the management process. The way forward lies in self-management.[3]

According to Aganbegyan's vision, the central aim of economic reform was to find an answer to 'the problem of inculcating in every worker the feeling of being a co-owner'.[4] To this end the Law on Enterprise provided for the election of managers by the workers themselves. The first result of this blow to the 'administrative-command system' was that popularly elected managers cemented their popularity, and defended their jobs, by abolishing existing wage ceilings and yielding substantial wage increases, without any corresponding increase in productivity. This only fuelled inflation.

How should one interpret the thinking behind this ill-fated reform? Is the worker's feeling that he is a co-owner the solution to the problem? Do workers work hard because they identify with the firm or do they do so in the hope of making money that will buy the things (or the opportunities and experiences) that they and their families want? If the latter is the answer, it would have been less important to increase democracy on the shop floor in the absence of consumer goods than to produce the consumer goods first, in order to provide some incentive to work harder. The solution preferred by economists such as Aganbegyan was in reality just another go at the attempt to produce a *homo sovieticus* as a substiute for the old economic man. In fact the self-management measures against the 'administrative-command' structure in industry would prove to be so destructive as to prompt the suspicion that their promised economic effects were not the only reason they were adopted. Gorbachev must also have been attracted by the idea of undermining a section of the apparatus that was opposed to him and giving succour to his other party enemies.

THE NATIONALITIES

This may also explain his bizarre nationalities policy. In 1988 the subject peoples in the constituent republics rose up with a vengeance in response to Gorbachev's calls for *perestroika* in the party and state apparatus. These calls had an ideological rationale, with a presumed Leninist (by which *perestroichiki* usually meant more lenient and sensitive) nationality policy now taking the place of the Stalinist one, the repository of all the past sins. Gorbachev's opponents had shown no particular disposition to oppose small changes that might permit the Union republics somewhat more leeway in dealings with the centre, but they were extremely unhappy about the various organizations mobilizing grass-roots initiative, such as the ones that met in Moscow in August 1987.[5] They did not much like local initiative in general, but they especially feared grass-roots nationalism taking hold in the republics.

Gorbachev's supporters, on the other hand, were wont to put a minus where his opponents put a plus. In this spirit Boris Kurashvili of the Academy of Sciences' Institute of State and Law actively promoted and encouraged mass organizations of non-party people, Popular Fronts and National Fronts – and even their combination into a grand 'Democratic Union for the Renewal of Socialism'. The idea was clearly to mobilize the 'informals', who had already proclaimed that they did not want an opposition role but one as a social support for reform, to do battle with Gorbachev's opponents and critics.[6] At the republic level, this meant an encouragement from Moscow to form nationalist organizations like the Lithuanian *Sajudis* which, to Moscow's horror, would soon be calling for secession from the USSR. The spring elections of delegates to the Nineteenth Party Conference provided the occasion for this Gorbachevist apparatus mobilization against the 'conservative' apparatus. It began in Estonia, where a nationalist faction won control of the republic Communist party organization and set up a Popular Front with non-party groups to support *perestroika*. It was in effect a model for the national Communists of the East European bloc. No sooner had it been founded than it turned the cutting edge of its appeals against the Russian language and the Russian population, whom it tended to regard as immigrants. The

Latvian Front, organized shortly afterward, was even more hostile to the Russian settlers.

In encouraging this endorsement of grass-roots popular initiative against his apparatus opponents, Gorbachev found himself in the position of the sorcerer's apprentice. No sooner had he established his superiority over his rivals, than he was faced with a wave of demands for secession from the Union. As these demands grew, the spirit of localism grew with them, threatening a breakdown in the Union's delicate international division of labour. His economic reforms, which ironically had taken root best in refractory areas such as the Baltic republics, were counteracted by the bumptious mood of nationalism. The net effect was to help drive the economy into what would turn out to be a catastrophic decline, heading toward the 'little famine' (*golodukha*) of the winter of 1990–91. It seemed that every broad and generous liberating impulse caused a short circuit in the functioning of the system. *Perestroika* was a hard course. Its ways and means would have presented formidable challenges even if the Politburo had been united. But in an atmosphere of factional combat that promoted a passion for radical measures, a good deal of furniture looked likely to be broken. Gorbachev's accumulation of unlimited power continued, but it was oddly accompanied by harbingers of genuine disaster.

GLASNOST AGAINST GROMYKO

The Washington summit was the occasion for another Ligachev coup. Bukharin was expected to be officially rehabilitated according to a decision agreed to before Gorbachev left, but while he was gone, Ligachev succeeded in blocking the action. On Gorbachev's return the way had to be cleared of obstacles laid by the ideology secretary. *Moscow News* aided the cause with the publication of an interview with Natalia Krestinskaya, Nina Lomova, Tamara Medvedeva, Elena Rukhimovich, and Tatiana Smilga – daughters of party leaders shot in the thirties, of whom Krestinsky and Rakovsky were Bukharin's co-defendants in the 1938 trial.[7] This prodded a plenum of the Supreme Court to announce on 4 February that Bukharin and the other defendants in the March 1938 show trial had been falsely accused and convicted. The case against them was dismissed 'for lack of evidence of a crime'.[8]

Up to this point the mythopoeic importance of the figure of Bukharin was as a defender of a humane system with prominent market elements. Western newspaper commentary concluded only that Gorbachev was trying to use Bukharin to make respectable the introduction of private enterprise into retail trade, light industry, and parts of agriculture. Not much had been said in the Soviet press at first about Bukharin's foreign policy.

ideas.[9] But historian Anatoli Latyshev sounded a new note in an article in *Nedelya*, stressing Bukharin's warnings in the thirties about the vitality of world capitalism, in contrast to Trotsky, Zinoviev, and Stalin, who expected its imminent collapse, and his recognition of the need to join with the Western democracies in the struggle against fascism. The others were said to take the view that there was no essential difference between the capitalist countries, nor a real difference between the Western social-democratic parties and the fascists. Latyshev correctly noted the position of Stalin in 1924 that the two tendencies were equivalent. Stalin was faulted for carrying this view into the period of Hitler's rise in 1928–33 (Latyshev ignored his 'Bukharinism' of 1924–28). This attitude was said to lead straight to the pact with Hitler in 1939. Stalin's viewpoint was depicted, as in the rest of the *glasnost* press, as one of consistent extreme leftism, actually rather more like the positions taken by Grigori Zinoviev, without any centrism or manoeuvres to the right.[10] But the distinction between the fascist powers and the democracies was certainly not lost on any of the above-mentioned personalities, not even those who advised war avoidance by a pact with Hitler's Germany. Our task here is not to dispute Latyshev's ludicrous historical assessment, although it is well worth disputing, but to assess its propaganda importance in the ongoing political struggle. The point seems to be that the foreign policy that Stalin finally turned to in making the pact with Hitler (that is, the foreign policy whose victory resulted in the shooting of Bukharin) was that of Molotov, Gromyko's patron of that period. Latyshev's remarks about Bukharin were bound and no doubt designed to make Gromyko squirm.

To attack the Third Period Comintern line of 1928–33, or the pact with Hitler, or anything else that can be ascribed to Molotov, was to attack Gromyko. Arguments for a renewal of detente with the West could now be given a politically important historical twist; for the first time Soviet readers were told that the pact with Hitler was not entirely forced by the treachery of the imperialists and their appeasement of Hitler at Munich. Diplomatic historian Vyacheslav Dashichev would by May be suggesting that Stalin's foreign and domestic policy, instead of preparing the country for war, actually made it more likely. The only chance to oppose Hitler, said Dashichev, lay in the 'classical configuration of power', which in the circumstances of the time, could only mean 'to revive the Triple Entente (Britain, France, and Tsarist Russia)'. British and French state interests demanded this; their anti-Communism and anti-Sovietism were troublesome but surmountable. However, said Dashichev,

as Stalin decapitated the Red Army, wiping out its best commanders, Britain and France could no longer view the Soviet Union as a dependable war ally. It was as well hard for them to do business with a

supreme ruler who trampled on human morality, consolidating his authoritarian power with unprecedented crimes and brutal methods.[11]

Criticism of Stalin's war leadership had been prominent in Khrushchev's
secret speech of 1956. And *glasnost* writers were following his lead in
connecting the failures at the front with the army purges of 1937–39. But
Khrushchev had defended the rectitude of Stalin's foreign policy, at least
in public. *Glasnost* writers such as Aleksandr Samsonov would be criticizing Stalin's generalship in the war years, with special reference to the
effects of the military purges on the leadership of the armed forces.[12]
Some noted the excessive and opportunistic nature of the discussion. The
controversy moved historian Mikhail Gefter to wonder at the efficacy of
pulling down Stalin from a pedestal in order to replace him with . . .
Khrushchev. He concluded that it would not do. Latyshev and Dashichev,
however, were taking things further, to the point of suggesting that the
Litvinov policy of urging Collective Security against Hitler was destroyed
by the purges and the turn to the Hitler–Stalin pact. A biography of
Litvinov appearing in 1989 would argue the same case. The year of the
50th anniversary of the Hitler–Stalin pact of 1939 was looming, and for
many the Stalin question had now become the question of August 1939.

The relative paucity of *glasnost* on current questions of foreign policy
was offset by an explosion of discussion about the past. There was a general fear of talking about 1939 because of the fact that the Baltic republics' annexation had been arranged by the Nazi–Soviet pact. No one thought
at that time that the Baltic republics had the perspective of revolt against
Moscow, but the sensitive nature of the question was understood. The
discussion of the Stalin era was bound to spill over into these national
questions. As *Ogonyok* and *Moscow News* published material on the purges,
showing photos of the sites of some of the shootings, such as Kuropaty
in Belorussia, it became more evident that the massive repressions against
the intelligentsia in the republics had amounted to a kind of genocide by
Russians against subject peoples.

None of the tender issues involving the nationalities particularly bothered
Gromyko, however, nor was there any problem in his mind about the
'white spots' in the past of concern to other bloc countries. When the Polish
ideology secretary Mieczyslaw Rakowski visited Moscow in January, he
assumed that Gorbachev's call for discussion of hitherto proscribed questions would entail a few answers for Poles. A joint commission of historians had been set up by Gorbachev and General Jaruzelski with the
idea of examining three matters sensitive to the Poles: the dissolution of
the Polish Communist Party in 1938 and the shooting of most of its leaders, the partition of Poland with Nazi Germany in 1939, and the deportation
of many thousands of Poles into Russia in the first years of the war.

The last point entailed re-examination of the Katyn forest massacre announced by the Nazis in 1943, on which occasion Stalin had taken the opportunity to break relations with the London exiles and form his own future Polish government. Yuri Afanasyev had given an interview to the Polish party weekly *Polityka* in the previous September, in which he had said that the time had come to tell the truth about Katyn. Rakowski happened to mention some of this history to Gromyko, whereupon he was told flatly that Poland should be grateful to Stalin for socialism. Stalin had 'fought like a lion' for the victory of the Communist party in Poland after the War. Of course that is perfectly true; Rakowski and his party could not have taken power without Stalin. But Rakowski, like his patron General Jaruzelski, was a supporter of *perestroika* and therefore not as consistent as Gromyko. In fact he told *Moscow News* that his party must of necessity view past relations with Russia differently: 'Suvorov (the general who led Russian forces into Poland at the time of the partitions of the eighteenth century) is a hero in your eyes, but not in the eyes of the Poles.'[13]

Gorbachev himself did not take positions on the historical and ideological questions, but only called for redoubled efforts to improve discussion of them. Since he was, according to his own analysis, fighting on two fronts, he found ways to agree with the *glasnost* writers and also with their critics. In a Central Committee meeting in January with the editors of major media, he urged a continuation of the struggle for *perestroika*, while of course avoiding the excesses of ultra-*perestroika* extremists. Vitali Korotich of *Ogonyok*, along with Ivan Laptev of *Izvestiia*, *Novy mir*'s S. P. Zalygin, and G. Ya. Baklanov of *Znamya*, called for a continuation of *perestroika* along with an intensification of efforts to, as Korotich put it, study the complex biographies and fates of the creators of our country'.[14] Valentin Chikin, editor of the Russian Federation daily, *Sovietskaya Rossiia*, who had held his post since the 27th Congress and published many articles harshly critical of those who might review party history, sounded a different note, complaining about the 'blasphemous tone' of *Moskovskaya pravda* under the editorship of Yeltsin's close supporter Mikhail Poltoranin. Gorbachev quickly agreed.

Valentin Falin, of *Novosti* press agency, was eager to stem any potential euphoria over the INF agreement. He argued that the West had actually lost enthusiasm for the Gorbachev reforms, and that, despite the favourable attitude of the American people toward the new USSR, 'official circles' there remained capable of manipulating public opinion, as they had in 1946–47 when they swung policy away from alliance with the Soviet Union to the Cold War.[15] Gorbachev again seemed to agree. Viktor Afanasyev of *Pravda* decried the irresponsibility that had been in evidence in some press discussions of the historical 'white spots', especially as presented in Mikhail Shatrov's play *Onward, Onward, Onward*, then being serialized in *Znamya*. Gorbachev did not object. Ever the man of the golden mean, he viewed with equanimity the opposed positions struggling for his soul, each of them claiming to be in harmony with the brilliant formulations in his November speech.

LIGACHEV RIDING HIGH

As ideology secretary, Ligachev was at this point the most prominent person in shaping the ideological position of *perestroika*. Yakovlev had seemed to be his rival since being appointed to the Politburo the previous spring, but Yakovlev vanished from view for several weeks beginning in mid-December, perhaps as a result of fallout from the Yeltsin affair. Ligachev, on the other hand, seemed to be riding high at the February plenum, the 'Educational plenum', at which he put forward grandiose but empty projections for reshaping Soviet schools at every level. The plenum's personnel actions strengthened him somewhat, but seemed the result of compromise. Yeltsin was removed from the Politburo, as a reflection of the current consensus. Yuri Maslyukov, who took the position vacated by the removal on 5 February of Nikolai Talyzin as head of Gosplan, became a candidate member of the Politburo. Also promoted to candidate was Georgy Razumovsky, who was thought to be a Gorbachev man; he had run agriculture in Krasnodar, next door to Gorbachev's native Stavropol, before being made first secretary by Andropov in 1983. Razumovsky had helped the General Secretary replace 41 per cent of the Central Committee for the 27th Congress in 1986, and was now thought to be involved in doing something similar for the upcoming June Party conference. Oleg Baklanov came into the Secretariat to run the armaments industry; Valery Boldin of the General Department, which controls the administration of the Commissar apparatus in the armed forces and the KGB, came into the Central Committee, as did the rather dispirited Viktor Mironenko of the Komsomol (who had admitted that he could not get his own son to join). Baklanov and Boldin would be among those attempting to overthrow Gorbachev in August 1991. Natalia Gellert, a tractor driver, was brought into the Politburo as a candidate, ostensibly to advise on nationality questions. The two Republic leaders most under fire by local advocates of *perestroika*, Ukraine's Shcherbitsky and Armenia's Karen Demirchyan, survived for the moment. Shcherbitsky was even awarded an Order of Lenin.

The plenum displayed two sharply opposed positions in discussions that V. Sitnikov, secretary for Irkutsk, described as 'fast and furious'. There were calls for further *glasnost* on party history. On the other hand Dzhumber

Patiashvili, head of the Georgian apparatus, decried 'certain publications who commit great sins trying to blacken the names of certain revolutionary and war heroes'.[16] Reporting the debate, a *Pravda* editorial also called for 'more careful comparisons of the path we have travelled and the path ahead with the precepts of Lenin'. 'Haste in assessments', it warned, was unacceptable, as were 'substandard works that obscure the truth'.[17] This was only two days after Gorbachev had told the plenum in a contrary spirit that

> It would be impossible to set the tasks of *perestroika* in the economy and in the political, social, and cultural spheres without reliance on theory and without ideological substantiation.[18]

The party must lead ideologically, said Gorbachev, of course avoiding 'hasty statements' and 'superficial conclusions', while at the same time continuing the struggle for 'a truthful and complete history'. He continued to stress the importance of 'a new role for universally shared values', consistent, he added, with the promotion of materialist dialectics.

Where did Gorbachev stand in relation to the two positions? Was the Western press correct to characterize the struggle as one between Gorbachev and the 'conservative' Stalinists? In a sense that was exactly what was taking place. But Gorbachev's way was more indirect than was generally recognized. True, he was responsible for the *glasnost* movement in the first place. Yet he did not consider himself responsible for any political or ideological position that the *glasnost* press might put forward, and indeed now usually agreed with its critics. It was almost as if Gorbachev had called the literature up in order to create a struggle in which he could play the role of *tertium gaudens*, the 'third who enjoys', making both sides more and more reliant on him and at the same time weakening both sides to his own advantage in the quest to rise above the collective leadership.

'ONWARD, ONWARD, ONWARD'

A second round in the re-examination of party history was unfolding, in some ways a more meaningful round than the first. As Rybakov's *Children of the Arbat* had served as a point of departure in 1987, Shatrov's play, *Onward, Onward, Onward* was now in that role. It was not seen on the stage until 13 March when it opened in Tomsk, but it could be read in serial form in Baklanov's *Znamya*. A sprawling didactic drama of a type familiar to Soviet audiences, it had groups of actual figures from history on stage debating their roles in broad exchanges and soliloquies. Shatrov was asking the question most central in all the *glasnost* literature, why and how the naive revolutionism of 1917 ended up in the regime of Stalin. Rosa Luxemburg issued her well-known (that is, well-known in the West) warnings about bureaucracy and the perils of a regime that tramples democracy underfoot, and Lenin agreed with her: 'Bravo, Rosa!' Stalin debated with the other actors the reason and necessity for his most objectionable measures, including shooting some of the principals, but in various ways his arguments were refuted, usually with Lenin's participation and approval. It was made clear that Lenin did not want Stalin as his successor, and that the latter's succession was made possible only because the other Bolsheviks suppressed and ignored Lenin's Testament and its call to remove Stalin.

Shatrov had a harder time debunking the historical case that Stalin used to justify his actions, especially the case against the devil Trotsky. He repeated the charge made in 1923–24 by Zinoviev and Stalin, and codified in the *Short Course*, to the effect that Trotsky had tried to subvert the insurrection that took the Bolsheviks to power in October 1917. But Shatrov added an original wrinkle, adding Stalin as a culprit, in the suggestion that both Trotsky and Stalin had opposed the rising and tried to prevent Lenin from coming to Smolny to direct the preparations. At one point Stalin argues that Trotsky's claim to a leading role in the organization of the rising is false. It is left to Aleksandr Kerensky to defend Trotsky by reminding Stalin of an article he (Stalin) wrote in 1918 before the struggle with Trotsky for the succession had become acute, an article giving Trotsky full credit as the organizer on the spot of the insurrection. Trotsky himself catalogues the various issues on which he differed with Lenin after the Revolution, and ascribes them, in the fashion of his accusers of 1923, as 'my non-Bolshevism'. Trotsky also denies the charges made against him and others in the Moscow Trials of the thirties, to the effect that they are organizing assassination and wrecking, and are in the pay of foreign intelligence agencies.

Thus Shatrov was not really providing any revelations, nor even a consistent line of interpretation that could be supported or criticized. Indeed, he was in some ways perpetuating the myths of the Stalinist *Short Course* of 1938. Yet he was raising issues about the Stalin era that Soviet audiences were not accustomed to hearing debated openly, and presenting the historical figures such as the White General Denikin and the liberal Petr Struve, as people whose ideas could legitimately be known and discussed openly.

The specific bombshell in the play was the open discussion of Lenin's Testament, which Gorbachev had distorted in his November speech Shatrov's official critics were forced to justify the suppression of the Testament at the Thirteenth Party Congress in May 1924. In *Pravda*

V. Glagolev complained that there was 'only a semblance of historical objectivity' in Shatrov's play. He admitted that 'even time will not soften the tragedy of 1937', but insisted on 'the need for works that show real knowledge of the laws of historical development'.[19] M. Kim argued that the instructions to remove Stalin should be seen in terms of Lenin's last articles concerned with the building of socialism in the Soviet Union. True, the rest of the Bolshevik leadership had saved Stalin from Lenin's instructions, but they had done so because of the threat of Trotsky and the Trotskyists, who fought against the Lenininist idea of building socialism, and thus they were justified.[20] It was not pointed out that the theory of Socialism in One Country, which Kim was summoning up, was only advanced by Stalin in December 1924, six months after the Congress, and that Trotsky did not publicly oppose the idea until he made his bloc with Zinoviev in April 1926. It was Zinoviev who opposed Socialism in One Country, but he was the most influential of those who decided to suppress Lenin's Testament!

The official critics of Shatrov seemed to sense that they were not standing on firm ground. His supporters, on the other hand, were more confident. The Soviet theatre lent its support in a letter to *Pravda* from directors Oleg Efremov, Mark Zakharov, and Georgy Tovstonogov, actor Mikhail Ulyanov, and the head of the Theatre Workers Union Kirill Lavrov. In *Moscow News* Dmitri Kazutin argued that the play should serve as a case for reopening the study of the reasons for the rise of Stalin. The Marxism that Soviets had been taught was Stalin's, a simplistic creed with little of the sophistication of Lenin's views. 'Nothing', he said, 'can replace serious study of the records of the Politburo, central committee plenums, and correspondence.'[21] He reminded his readers that the decision to suppress Lenin's Testament, which turned out to be 'so fateful for the party', was agreed by almost all the major leaders: Zinoviev, Kamenev, Bukharin, Rykov, Dzherzhinsky, Ordzhonikidze, Kalinin, Kirov, and many others. Why had they taken this step? Out of fear of Trotsky. The question of Stalin was ultimately the question of Trotsky, or rather: the premise of Stalin's rise to supreme power was fear on the part of the major party leaders of the prospect of living under Trotsky.

Glasnost journalism was trailblazing without much guidance from Soviet professional historians or from party ideologists. Observing this painful controversy with the advantage of Western scholarly literature, one is tempted to ask of its participants: why not state the thing frankly? Why not say that all the party leaders, and not Stalin alone, falsified the history of the Revolution and the history of their party when they invented the heresy of 'Trotskyism'. This put them on a slippery slope toward the abyss of Stalinism because, in their zeal to vie for power, they did not hesitate to call each other 'Trotskyists', to seek Stalin's help against each other, and thus to permit Stalin the room to manoeuvre and defeat them. They went to their deaths at Stalin's hand, praising Stalin and accusing each other of 'Trotskyism'. Why not admit that the whole thing had been false from the beginning, an excess of the intense struggle of people like Zinoviev, Kamenev, and Bukharin, to succeed Lenin. This need not imply conversion to 'Trotskyism', whatever that might mean in present circumstances, nor even endorsement of all or any of his political ideas. It would at least remove the onerous superstitions about the devil Trotsky, the historical figure about which Soviets were least able to talk rationally.

It did appear that the process of re-examination might find its way to something like that, a rehabilitation of all the victims of Stalin and the granting of permission to publish, read and discuss their works. But the case of Trotsky seemed to hold it up. During the first wave of *glasnost* discussion in 1987, Iuri Afanasyev had urged that Trotsky be recognized in the various exhibits of the history of the Revolution in Leningrad, and had been answered by a phalanx of Stalinist resistance to the idea, led by historian F. M. Vaganov.[22] In September, *Ogonyok* had carried an interview with Ivan Vrachev, a ninety-year-old former Trotsky supporter expelled from the party in 1927, who referred to Trotsky as 'an outstanding activist of our party who was forced onto the path of isolation'. At the same time *Sovietskaya Rossiia* and the trade union paper *Trud* campaigned for the traditional Stalinist view. *Trud* ran an article on the trade union debate of 1920, in which Trotsky had advocated the 'militarization of labour', commenting with no little disingenuousness that this would have prevented the unions from protecting the rights of the workers. In February 1988 the Soviet Military Encyclopedia decided to return the name of Trotsky, the founder of the Red Army, to its pages, reasoning that since there was already an entry for Hitler, Göring, and 22 White generals, they could do no less.

The periodicals of the extreme Russian nationalist standpoint such as *Nash sovremennik* and *Molodaya gvardiia* conducted their own campaign against Trotsky as an illustration of their contention that the worst phases of Stalinism were due to the party having been captured by non-Russian elements. *Glasnost* discussion of the elimination of the major leaders of the Lenin era in the purges of the thirties for them revealed a bias toward Jews and other alien elements, a 'psuedo-intelligentsia' within the Soviet intelligentsia. The purge was part of a general reconstruction of Russia according to abstract ideals most congenial to the cosmopolitan Trotsky; it was, for example, the Jew Kaganovich who supervised the destruction of so many cultural sites under the slogan that 'the old Moscow must give way to the new Moscow'.[23] The traditional Stalinist resistance to rehabilitating the purge victims converged with a case against the Old Bolsheviks from the standpoint of 'national radicalism'. Not wanting to find themselves on the wrong side of the issue of Russian nationalism, Ligachev and his supporters were tempted to appeal to these potential allies.

A TRULY LENINIST NATIONALITY POLICY

However much the *glasnost* writers might stumble on the full rehabilitation of Trotsky, other victims would be cleared. At this point, in spring 1988, Bukharin and his co-thinkers (for example, A. Chayanov and other former 'neo-narodnik' economists such as Kondratyev, known widely in the West for his theory of 'long business cycles') were the only ones. Aganbegyan specifically cited Chayanov as a forerunner of his ideas on cooperatives.[24] By the summer the rest of the victims of the other two Moscow trials, including Zinoviev, Kamenev, Radek, Sokolnikov, and others (but not Trotsky) would be similarly cleared. In the meantime they went to work on de-Stalinization according to the path of least resistance: a revival of Lenin. They centred their attention on the 1922 controversy between Lenin and Stalin over the relation of the major nationalities in the Soviet state. Stalin had maintained the view that the Ukraine, Belorussia, Azerbaijan, Georgia, and Armenia should be brought into the Russian Federation as Autonomous Soviet Socialist Republics, with a status comparable to today's Tatar, Bashkir, or Udmurt ASSRs. At first Lenin had seemed to go along with these proposals, but as he came increasingly into conflict with Stalin on other matters, the differences on the nationalities sharpened up as well. Eventually Lenin supported the idea of a Soviet Union with formal equality for all the republics. Differences were exacerbated by a controversy over the treatment of the Georgian Communists by Stalin and his supporters. In the course of the resultant disputes, Lenin came to the conclusion that there could be seen in Stalin something of the Great Russian chauvinist, and Stalin in turn referred to Lenin at least once in terms of a presumed limitation of 'national liberalism'.

Historian Albert Nenorokov spoke of distortions of the Lenin line during the period of Stalin's rule. Lenin, in his view, had defended the real freedom of the nationalities within the Union against the 'schematic' views of Stalin, views that had resulted in a real oppression of the nationalities in the succeeding decades.[25] Timur Pulatov, a writer from Tashkent who reported with sensitivity and understanding of the problems of the Central Asian peoples, complained that the Kara-Kalpaks suffered greatly from the drying up of the Aral Sea, caused by the diversion of the waters feeding it to the cotton fields of Uzbekistan. Too much cotton was produced in Uzbekistan, because of administrative decisions taken in Moscow without proper study or consultation with the local people. This was a result, he said, of 'the legacy of Stalin's nationality policy', as was the terrible mafia-ridden regime of former Uzbek party chieftain Saraf Rashidov.[26] It will be remembered that Rashidov, along with Medunov in Krasnodar, had been targets of Andropov's anti-corruption campaign of 1982–83. Kazakhstan's Kunaev, whose removal in 1986 had started the whole rash of local rebellions against Moscow that would soon be so dangerous, had been part of the Brezhnev racketeering and patronage machine.

One could say then that Gorbachev's apparent sympathy for the cause of the nationalities had a certain momentum from Andropov's anti-corruption campaign, a campaign which in retrospect could now be described by the *glasnost* enthusiasts as having been based on Leninist nationality policy as opposed to entrenched Stalinist-Brezhnevist practice. To present things this way was certainly no disadvantage to Gorbachev in his struggle against current Politburo opponents.

Moreover, nationality policy did not seem at the start of 1988 to be the minefield that it eventually turned out to be. The demonstrations in the Baltic republics in the summer of 1987 had been troubling, as were the attendant demands and pleas from variegated groups urging protection of the local ecology, the culture or the language. Usually these were accompanied by demands for publication of the Molotov–Ribbentrop pact of 1939. An Open Letter had been sent to Gorbachev from Belorussia urging efforts to save its language and a return to Lenin's nationality policy. Estonian non-party papers criticized the increasing Russification in that province. In the Soviet arms buildup of the seventies, new submarine bases and other military installations had been built in Estonia and many Russians had come into the area to work. Yet the 1987 demonstrations had not been very impressive in their numbers, not comparable to the tens of thousands and even the 100 000 who would be protesting in the summer of 1988. Nevertheless in January 1988 an Estonian National Independence Party actually called for a referendum on secession.

At the beginning of 1988 the response to these manifestations did not show any great fear about their potential. Historian V. Maamagi of the Estonian Academy of Sciences ascribed them to 'ideologically immature and even anti-socialist elements', who had been spurred on by 'western radio voices'. He thought the problem could be addressed by the introduction of a convertible ruble and more emphasis on the teaching of the Estonian language.[27] Dmitri Kazutin thought that Estonians, 90 per cent of whom knew Russian and 30 per cent with fluency, must surely recognize that this knowledge gave them access to world culture.[28] Soviet social science could not decide whether its traditional view of the progressive nature of the assimilation of the non-Russian languages and cultures was consistent with Lenin, or just something that had grown up under Stalin

And there was a pervasive suspicion that Academician S. A. Arutyunov was correct in supposing that Lenin's conception was compatible with the idea that 'The disappearance of any ethnos is a tragedy ... The conception of ethnic pluralism must have a Communist variant.'[29] Inherent in the generosity of that opinion is an assumption of the historical victory of the Russian culture and language in a multinational Soviet Union, a fact that made possible a paternal cultivation of ethnic minorities within the framework of borders and rules of interaction long since established.

Suddenly the problem acquired new dimensions with the decision of the Nagorno-Karabakh regional soviet on 20 February to transfer the region administratively from Azerbaijan to Armenia. Karabakh is entirely enclosed by the territory of Azerbaijan, despite its large Armenian majority. Its position would seem to be analogous to that of Nakhichevan (now having an Azeri majority), which lies within Armenian territory along the river Araxes forming the border with Iran. Both districts were, however, under Azeri administration, a fact that grated on the Armenian national sensibility. This was the result of the successful campaign of the Turks in 1920, occupying the Kars district previously claimed by the newly independent Armenia. The assignment of Karabakh and Nakhichevan to Azerbaijan was the result of the Treaty of Kars (13 October 1920), arranging a partition of Armenia between Turkey and Soviet Russia. When the dispute arose in 1987–88, *glasnost* writers stood on their heads to present the injurious territorial anomalies as the work of Stalin as Commissar of Nationalities, so much at variance were these, they claimed, with the spirit of Lenin's nationality policy. But this was all nonsense. The border arrangements were originally the work of Lenin and reflected the cordial relationship that had developed by then between the Turkey of Kemal Pasha and the new Soviet state, a relationship that had as an incidental by-product Kemal's liquidation of the Turkish Communist party.

From the standpoint of linguistic self-determination it would seem that if Nakhichevan were administered by Azerbaijan, then Karabakh should be ruled from Erevan. That was the opinion of the delegation of Armenians from Karabakh that visited Moscow three times between November 1987 and February 1988 to lobby for the transfer to Armenia. The case they made was essentially that of Souren Aivazyan in his letter to Gorbachev in March 1987.[30] They complained that Gaidar Aliev, head of the Azerbaijan party from 1969 to 1982, had actively worked to settle Azeris in Karabakh in order to reverse the linguistic balance, something that had earlier occurred in his native Nakhichevan. Aliev had some credence on this issue, having already convinced the Brezhnev Politburo that Soviet Moslems could play a role in the extension of Communist influence in the Middle East.[31] Aliev took advantage of their temptation to think that Shia Azeris could have some influence in Iran, at a time when the Iranian Tudeh party was supporting Khomeini. Aliev's successor as First Secretary in Azerbaijan, Kyamran Bagirov, had continued in the same vein, the Armenians claimed, even to the point of permitting Shia fundamentalist propaganda to be disseminated. In the Autumn of 1987, *Literaturnaya gazeta* had claimed that there was an active Khomeinist movement in Azerbaijan. In another venue of Persian cultural influence, Tadzhikistan, there was a report in January of a spillover from the Afghanistan war, in the form of a Moslem guerrilla movement led by the Sufi brotherhood. The Armenians had reason to think that Gorbachev heard them with a sympathetic ear. The Tudeh party had been ruthlessly purged in Iran in the early eighties and Khomeini spoke confidently of converting the godless Soviets. Aliev's star was in eclipse because of his rivalry with Ligachev (he was removed from the Politburo in October 1987) and his line was no longer in vogue.

In fact, if anything, the atmosphere of *glasnost* was promoting the opposite mood. Preparations were going forward for the celebration later that year of the millennium of the Baptism of Rus'. *Moscow News* carried a cordial interview three weeks before the crisis with Catholicos Vazgen 1, the Patriarch of the Armenian church, in which the interviewer referred to the 'permanent and cruel torture of Armenia by Byzantium and Arabs, Timur and Turks'.[32] There seemed to be some unconscious temptation to feel that Christianity was after all a more progressive thing than Islam and, moreover, a link to the West, while on the other side, promotion of the Moslem cause in the end only added to the anarchy in the world.

An additional irony was provided by the fact that it was the Nagorno-Karabakh *oblast* soviet that had passed the controversial vote only days after Boris Kurashvili, in an *Izvestiia* interview, proclaimed the necessity for greater independence on the part of the smaller soviets, in the spirit as he described it, of Gorbachev's suggestions to the January 1987 plenum about a greater role for elected bodies. As with Kurashvili's idea on an increased role for Popular Fronts and National Fronts, it was intended that these measures 'would make it possible to realize more fully Lenin's concept – to govern not only "for the workers" but "through the workers"'.[33] Gorbachev had made a speech to the same effect only days before the Karabakh vote.

It was convenient for Gorbachev that the agitated situation in Armenia gave new urgency to the campaign he had been waging for some months against Armenian First Secretary Demirchyan. In a stormy meeting in December, more than 20 local Communists had denounced Demirchyan, including some of his favourite protégés. Demirchyan was not a special case; similar campaigns were under way in Uzbekistan, Kazakhstan, Georgia, and Latvia. When large demonstrations broke out in Erevan, Gorbachev sent Razumovsky, the secretary responsible for party organization (and a Gorbachev man), to the scene. Within days heads began to roll. The First Secretary of the Azerbaijan party in Karabakh, Boris Kevorkov, an Armenian with a Russian name, was replaced at the head of the regional soviet by an Armenian with at least an Armenian surname, Genrikh Pogosyan.

But that did not keep matters from getting further out of control. On 28–29 February, a virtual pogrom against Armenians broke out in the Azeri port town of Sumgait, with over thirty killed and hundreds injured. After that Armenians began to sense that Moscow was no longer in their corner on Karabakh, but now just hoping for damage control. By May both first secretaries Demirchyan and Bagirov would be removed. The local problem was by no means solved. The atmosphere between Armenia and Azerbaijan would in fact remain on the verge of civil war from this point on, but even in the spreading chaos Gorbachev continued to make giant steps in the struggle against his apparatus opponents in the republics.

THE LETTER OF MME. ANDREYEVA: MOSCOW AND LENINGRAD

When the Karabakh problem erupted, the Ligachev group in the Politburo was still struggling to find its voice. Since January 1987 calls for restraint in cadre policy had been issued from several members, and Ligachev periodically urged that the leaders avoid 'haste', something that he had apparently learned to do from reading the documents on Lenin's struggle with Stalin in 1922, in which excessive haste is Ilyich's recurrent criticism. Gorbachev found it difficult to contend with an opponent who was only advising against undue haste. In March, however, while Gorbachev was in Yugoslavia, explaining his idea of the coming House of Europe, the critics of *glasnost* found a champion in a Leningrad chemistry teacher, Nina Andreyeva, who published an article in *Sovetskaya Rossiia* with the title 'I Cannot Forsake Principles'. Gorbachev was at the time courting the Yugoslavs by reaffirming the wisdom of the declarations of 1955 and 1956 in which Khrushchev had buried the hatchet with Tito, and assuring the Communist parties that they could henceforth do their work free of hindrance from Moscow, a position that in retrospect appears to have been a fatal blow to the Brezhnev Doctrine and a green light for the historic innovations of the Communist parties themselves in 1989.

Andreyeva's letter had apparently been sent to a number of periodicals before Valentin Chikin, editor of *Sovetskaya rossiia*, took it up. Before publishing it, he and others decided to rewrite it into a kind of anti-*glasnost* manifesto.[34] The public was shocked to see a platform document seemingly announcing a change of course. But the text of the article was not so sweeping as the supporters of *glasnost* and *perestroika* were to charge. It was not, for example, an anti-Gorbachev platform. Andreyeva stated explicitly that, in rejecting the 'excesses' of *glasnost*, she nevertheless endorsed the party's position of 1956 on Stalin's 'personality cult', and Gorbachev's 1987 speech on the 70th anniversary of October which, she said, 'remain the scientific guidelines to this day'.[35] She cited Gorbachev's words urging recognition of the contradictory nature of the Stalin era in which 'victories and setbacks, discoveries and mistakes, the radiant and the tragic . . . were combined'.[36] Yet the first authority that she quoted was Ligachev, for his understanding of the relation between general-human and class interests, which included recognition of the 'class essence' of the changes currently taking place. She complained, however, that, in the manifestations of *glasnost* were to be found 'some promptings of western radio voices', causing Soviet students and intellectuals to come to 'hasty' conclusions about the Stalin era.

In Shatrov's plays, for example, the Andreyeva letter found 'dramatic fantasies' bearing little relation to scientific history. In *The Brest Peace*, 'Lenin kneels before Trotsky', and *Onward, Onward, Onward*, said Andreyeva, showed the influence of Boris Souvarine's Stalin biography of 1935.[37] The worst of this was that Shatrov appeared to give credence to the views of some members of the Central Committee ('opponents of Leninism') in the inner-party struggle of the twenties. The letter simply restated Gorbachev's conclusion, as related in the 1987 speech, that in the succession struggle after Lenin's death, Stalin defended Leninism against Trotsky, Zinoviev, Bukharin, and the others. Andreyeva stressed, with Gorbachev, her unhappiness about the purge era: 'I share the indignation of the Soviet people over the massive repressions that took place in the thirties and forties and with the party-state leadership of the time, which was at fault.'[38] But she rejected any attempt to portray these shortcomings as a kind of counter-revolution, or as a cause of spiritual slavery; nor, she thought, should Stalin be charged with responsibility for Hitler's coming to power in Germany. The deviations of the *glasnost* press, enthusiasm for rock and roll music or frank discussions of sex in the press were avoidable excesses, in her view, as were other occasional slips that she had spotted, such as references to a supernatural intelligence in nature, or claims that culture is not learned but inherited. She bristled at the suggestion accompanying discussions of universal human values that relations between states with different economic systems could be devoid of class content.

The Andreyeva letter urged a middle course between the twin evils of 'left-liberal socialism', and 'neo-Slavophilism'. It criticized the latter for failure to recognize the historical importance of the Revolution and for its characterization of the collectivization of agriculture as an unalloyed atrocity against the peasantry. Andreyeva was, however, not such a fighter against nativist prejudices that she could forgo an opportunity to attack 'cosmopolitanism', as historically embodied, of course, by Trotsky. She quoted a statement of Trotsky's to the effect that he did not regard himself as a Jew but as an internationalist, something she seemed to doubt. Trotsky was expressing a sentiment that in itself must have been perfectly in tune with the predispositions of Lenin and all the other Bolsheviks of the time. But Andreyeva found in it a kind of treason to Great Russian national pride. It is, moreover, clear enough from the text that Andreyeva only raised the issue to juxtapose Trotsky's 'Jewish internationalism' alongside contemporary Jewish refuseniks and other cosmopolitans.

It is not clear whether the article's anti-semitism was the work of Andreyeva herself or the team of Gorbachev opponents and Politburo members, including Ligachev and Nikonov, who helped to rewrite it. But at any rate, the anti-semitism was certainly no slip of the pen. Politically, most of the article restated the Gorbachev–Ligachev consensus cemented by the cooperation of the two at the time of the 70th anniversary speech and the fall of Yeltsin. The anti-semitic note, however, reflected a choice of the Ligachev–Chebrikov–Shcherbitsky–Nikonov group in the Politburo for an ideological stance harking back to the views of Suslov in the seventies, a kind of orthodox Stalinism which, in the name of Soviet patriotism, had made its peace with an unreconstructed Russian nationalism. To be sure, the Andreyeva letter made a feint in its brief suggestion of criticism of modern neo-Slavophilism (by which it referred to proponents of 'peasant socialism', and to Astafiev, Rasputin, and other 'village school' writers) but it emphasized that the deepest deviation of *glasnost* lay in treason to national pride and failure to recognize the historic role of the Slavic peoples with the Russians at their head. In this respect, the authors of the letter were slipping away from the course originally set by Andropov, who had brought most of them to prominence, and returning to that of Suslov.

Ligachev and his group must have thought themselves exceedingly shrewd to take advantage of the backlash inside the Russian republic to the rise of bumptious nationalist moods among the non-Russian peoples. It had become a reflex among some Russians to match calls made by non-Russians for secession from the Union with their own calls for *Russian* secession, as if to say: 'very well, let's see how you get along without us, our oil, etc.' In their desire to put a minus where Gorbachev put a plus, the Ligachev group instinctively reached out for this potential constituency. But in coquetting with Great Russian national Communism, they were in effect weakening the cause of the Soviet Union, the crucial idea that permitted them a rationale for Russian leadership of the non-Russian nationalities. Ultimately the Great Russian national cause, insofar as it implied the domination of non-Russian peoples in the name of nothing more than the 'Russian Idea', an inheritance from tsardom, made more difficult the Union of the peoples and more natural the nationalism of the subject peoples. Great Russian nationalism tends to serve the cause of a little Russia, that is, a sovereign and isolated Russian Federation, the ultimate product of half-baked ideas about Russian secession.

The Andreyeva letter quickly provoked a showdown between Gorbachev and Ligachev. Ligachev held a meeting with editors several days after the article's appearance, pointedly excluding the editors of *Moscow News* and *Ogonyok*, to encourage the reprinting of the article. All over the Soviet Union there was a kind of shock and paralysis. Most people simply thought that the line had changed, and wondered if arrests would soon follow. When Gorbachev returned from Yugoslavia, he immediately began a whirlwind tour of meetings to shore up support, the culmination coming at a Politburo session in the first week of April, with Ligachev conveniently away on a trip to the Vologda area.[39] Gorbachev made his case with customary eloquence and followed with a threat to resign. Lenin had once remarked that the most difficult disputes in the party were usually settled this way.[40] And the point here was that Ligachev was not prepared to take the leadership in the event of Gorbachev's resignation. None of Gorbachev's opponents had the perspective of alternative leadership, but only that of limiting the General Secretary by vigorous exercise of the functions of the Second Secretary in his capacity as guardian of ideology, as Suslov had done. It was later revealed that when Gorbachev made his threat to resign, even Ligachev had agreed that he must continue at the helm.

Gorbachev's move carried the day. Chikin was criticized, Ligachev warned, and Gorbachev ally Aleksandr Yakovlev was commissioned to write a reply in *Pravda* to the letter. This editorial, which appeared on 5 April, was taken throughout the apparatus and the government as a sign that the line of the Andreyeva letter was not in fact the party's line. That was the key thing for most people. But the editorial itself was an odd document with rather few articulated differences of 'principle' with the Andreyeva letter. It stated firmly that there was no alternative to *perestroika*, but Ligachev had said that a number of times. The reply admitted that there had been considerable excess in the *glasnost* campaign: debates had been 'lacking in sophistication', with 'a lack of knowledge and cogent arguments'.[41] It spoke of broad recognition of 'Stalin's indisputable contribution to the struggle for socialism and defence of its gains'. It put the excesses of *glasnost* alongside two other categories of error, first the view of *perestroika* as a mere cosmetic adjustment, and second, the attempt to skip stages. The first clearly referred to 'conservatives', and the second to Yeltsin. The Yakovlev reply mentioned the reference in the Andreyeva letter to the threat from the descendants of Dan, Martov, Trotsky, Yagoda, NEPmen, and Basmachi by calling this a kind of genetic determinism.[42] The letter's anti-semitism was briefly and meekly noted as a naive confidence in determining political positions by national origin. Yakovlev passed up an opportunity to criticize the Andreyeva letter's flirtation with Russian nationalism. His previous criticism of Russian nationalism in 1979 had resulted in his being sharply censured and reassigned to the Soviet Embassy in Canada.

The reply was forced to take note of the fact that Andreyeva accepted the line of 1956 and Gorbachev's history speech of 1987, but it claimed that despite this, the Andreyeva authors in fact were trying to overturn the reform course. Why? Because they saw in it a threat to the administrative-command system and its 'selfish interest'.[43] Their implicit defence of bureaucracy and corruption, said the reply, was undermining unity of the supporters of *perestroika* 'of all shades of opinion'. This could be read as a kind of appeal to unite the forces (including the Ligachev opposition) that had marched under the banner of Andropov's anti-corruption campaign, forces that had only come into conflict since January 1987. The reply tried to use polemical devices of the Andreyeva letter against it. Where Andreyeva had accused Shatrov of depending on Western historical writings such as those of Boris Souvarine (expelled from the French Communist party as a 'Trotskyist' in 1924), the reply accused Andreyeva of putting into Churchill's voice the famous remark of 'the British Trotskyist Isaac Deutscher' – that Stalin had 'found Russia with a wooden plough, but left it with atomic weapons'. That was a way of saying that there were no real differences about Trotsky or Western 'cosmopolitan' historians. Where Ligachev and Andreyeva had warned about 'haste', the reply also quoted a warning of Lenin about 'haste' in making assessments of cultural topics. The reply chose to end by repeating Ligachev's slogan of 1986, designating the main desideratum of reform: 'more socialism'.

The exchange between the two documents had not revealed huge ideological differences, yet the struggle between the two factions making their stand on those documents was nevertheless passionately intense. One could not know how things would turn out from one moment to the next. Ligachev was reported to have been 'vacationed' on 21 April; on 22 April he and Gorbachev were photographed in amiable conversation. No one was fooled by this show into thinking that the fight was over. In fact, all eyes were on the upcoming Nineteenth Party Conference, the first meeting of its kind to be staged since 1941. The expectation was that Gorbachev would gain another march on his rivals by setting new party rules permitting the ranks of the party to take a share in major decisions. The assumption was that the most influential members committed to reform would be present among the 4,991 delegates. The party apparatus naturally tried to influence elections to delegate status as much as possible. Demonstrations against unfair election procedures rose up all over the country, with party secretaries forced out of office in protests in Astrakhan, Kuibyshev, and Yaroslavl. Yet many *perestroika* supporters were denied delegate status. The city of Leningrad elected 176 out of 176 apparatus nominees, in the face of large popular protests.

There were some who felt it to be significant that Leningrad should be in effect the rallying ground for the fight against Gorbachev's reforms. Yuri Afanasyev insinuated that it was 'understandable' that Andreyeva should come from there. Another commentator suggested that Leningrad was a 'suspicious city'. 'The opposition [to Stalin] rallied there', said Mikhail Chulaki, 'The 14th Party Congress [of 1925] was a trial of the Leningrad delegation.'[44] It was the place where Kirov was killed, the site of the persecutions of the 'Leningrad Case' of 1949,[45] and more recently the stronghold of Gorbachev's most bitter opponent, Romanov. Afanasyev thought he saw 'an approving response to Andreyeva', in the city whose major university was still named after Zhdanov. He noted the 'heavy Zhdanov–Romanov heritage and the city's reputation for being more rightist than Moscow'.[46]

True enough, one can see in Soviet party history a recurrent opposition between the Moscow and Leningrad party organizations, from the time of the 14th Congress, when Zinoviev and Kamenev led the critique of the moderate and, to their mind, 'Thermidorian' policies of Bukharin and Stalin. But the Leningraders were usually considered the left of the party, the repository of its proletarian symbolism and revolutionary images, hence the primary constituency for protests against a perceived loss of militancy and internationalism. In view of their protest against the loss of class perspective and international militancy among the *glasnost* writers, it made sense to see Andreyeva and the Ligachev group in this light. The accompanying idea, that party policy would eventually be swinging back to their perspectives when the current reform line was exhausted, was also quite thinkable.

But the line of *perestroika* was far from exhausted, with a feverish enthusiasm for democratization sweeping the country. Even among apparatus oppositionists who were now much more clearly visible to the public because of the Andreyeva episode, there was really no sentiment in favour of attacking Gorbachev himself. The major reason, as has already been suggested, was that Ligachev and his group did not see themselves as a potential ruling faction in the Politburo. They only wanted to hem in the leader in the spirit of collective leadership inherited by them from the experience of the last quarter of a century. Moreover, the foreign policy of Gorbachev still held great attraction even for people like Ligachev, who, whatever they might think of Universal Human Values, had to admire Gorbachev's defusing the more explosive aspects of the superpower relationship. Each summit meeting with American leaders added to his aura of mastery in their eyes and reinforced their sense of his indispensability.

So, with the summit that was held in Moscow at the end of May and beginning of June, Gorbachev's prestige received another boost. By comparison with Geneva, Reykjavik, and Washington, this was an uneventful summit. The earlier goals for START negotiations were reaffirmed. 'Some progress' was reported on difficult verification problems such as mobile ICBMs and sea-launched cruise missiles. The two sides were said to have some 'common ground' on regional conflicts such as Namibia and Angola. By far the most significant event occurred on a walking tour of Red Square, when Reagan told reporters what he would later tell the British parliament: that the Soviet reforms meant a 'lasting change' and a major turn in history. As Gorbachev told assembled reporters at his first press conference after the summit,

> Somebody asked the President whether he still considered the Soviet Union to be the 'evil empire'. He said no, and he said it within the walls of the Kremlin, next to the Tsar cannon, right in the heart of the 'evil empire'. We take note of that. As the ancient Greeks say, 'everything flows, everything changes. Everything is in a state of flux.'

Gorbachev advertised himself as having worked his magic on the entire foreign policy elite of the United States, and in particular as having tamed the man he now disparagingly called 'grandfather Reagan,' the former leading militant of world anti-Communism. Things that had never been achieved by means of competition in the nuclear arms race were now being achieved by 'political means'. And Gorbachev appeared to be the only politician who could achieve them. Even his most entrenched opponents had to give him his due. One can forgive some domestic errors of a leader who exhibits such mastery in foreign policy. Even Ronald Reagan had read his book and, he said, 'found much to agree with'.

9 Dropping the Pilot: Gorbachev Retires Gromyko

> Usually, when western sources discuss the non-aggression pact they raise the question of a secret protocol, supposedly signed by Molotov and Ribbentrop ... on the need for certain territorial changes to be made in the countries lying between Germany and the USSR ... The Soviet chief prosecutor at Nuremburg labelled it a forgery, and correctly so, since no such 'protocol' has ever been found, either in the USSR or in any other country – nor could it be.
>
> Andrei Gromyko

> We worked him too hard. But today it is said that Comrade Gromyko is out of touch with life. He has done his work, and his noble deeds are in the people's memory.
>
> V. V. Shcherbitsky

By spring 1988 Gorbachev was using the word 'democracy' routinely, almost casually – democracy, the quality that would alone, in his view, give genuine substance to *perestroika*. Gorbachev unequivocally called his reforms a 'revolution', and himself a revolutionary. And the Soviet press seemed to receive the news without alarm, treating it as another slogan in the ongoing campaign, what the French press sometimes called a 'parachutage des mots d'ordres'. At the time, I could not understand the extraordinary equanimity. I was in the habit of thinking of the Soviet system as fundamentally a system of Stalinism, one which had, to be sure, undergone a far-reaching process of rationalization in the successive post-Stalin regimes, but one which nevertheless remained a grinding dictatorship. It was capable of many things: broad and open-ended detente with the West, increased press freedom, inner-party discussion about party history, far-reaching reassessments in culture and media policy, every kind of economic reform, including the development of a consumer goods and light industry sector governed according to market principles. But not democracy. Not, that is, if the term were meant literally. That was the main rub in analyzing the matter from western perspectives. Soviet propaganda had long maintained that the system was the most democratic in the world. This was not entirely cynical. There was a sense in which the

Soviet Communists thought they exercised a 'Democratic Dictatorship' of the type of the French Jacobins of 1793–94, in which there were no elections to give the government a mandate, but populist measures or pretensions, or more often a willingness to compel action in the national interest, nevertheless prompted a claim to rule in the name of the great democracy, that is, the peasantry and urban plebs. Was this what Gorbachev meant by democracy? If so, some Communists could tell themselves that they were not against a certain broadening of the 'democratic' character of the existing regime. They would soon learn that Gorbachev meant something more than this: that he meant to take advantage of the radical desire of many Communists to make their socialism compatible with democracy – as Social Democrats in the West might understand the term.

It was not so far-fetched that some might cultivate this hope, in view of the widely held assumption that the Soviet public was deeply committed to socialism. In his famous tract, *Will the Soviet Union Survive until 1984?*, Andrei Amalrik had assumed that the democratic opposition that looked to leaders like Roy Medvedev or Andrei Sakharov, even in seeking to create a democratic society of the western kind, would not want to tamper with the foundations of governmental ownership of industry.[1] Truly bizarre was the idea that Communism itself could continue its control over affairs under the legitimation of democracy. The people would be so grateful to their jailers for setting them free as to entrust their future to them by a democratic mandate. No one seemed to have considered that any newly enfranchised electorate could never be confident of having demonstrated its new electoral weapon until it had turned out the Communist Party – with thanks, of course.

Communists, however, were in the habit of speaking and writing about their intimacy with the masses, who required their indispensable tutelage. They had all come up in the party addressing large meetings of workers lined up behind banners and placards, workers whom they could easily imagine genuinely to need their leadership. There was a temptation to believe their own ritually repeated propaganda. Some thoughtful Communists might have harboured premonitions and forebodings about the masses and might have advised instead an advance blueprint for a measured long-term transition for the regime. But democratization did not arise according to plan; it was in fact a hastily improvised strategy for undermining Gorbachev's opposition.

After the bitter fight in the party and its local organizations over the letter of Mme Andreyeva, it became clear the entire country would be the arena for playing out the two interrelated struggles in concentric circles, the fierce fight between the *glasnost* intellectuals and their more traditionalist critics, and the rather more controlled antagonism between the Ligachev and Yakovlev forces in the leadership, who were in effect contending for the ownership of Gorbachev's soul. The idea of a historic rendezvous between democracy and socialism elevated this struggle for power in many minds to a war having for its stakes the very fate of civilization. Since the meeting of the Communist parties at the time of the 70th anniversary celebration in Moscow, Gorbachev had been describing his efforts at reform as having special significance as a model for the other Communist parties. He left no doubt that he considered his advent to power as a signal for the reinvigoration of 'socialist internationalism', and his remarks were taken seriously by many foreign Communists, including the South African Communist leader Joe Slovo, comrades who would soon be reassessing their entire approach to the relationship between socialism and democracy. One can assume that Gorbachev was sincere in thinking that the international Communist movement was capable of substantial changes in outlook and a reassessment of its history in the spirit of socialist renewal.

No special attempt was made at this time to force the East European parties to imitate a Gorbachev model of reform, but the Soviet leader permitted the impression that the East European parties might gain from being attentive to the processes unfolding in the USSR. On Gorbachev's trip through the area in the Spring of 1987, it was precisely this implicit suggestion that caused his reforms to be received so coldly by the leaders such as Honecker, Ceauşescu, Grosz, and the others.

At least one Soviet official, Novosti's Valentin Falin, who would rise rapidly in influence in the next months, seemed to say that the day of the Brezhnev Doctrine, the idea describing a Soviet mission to 'defend the gains of socialism' by force, belonged to the past. In an interview with the West German weekly *Stern*, Falin was asked a question referring to 'Brezhnev's doctrine which served to justify Moscow's intervention in Czechoslovakia in 1968', to which he answered that the Soviet Union had 'long since outgrown the inclination to play censor to other socialist countries'.[2] From the perspective of post-1989 that seems to be an augury of the revolutions that tore the countries of the east from Soviet control in 1989, but at that time it may actually have been interpreted by East European Communists in a reassuring way, as Soviet reluctance to impose a reform course on the satellite regimes.

The international message of the Gorbachev reforms inevitably affected the eastern countries because of Gorbachev's promotion of the idea of a Common House of Europe. If a Europe from the Atlantic to the Urals was to be seen as having a certain shared interest, Eastern Europe must therefore draw itself into some kind of relation and dialogue with the Western states. The economic side of this vision was dramatized by the beginning of contacts between the EC and COMECON in June 1988. A certain 'interdependence' between the two seemed to be natural.

Once again one can see the familiar convention of 'infusing content' into new conceptions. If a Suslov-era Soviet ideologist had been asked to define 'interdependence', he would no doubt have replied that this was a Western imperialist propaganda term describing the class ideal of a perfectly functioning capitalist world economy according to the bourgeois law of comparative advantage. But under the New Thinking, it became fashionable to use the word 'interdependence' to depict the community of fate shared by the superpowers, who stood to gain from preventing nuclear war. Then Soviet reform economists could use the word to argue for a slightly more intimate interface with the capitalist world, and after that for a reconciliation with the original definition: an end to autarchy and full integration with the world economy. And what more felicitous place to start the process than with steps toward the integration of east and west Europe? Such was the erratic pattern of discourse under *glasnost*. The idea of de-ideologizing foreign policy had arisen as a critique of the Reagan doctrine and US 'neo-globalism'. Once it became accepted as a formula, one found that it was being applied to *Soviet* foreign policy. One can trace a similar evolution in *glasnost* discourse with 'pluralism' and 'market economy'.

The Soviet political mind was already a battleground in which the Ligachevist forces fought with the *glasnost* writers to define these new terms. As the struggle intensified, it became only natural to seek succour from the leadership of the East European Communist leaders, especially for Ligachev and his supporters, who could not fail to mark that most of these leaders saw things their way, with only the Polish General Jaruzelski giving full support to the Soviet reforms and to Gorbachev personally. So the Soviet inner-party conflict broadened to include the East Europeans. Yet even were this not the case, there would still have been a sort of logic driving the east European intellectuals to join in the discussion of the sins of Stalinism, since any Soviet recognition of past offences would automatically suggest changes in their regimes. The Hitler–Stalin pact thus had special meaning for the Poles and peoples of the Baltic Republics, as the crushing of the Hungarian revolution in 1956 had for the Hungarians and the Prague Spring of 1968 for the Czechoslovaks.

For all these reasons Gorbachev's struggle to rise above the collective leadership, already a struggle against the 'enemies of *perestroika*' in the USSR, necessarily expanded into a struggle against the traditionalist leaders of the Red Rejection Front in East Europe.

The campaign to weaken Ligachev and his co-thinkers in the USSR was now multiplying in intensity and extending to the farthest reach of party press organs. An explosion of protest against a rigged vote for delegates to the upcoming Nineteenth Party Conference removed the First Secretary on Sakhalin island. Outside Moscow it was hard to avoid choosing sides, yet Gorbachev and Ligachev had more room to manoeuvre than the armies of 'conservatives' and *perestroika* supporters at the Republic level. The two men could stake out various centrist positions and control the level of antagonism in the highest councils of the government. Moreover, they had between them the conciliatory figure of Andrei Gromyko, the last agent of the successions from Brezhnev to Andropov to Chernenko to Gorbachev, the man who had cast the vote that brought Gorbachev to power. Gromyko no longer made foreign policy, but as a President, he was not yet a mere figurehead, as for example Kalinin had been for Stalin. Since the struggle had broken out in January 1987, he had tried to round off the hard edges and to counsel moderation to the two factions. He thus stood between the Ligachevs and the Shcherbitskys on the one side and the Shevardnadzes and Yakovlevs on the other. It is likely that he himself commanded a certain support in this position, perhaps from Zaikov and Ryzhkov. As long as Gromyko still appeared at meetings of the Politburo the two factions were buffered from one another. And, since Ligachev sought nothing more than limits on Gorbachev's prerogative, that worked in his favour. So, it followed that if Gorbachev's fight was to be taken to a finish and Gorbachev to rise to a position independent of a second secretary and collective leadership, the Gromyko buffer itself would have to be stormed. Once Gromyko was out of the way, the last remnants of the Ligachev resistance and the Rejectionists in Eastern Europe could be confronted: for renewal or not! Gorbachev, a man of seemingly inexhaustible resources, would prove himself to be equal to this task as he had the others.

A PIVOTAL PARTY CONFERENCE

The next phase of the struggle opened with the Nineteenth All-Union Party Conference at the end of June, with 4991 delegates in attendance. Anticipating the event, Western commentators were divided between a large number who expected it to be a showdown between forces supporting and opposing reform, and a rather smaller number who expected what Michel Tatu called a 'Brezhnev era non-event'. But it proved to be neither. There was in evidence roughly the same balance of forces between the contending Politburo factions that had been on display since the Andreyeva affair. These factions showed some sign of being willing to chip away at the Gromyko buffer between them, and at the same time defended their own *modus vivendi* from Boris Yeltsin's highly threatening request 'to be rehabilitated in my own lifetime'. One might conclude that the conference simply ratified stalemate, except for the fact that it approved changes in the Soviet political system that were capable of shaking it to its foundations.

Some indications of the changes were given in the Ten Theses approved by the Central Committee plenum that convened at the end of May, just prior to the conference. They approved the ongoing ideological renewal, asserting at the same time that 'debates are only fruitful if they are based on socialism'. Within the context of a one-party system, there was ample encouragement for input from below and a seemingly earnest attempt to embark on 'the democratization of Soviet society and socialist self-government'. This was to be accomplished, according to the Theses, by a strengthening of the elective principle with regard to inner-party matters and turning the existing soviets into really effective and democratically elected organs of local government, 'with primacy over local executive bodies'.[3] The latter provision amounted to abandoning the role of the party in supervising the economy.

The Theses were provided as a document for pre-conference discussion. But Gorbachev's opening speech to the conference a month later took the Theses only as a point of departure and added his own proposal for all-Union elections to a new Congress of Peoples' Deputies, to be henceforth a 'representative supreme body of state power'.[4] Fifteen hundred deputies were to be directly elected to five-year terms, with 750 reserved for the party and related organizations. These would in turn elect from their number a small Supreme Soviet in two houses, a permanent body accountable to the Congress. The Congress would elect an executive President (a post that Gorbachev envisioned for himself) who would preside over its sessions.

The proposals amounted to a Gorbachev *coup* against the party Central Committee, which no doubt thought that it was going a long way in advocating its own programme to revive democracy in the party and to promote initiative in the manner of the earliest days of the revolutionary 'soviet power'. The Central Committee theses showed that the consensus between the opposed party factions was still a quite radical one by Soviet standards. On the other hand, Gorbachev in effect wanted to create a parliamentary power recognizable to the democracies of the West, with only a minority position in the parliament assured to the Communist party. The President was to have a mantle of democratic legitimacy that the general secretary had not enjoyed in recent history. The party itself was now driven onto the path of becoming a parliamentary party. As it reluctantly peered out into this brave new world the party was being given a push from behind by the confident helmsman (or rather, herdsman) Gorbachev.

GORBACHEV AND LIGACHEV IN THE CENTRE

No doubt could be left in the mind of any delegate to the conference that Gorbachev was the indispensable leader of genius. He presided over the sessions as if he were consulting with a kitchen cabinet, gently admonishing comrades to stick to their main thoughts and to eschew the old style of reciting accomplishments in their own bailiwick. He deferred the thorniest questions, cut short the long-winded, and gently prodded those who were too vague for more specifics, occasionally asking the whole body of thousands if they thought the speaker was right on this or that minor point. Comrades went through the difficult process of making a real speech and determining what they would say in the absence of a set role as in the Congresses of the past. Delegates clearly found the process exhilarating.

They wanted to talk about broad questions of history and theory. In the wake of the 30 May decision to cancel secondary school exams in history for want of a reliable text, the delegates focused on the question of the future (and the excesses) of *glasnost*. The writer Yuri Bondarev, no friend of *glasnost*, charged the the editors of 'certain periodicals' with having adopted the morality of Ivan Karamazov. *Znamya*'s Baklanov, on the other side, later wondered why none of those present was willing to answer 'Nina Andreyeva and Yuri Bondarev?' He cautioned the delegates to remember the position of Soviet literature under the ideological procuracy of Suslov. 'Suslov promised [Vasili] Grossman that his novel [*Life and Fate*, currently being published in *Oktyabr*] would not be published even in 200 years'.[5]

Despite the prior agreement that no personnel matters would be decided at the Conference, V. I. Melnikov, First Secretary from Komi ASSR, nevertheless managed to insert an attack on 'those who actively pursued the policies of stagnation in the past'. From the chair Gorbachev prompted him to name names. 'I address this to Comrade M. S. Solomentsev primarily', said Melnikov, 'and to Comrades A. A. Gromyko, V. G. Afanasyev, G. A. Arbatov, and others'.[6] Gorbachev's promptings understandably convinced many that Melnikov's statement was not entirely spontaneous. And none other than Shcherbitsky, himself under fire since the spring of 1987, read a note from a 'Comrade Mamayev', referring to Gromyko as being 'out of touch with life. He has done his work and his noble deeds are in the peoples' memory.'[7] Gromyko, on the dais and neither yet dead nor retired, had to sit and listen to these funereal tributes.

They may have been laid on in advance by both Gorbachev and Ligachev. In the pre-conference discussion period, numerous party officials had carried on a detailed critique of the foreign policy of the Brezhnev era, when Gromyko was foreign minister. Aleksandr Bovin argued for *perestroika* in foreign policy, long burdened by what the Ten Theses had called 'dogmatism and subjective approaches', and responsible for the decision to invade Afghanistan and to install SS-20s in Eastern Europe.[8] 'Dogmatism' had been Khrushchev's designation for the foreign policy outlook of Molotov, Gromyko's political godfather. In a speech at Togliatti in the beginning of June Ligachev got into the act, referring to those who charged that opposition to *perestroika* was growing as having repeated the error of the 1930s, when it was said that 'the class struggle intensifies' with the approach of socialism.[9] This was the theory on which the great purges had been ostensibly based, argued by Molotov and Kaganovich against the objections of Kirov as early as 1932. In his speech to the conference, Ligachev similarly blamed 'some newspaper editors' for trying to frighten the party with the idea that opposition to *perestroika* was growing.

On the last day, when everything seemed to have been said, Boris Yeltsin made his case 'for rehabilitation in my own lifetime' before the assembled delegates. He justified an interview with Western media in which he admitted his differences with Ligachev on 'the tactics of *perestroika*, on questions of social justice and on the style of his work'. Solomentsev was also cited for his 'liberalism toward millionaires'. Yeltsin seemed to attribute these faults to the legacy of corruption from the Brezhnev era. Ligachev answered by citing his work in Tomsk through the Brezhnev years, far from the centres of corruption. He asserted that the work of cleaning up the nests of privilege and misuse of state property was begun not under Gorbachev, but Andropov. And with regard to the apparent attacks on senior party personalities, Melnikov's 'those who actively pursued policies of stagnation in the past', he reminded the delegates of the 'anxious days' at the time of Gorbachev's election in 1985, and of the 'firm stand by Politburo members comrades Chebrikov, Solomentsev, and Gromyko and a large group of *obkom* first secretaries' on the choice of Gorbachev and the setting of a new course.[10]

On the subject of *glasnost*, Ligachev endorsed to the full the harsh criticisms made by Bondarev, scoring in particular the 'ersatz data' on corruption provided by *Moscow News*,[11] and its calls, voiced by 'unworthy people', for the hunting down of the enemies of *perestroika*. Gorbachev's concluding speech gave him support, and seemed to belie the talk of a split in the leadership, which Ligachev had already been at such great pains to deny. The two men were absolutely united in their unyielding attitude toward Yeltsin, and in their desire to ratify the decision made on him in October 1987.

GLASNOST AGAINST LIGACHEV

One could sense differences on the questions concerning 'white spots' in party history, but these did not stop the Politburo's special commission from rehabilitating the victims of the second and third Moscow trials of 1936–37, including Zinoviev, Kamenev, Radek, Pyatikov, and others. In addition to this, the Supreme Court cleared the accused and published interesting details about the 'Ryutin affair' of 1932, which had served as the point of departure for Stalin's suspicions about the vast conspiracy he thought he confronted.[12] *Glasnost* now having come close to its apogee, it only remained to rehabilitate Trotsky, but this was not done. True, the CPSU had now rehabilitated everyone else among Stalin's Bolshevik victims, but the figure of Trotsky, whom Soviets had been taught for generations to see as the mastermind of the anti-Stalin conspiracy, still induced hesitation and confusion, perhaps among other reasons because the current *glasnost* mythology had cast him in the role of mastermind of Stalinism.[13]

Nevertheless this version of party history was closer to the truth than the regime had come since Stalin's time, and most felt that the revelations and investigations were only beginning. There was a sense of elation at having defended so ably against the impositions of the 'conservatives'. In this mood Foreign Minister Shevardnadze was moved to expand on some criticisms of Stalin's foreign policy that, in view of the campaign against Gromyko, had a contemporary ring. Vyacheslav Dashichev, of the Institute of the Economics of the World Socialist System, had offered the view that

> During Stalin's time, there developed serious deformations not only in domestic policy but in foreign policy as well. In essence the foreign policy practice of Stalin and those around him (*okruzheniia*) were of a character alien to socialism and based instead on the ultra-left ideas of Blanquism and Trotskyism.[14]

At the end of July, Shevardnadze told a large foreign policy conference held under the auspices of the Foreign Ministry that, just as New Thinking depended on *perestroika*, so it had been with Stalin. The deformations in his foreign policy were an outgrowth of those in his domestic policy, specifically in his failure, because of his ideological prejudices, to use political methods to prevent the onset of the Cold War.[15] By contrast to the Stalin era, foreign policy guided by the New Thinking 'cannot be equated with the class struggle'. Moreover: 'The struggle between the two social systems is no longer the defining tendency today.'

What would Ligachev, the Second Secretary for Ideology, say to that? The answer came in the beginning of August, at a meeting with the local party people in the city of Gorky. Ligachev first tried to roll back some of the more dangerous deviations in the press. 'Commodity–money relations' were well established in the country, he said, but they were no panacea, and could not serve to regulate the economy; rather the reverse was the case – the party must always regulate them. And they must not impinge on public property, the basis of the Soviet system. The party would actually have *greater* control over the economy for not being involved in its everyday workings, he claimed, in a statement whose logic may have been grasped only by those in the state of grace. As to foreign policy, Communists could not remain on the sidelines of global struggles. To deny the class nature of international relations only put confusion into the minds of the Soviet people and of friends abroad. Moreover, he concluded, Gorbachev's speeches to the Nineteenth conference had made all this abundantly clear.

Gorbachev had left to take a vacation in the south on 1 August and Shevardnadze had left on a trip to Afghanistan; so, as he had done in the Andreyeva affair, Ligachev had again turned up the volume when his opponents were out of town. He even got some support from abroad. The leader of the US Communist party, Gus Hall, saluted Ligachev for refuting the 'slander' against Gromyko and Solomentsev, and criticized supporters of reform: Aleksandr Bovin for presumably allowing that capitalism had a right to exist, Dmitri Likhachev for various 'god-building' tendencies such as referring to an 'extraordinary intelligence at work for millions of years', or to genetic heredity as a source of the ethical properties of the intelligentsia, and Yevgeni Ambartsumov for an endorsement of Universal Human Values as the more inclusive category under which class struggle must be subsumed.[16] Not being under the same constraints as Ligachev, Hall objected to the idea of the same person (Gorbachev) being both General Secretary and President.

A RESPONSE AT HOME AND IN THE BLOC

Even in hailing the work of the Nineteenth Conference, other parties gave indications of reluctance to follow Gorbachev's apparent course. The East German paper *Neues Deutschland*, which had printed with apparent approval the letter of Mme Andreyeva, chose the discussion period prior to the conference, set aside for a debate of the Ten Theses, to publish a laudatory article about former party leader Walter Ulbricht, who had been so adamant in demanding Soviet intervention in Czechoslovakia in 1968. GDR chief Erich Honecker flatly stated (and apparently with a straight face) that 'German Communists never aimed to apply the Soviet system in their country.' 'We intend', he said, 'to continue in the proven way.'[17] In Romania, Nicolae Ceauşescu was in no mood to try experiments with Lenin's ideas about a more lenient nationality policy. He announced instead his plan to raze 7000 villages in Transylvania, most of them inhabited by Hungarians. Accordingly, he decided as well to close the Romanian consulates in Bucharest and Cluj.

An international bloc of opponents of the Soviet reforms seemed to be gathering steam and becoming more aggressive. The Romanian move, however, made things extremely difficult for the anti-*perestroika* forces in Hungary, led by Karolyi Grosz, to consolidate their position in the wake of the forced retirement of János Kádár. Rather than play its part in the solid phalanx of opponents to reform, Hungary, faced with this threat from its most bitterly hostile neighbour, was liable to reverse its field and cast in its lot with Gorbachev and radical reform of the entire Rejection Front. This was to be crucial for the revolutions that swept East Central Europe a year later.

In some ways the most important statement against the Soviet reformers came in the form of an article by Valentin Falin and Lev Bezymensky, 'Who Unleashed the Cold War?' that appeared in *Pravda* at the end of August. The authors clearly wanted to counter the impression made in articles by Dashichev and others that Stalin and Stalinism were responsible for beginning the post-war conflict and arms race with the West, in the language of New Thinking, for 'not exhausting all the political means' to avoid a hostile confrontation. They insisted instead that 'The Cold War was not our choice.' The article was a careful, if tendentious, tour through a series of recently released documents on US foreign policy, with special attention to the various contingency plans of 1947–50 ('Pincher', 'Grabber', 'Fleetwood', 'Intermezzo') for an American atomic attack on the Soviet Union in the event of the failure of diplomacy.[18] A map showing the bomber ranges for plan 'Pincher', extending over most of the USSR, accompanied the article. The authors gave a highly selective analysis of steps taken by American policy toward confrontation with the Soviets, all in their view tending dangerously and almost inexorably toward a 'nuclear bacchanalia'.[19]

The article was intended as a correction to the writings of Dashichev and others, as well as a counter to the Universal Human Values theory of Dobrynin, at the time head of the International Department of the Central Committee. It was no coincidence that Falin was to take Dobrynin's job in the major shuffle of positions at the end of September. The editors of *Pravda*, undoubtedly including Viktor Afanasyev, introduced the article by drawing attention to its thesis of a US decision, as early as 1943, for a crusade against 'totalitarian state aggression'. They called for balance in the discussion of the foreign policy of Stalin:

Here it is necessary to pause and think rather than give oneself up to the temptation to find the truth 'somewhere in between', that is, to apportion metaphysically a vague guilt for all the pre- and post-war complications, difficulties, tragedies, as if for the sake of some 'demarcation' from the past, and thus in effect to admit only the sins of Stalin and Stalinism. Such a method would increase neither knowledge nor political wisdom.[20]

Western officials supposed that the controversy was part of a struggle to reorient foreign policy in a fundamental way, another mystifying Soviet zigzag. But no one in the Soviet debate wanted anything so extreme as, for example, stopping the destruction of missiles that was going forward in accordance with the INF treaty signed the previous December. Nor was there any real alternative presented to the policies being carried out. The quarrel concerned abstract questions relating to the appropriate labels for the New Thinking and, just as the struggle between Ligachev and Yakovlev was up to this point a struggle for Gorbachev's soul, this was a struggle for the soul of the New Thinking, which no one sought to renounce. It was really a struggle over the prerogatives of the General Secretary as opposed to the Second Secretary, the latter clinging to the tattered banner of collective leadership. Positions on issues had to be chosen in the heat of action. Falin was no 'conservative' fanatic, nor would he adhere insensately to his historical thesis about the Cold War once he had gained the leadership of the International Department. In fact, in an interview on the television programme 'Studio Nine' on 15 October, he would actually speak of Soviet leadership having erred in permitting the country to be dragged against its will into the arms race, for want of recourse to political means. In his view as stated at that time, 'Stalin showed too much enthusiasm for the military factor.'[21] But back in August, it was a different story. In the atmosphere of approaching conflict between Ligachev's group and Yakovlev–Shevardnadze–Dobrynin, Falin had opened up a new front.

GORBACHEV DEALS A BLOW

The showdown came at the end of September, in a series of Politburo meetings, in which Gorbachev succeeded in reshuffling its composition. The personnel changes amounted to the greatest single step he had been able to make in the direction of freedom from collective leadership. These were rumoured to have been the result of a 'counter-coup' that he had to mount against some sudden threat to his leadership. According to *L'Unità*'s correspondent Giulietto Chiesa, the Moscow military district was put on alert, with military aircraft and helicopters flying over the central sections of the city.[22] But no further evidence of a coup attempt by Gorbachev's opponents has surfaced, and indeed it seems unlikely in view of the fact that the major figures, including Ligachev, were out of town. Moreover, the sweeping reorganization that Gorbachev installed does not give the appearance of a hasty improvisation.[23]

The changes were a consequence of the campaign against Gromyko visible to all from the time of the Nineteenth Conference. Up to this time, Gorbachev must only have been able to count on a minority of votes, Yakovlev, Shevardnadze, Slyunkov, and perhaps Ryzhkov and Zaikov. His opponents would have included Ligachev, Chebrikov, Vorotnikov, Solomentsev, Nikonov, and Shcherbitsky, with Gromyko in the position of a buffer and a conciliatory vote, especially if the lineup were as just described, six against six. On the key days, 25 and 26 September, Ligachev was on vacation in the south, Vorotnikov and Solomentsev were out of the country, Shcherbitsky was in Kiev, and Shevardnadze was in New York addressing the United Nations. Chebrikov and Nikonov were the only representatives of the Ligachev group, with Yakovlev and Slyunkov undoubtedly in support of Gorbachev, and Ryzhkov and Zaikov probably similarly inclined.

At this point the question of Gromyko's retirement, the central point in the reorganization, was raised. Of the seven, aside from Gromyko, only Chebrikov and Nikonov ended up by having their positions even slightly reduced. The rest came through unscathed. Until further evidence is produced, one must surmise that this is the way the vote went, assuming of course that, under the circumstances, Chebrikov and Nikonov did not also decide to go along cheerfully.

The result was that Gromyko was retired with thanks, along with Solomentsev. A series of Politburo commissions was set up, with some who lost ground in the Politburo compensated by being given a commission to head. Ligachev lost his position as second ideology secretary and was given a commission on agriculture (Nikonov had up to this time also had responsibilities for agriculture); and Chebrikov lost the KGB, but gained the chair of a commission on legal policy. Yakovlev, however, was not made the ideology secretary, as that position was given to Vadim Medvedev, who came into the Politburo as a full member without having to go through a period as a candidate. Yakovlev was, however given the chair of the commission on international policy. Dobrynin had to surrender the International Department to Valentin Falin. Vladimir Kryuchkov, a specialist on foreign intelligence, got the KGB. Dolgikh and Demichev were removed from the candidate members and replaced by Nikolai Talyzin and Aleksandra Biryukova. Gorbachev's old friend, Anatoli Lukyanov, was made first deputy chairman of the Presidium of the Supreme Soviet and a candidate member of the Politburo.

Most dramatic was the apparent reduction of Ligachev, deprived of the ideology portfolio and reduced to sharing agriculture responsibilities with Nikonov. Chebrikov was at first thought to have benefited, despite losing the KGB, because he was added to the Secretariat. But in the days that followed, it became clear that the positions and offices themselves had been reduced in importance, and that in fact the Secretariat itself was for the moment defunct. The party, that is, the collective leadership, had lost control over its day-to-day affairs.

Medvedev, in a press conference immediately after the Central Committee plenum that ratified the changes, said that Ligachev no longer chaired sessions of the Secretariat, and more than that: there was no longer a position of second secretary at all. As the new chief in the somewhat truncated field of ideology, Medvedev announced to the press what the fight had been about. Talks with ordinary citizens, such as the talk Gorbachev had had with people in Krasnoyarsk, had caused the General Secretary to realize the need to reshape the apparatus. Ligachev had been wrong, said Medvedev, to affirm the primacy of class struggle perspectives in foreign policy. Universal Human Values (*obshchechelovechestvo*) must be the keynote of renewal in international relations. To judge by this, Soviet foreign policy had been freed from what Soviet UN representative Aleksei Belonogov had called Gromyko's 'excessive dedication to ideology'.

In that event, what would the new role of ideology be in a Politburo without a second secretary? Medvedev's views were not entirely clear. On the one side, he stressed the historic challenge of the present period, one that demanded a 'deep and qualitative renewal' and a 'stronger dynamism'. He argued that the reform ideas of the sixties had not taken root because the perspective of the leadership was that of overtaking the United States in 20 years, something that had proved to be unrealistic. In truth, he argued, the socialist countries could not develop as a separate system in isolation from the world economy; there must be, if not a full convergence, then at least some sort of interface between capitalism and socialism. On the other side, Medvedev spoke of the need for greater vigilance in defending Leninism. Too many were under the influence of Stalin's 'distortions of theory' and his 'deviations from Leninist concepts'.[24] In view of what had been said about these things in the last two years, this might have given an indication that Medvedev was going to try to contribute something on Stalin's 'deviations'. But this was just a pretentious phrase. Medvedev's reign as ideologist was to be a dull and uneventful one. In fact, ideology was now represented in the Politburo by no less than three persons, Medvedev, Yakovlev, and Ligachev, a sign as it proved, of its rapidly decreasing importance.

In breaking the authority of the party ideologists, Gorbachev had removed a barrier between himself and the *glasnost* intellectuals, who were now coming into a position from which their thinking could have a direct impact on the leader. Readers of the Soviet press had since 1987 slowly got used to the idea that not everything they read was policy. Now they had come to realize as well that a given article was likely to shape policy immediately. Among bloc Communists the change was important. They were used to supposing that a statement made by a Soviet official could not have been made without Politburo approval. Now that the central discipline over ideology was gone, a Soviet official or an institute intellectual did not have to fear censure if he were to urge a new line of conduct on a bloc party, which in turn did not have to fear censure if it followed the advice. So domestic and foreign policy began to evolve spontaneously. Such was the spirit of initiative in Communist intellectual and political life introduced by Gorbachev.

Not that Medvedev ever understood this. He was prepared to make a series of far-ranging edicts, if not on Leninism itself, then at least on what the Soviet press ought not to propagate as Leninism. He was sure, for example, that Solzhenitsyn's *Gulag Archipelago* should not be published, and he was able to dissuade *Novy mir* from printing the first instalment of its planned serialization in October 1988, although several did appear by the end of 1989.

He was incensed at the same journal's publication in May of Vasili Selyunin's article, 'Sources'.[25] Even before the article appeared, Selyunin had become known for advocating a strategy of 'acceleration' different from the official one. As he saw it, the Gorbachev economic reforms began with the idea of compensating for the manpower shortage with an increase in productivity, but they wanted to continue to direct the lion's share of resources into production of producer goods, in many sectors of which the Soviet Union was already the world's leader. The meagre actual increase in the standard of living by the ordinary worker was scarcely sufficient to motivate him to work harder. So, said Selyunin, the economy should actually plan a decline in the growth rate of heavy industry and an increase in consumer goods, so as to raise the standard of living.[26] Thus one could compensate for the Soviet economy not having a capitalist 'braking mechanism', such as crises and recessions, to halt heavy industry overproduction.

Selyunin's advice for the first steps of 'acceleration' seems to me superior to that of Aganbegyan. In the final analysis, Aganbegyan's conception depended on non-economic incentives connected with increased power over the process of production, whereas Selyunin assumed the primacy of economic ones. Selyunin was not content to leave it at that, however. In 'Sources', he took the reader on a tour of Russian and Soviet history to suggest that the Russian political culture, with its powerful state dominance over market forces, was itself the cause of Russian economic backwardness by contrast with western Europe. Some Western historians of

Imperial Russia might regard this as an acceptable proposition. Yet one could also take into consideration that Russia's position in the international division of labour, after it was opened to trade with the western countries in the sixteenth century, was as a supplier of raw materials and foodstuffs for the western market. It was in part to serve this market that Russian governments used compulsion to prevent peasant flight to the frontier land and eventually to impose serfdom. It is at least arguable that autocracy prevented the disintegration and partition of the Russian lands, that it prevented Russia from going the way of partitioned Poland, a state that also relied on serfdom, but tried unsuccessfully to survive in a fiercely constitutional polity under a weak monarch. Forests have been felled to provide the paper to discuss this question, and it cannot be taken up at greater length here. The point for purposes of our inquiry is that Selyunin wanted to say that Lenin continued in the old oppressive pre-revolutionary way when he approved various measures to suppress the peasant market during the civil war of 1918–21.

Selyunin adhered to the fashionable rendering of Bukharin's 'alternative', and the glorification of the NEP as opposed to what preceded, War Communism, and what followed, the collectivization of agriculture and five-year plans. He made Trotsky a straw man as the advocate of barracks communism and levelling poverty, and Stalin as the leader who realized these prejudices in action. His unique twist was to remind the reader of Lenin's suspicions about the market and the peasantry, and thus in effect to make Communism the continuator of the tsarist autocracy's suppression of markets, the obstacle to a radical restructuring that would entail dismantling, not reforming the economic bureaucracy. But Selyunin also presented a political manifesto, indicting Bolshevism, but not always Lenin himself, for excessive deference toward the experience of the French Revolution and the heroic example of the Jacobins. Selyunin found it significant that French Jacobinism also punished the people in the name of a utopia that suppressed the market.

> In his search for practical answers to real problems, Ilyich more than once recalled their example and compared the French Revolution with ours, reflecting on the limits of brute force in economic construction.[27]

Selyunin was probably the most clever and striking of the economists, as well as the most courageous, since he was more frank than any of the others in providing a well-reasoned intellectual and political exit from Marxism. His arguments were not the last word in historical erudition, but others showed that the whole Marxist and Bolshevik tradition could be re-examined in such a way as to arrive at similar conclusions. The historian Mikhail Gefter, whose work had been suppressed in the Brezhnev

era, and who now gave slithery interviews capable of various interpreta
tions, nevertheless also provided a powerful argument for putting
Communism's historical dilemma in Russia into broader European terms

Gefter began with Russia's archaic peasant collectivist tradition, not
ing that Karl Marx had made a certain wager on it when he encouraged
the Populist revolutionaries to think that Russia might have a chance o
skipping the capitalist stage of development. The peasant repartitiona
commune might serve as a point of departure for socialist developmen
in backward Russia, Marx had allowed, if a Russian revolution agains
tsardom were to be accompanied by a proletarian socialist revolution i
a more advanced country that might take backward Russia under its wing
Marx was certainly not sanguine about the possibility. He had always
preferred those Russian writers who exposed the primitiveness and bar
barity of agrarian conditions.[28] Yet he was brought to the point of en
couraging the Populists in letters to N. F. Danielson (Nikolai-on) and
Vera Zasulich in 1881. Thus the curious fact that the founder of Russiar
Marxism, G. V. Plekhanov, espoused an idea of a 'semi-Asiatic Russia
repeating the French Revolution of 1789 in most particulars, while Marx
himself disdained Plekhanov and Russian Marxists to encourage revolu
tionaries who professed ideas having little in common with Marxism.

For historian Gefter, Lenin was closer to Marx's thinking than
Plekhanov.[29] Lenin had seen two paths for old Russia: a 'Prussian' sys
tem in which landlords and industrialists enjoyed the protection of a
powerful monarchy or an 'American' solution in which the peasantry es
tablished a smallholder's paradise. But in the end, Lenin put a new gloss
on Plekhanov's idea of Russia recapitulating the French Revolution of
1789 when he suggested that the 'Asiatic' peasantry might itself serve in
the revolutionary role of the bourgeoisie if, combining with the prole
tariat, it helped to provide a radical democracy with socialist goals.[30] Thus
said Gefter, 'the two paths became one'. Lenin became the advocate of a
Russian socialist Jacobinism, as shown in the slogan of a Revolutionary
Democratic Dictatorship of the Proletariat and the Peasantry.

The deficiencies of Lenin's formula became evident, argued Gefter, in
the period 1917–29, during which revolutionary Russia's socialist choice
dictated two further options: War Communism or NEP. And, said Gefter
each turn in the succession struggle after Lenin's death ('1923, 1928, or
even 1934'[31] – Gefter gave the symbolic dates for the defeat of Trotsky
Bukharin, and Kirov) saw a narrowing of the range of choices for the
creative development of a 'multiform (*mnogoukladnoi*) NEPist Russia'.[32]
Loss of this creative alternative meant that Russia was to have 'not an
idyll, but a tragedy of many acts', with Stalin, as 'a second Peter . . . a
revolutionary on the throne'.

This influential article had been originally penned in 1977, but suppressed

Appearing in the middle of 1988, it seemed to speak eloquently to a new constituency of intellectuals who drank deep of the myth of the Bukharin alternative and saw NEP as a synthesis of the native socialist traditions and a quest for a westernizing political culture. As had Boris Kagarlitsky and Fyodor Burlatsky, Gefter quoted Petr Chaadayev, the first philosopher of Russia's westernizing mission: 'Not to repeat, but to rebegin' (*Ne povtorit, a perenachat*). Of course, whatever was said about the dilemmas of modernizing Russia in 1917–29 was in effect also said about Soviet policy toward the Third World. The argument for the possibilities of 'multiform' development within NEP was an argument against the exaggerated pretensions of 'non-capitalist development' and 'socialist orientation' in Third World countries where in fact the example of NEP might better serve as a model. Gefter was recapitulating the literature of Soviet experts on the Third World who had been expressing doubts about support for radicalism and national liberation struggles for some time and whose work had been watched with interest by Western scholars. He was redefining socialism and Leninism, in the phrase of a co-thinker, as 'civil society based on multiform modes of human activity'.[33]

The same co-thinker, Vladimir Maksimenko, nevertheless cautioned against interpreting this synthesis as a repudiation of the revolutionary heritage of Communism as expressed, for example in the experience of War Communism. He knew that this was a pronounced tendency in the current literature, but he warned that 'People who speak of such a deformation should, to follow their own logic, also declare the program of the Russian Communist Party, numerous works of Lenin, and the entire Russian Revolution to have been "perversions".'[34]

At the end of 1988, a consultant to the Ideology Department of the Central Committee, Aleksandr Tsipko, took this advice and extended the ideas of Selyunin and Gefter to precisely this point. In a series of articles in *Nauka i zhizn* Tsipko seized on Selyunin's thesis about the compulsory grain requisitions of 1918: instead of these measures being the result of famine, they had been its cause.[35] For Tsipko, this served as the springboard for arguing that all the depredations and tragedies of Soviet history were the result of hubristic efforts to suppress markets. From there Tsipko was able to recapitulate the Western writing about Soviet totalitarianism of the last several decades. At first putting the critique of Soviet socialism in terms of deviations from the wisdom of Lenin and Marx, he soon concluded that the roots of Stalinism lay in the radical anti-peasant and anti-market prejudices of the nineteenth-century Russian intelligentsia and in Marxism as the convenient instrument of these passions.

Tsipko was striving for a more extreme and simple version of the Bukharin myth, which for him was only a starting point. He wanted to discard the idea that Stalin was some sort of Thermidorean who ruined the socialist experiment. For him, Stalin was the most consistent leftist, who realized the visions of the demon Trotsky, the latter a man who feared consumerism and hated the thought of the workers getting any material benefit from labour, and who was only happy when people were hungry.[36] Stalin was no gravedigger of Bolshevik revolutionism, thought Tsipko, but the heir of the most extreme Russian radicals, going back to the anarchist Bakunin, and was not satisfied until he had achieved a 'total Totalitarianism' consistent with the original premises of the intelligentsia's struggle against markets. Critics such as Anatoli Butenko were perfectly correct to accuse Tsipko of having repudiated the entire Russian Revolution.

A kind of platform for a democratic revolution against the Soviet *ancien régime* had thus gradually emerged from the ruminations of the Soviet intelligentsia on the question of the origins of Stalinism. Ideology Secretary Medvedev, not a great intellectual innovator, nevertheless recognized the obvious threats, and did his best to suppress publication of Selyunin, Solzhenytsin, and other anti-Communist publicists. He denounced the newly formed Democratic Union as a political party seeking the overthrow of Soviet institutions, and called for adherence to the traditions of October and the movement of *perestroika*. But he was at a crushing disadvantage. He stood on the ideological platform that had conquered at the September–October 1988 plenum, a victory of Universal Human Values over class struggle, an endorsement of a programme of de-ideologizing Soviet foreign policy. The ideas of the anti-Communists seemed to follow logically from these premises or, at any rate, large sections of the intelligentsia were bound to conclude as much. Resistance to the trend demanded more ingenuity than the new ideological secretary was able to muster.

CONFRONTING THE BREZHNEV DOCTRINE

Having liberated himself from the strictures and limitations of Gromyko's prestige on foreign policy matters, Gorbachev gave evidence of sensing that he had a free hand to make the kind of sweeping initiatives that could truly liquidate the decades-old European tensions. Yet he was not entirely satisfied with his impact on American public opinion. Studies coming out of his own institutes reassuringly claimed that his speeches had made a deep impression on the foreign policy elites, and that the United States had virtually abandoned the perspectives of the Reagan Doctrine and 'neo-globalism'.[37] Yet the feeling remained that enthusiasm for Gorbachev's measures had reached only a part of US opinion. There remained a large and variegated group who suspected that *perestroika* was designed to make the Soviet military more, not less effective, in line with the plans of General Ogarkov.[38] At an international conference on European conventional forces held in Moscow in October, Arnold Horelick, of the RAND/UCLA centre commented that the idea of the Cold War's having disappeared might be premature:

> Many of us in the West see the root of the Cold War in the Soviet occupation of Eastern Europe ... in effect making Eastern Europe a salient under Soviet control pointed at the heart of the West ... a fundamental reordering of the European and global security system is going to have to await fundamental systemic changes in the East.[39]

At a September conference on European conventional weapons held in Budapest, Falin and Helmut Sonnenfeldt seemed to agree that Hungary should be a testing ground for a unilateral removal of Soviet troops. Falin expressed the view that 'Hungary with its vast intellectual potential is capable of further contributions to solving the problem of European security.' Sonnenfeldt added that the Soviets 'could easily cut by half' their forces in Hungary: 'they have ample room for taking unilateral moves'.[40]

Gorbachev had not convinced everyone he sought to convince. Nor had he cemented his new alliance with his own military people who, in view of what they claimed was increased military aid to the Afghan rebels, now wanted to reconsider the decision to quit that country taken a year previously. Suspending the withdrawal, however, was no solution, as it signalled the bankruptcy of his entire policy. The break with the traditional foreign policy symbolized by the campaign against Gromyko would all go for nothing if the policy were not pursued to the end, and it was increasingly clear that this could only be done by another 'escape forward'.

Gorbachev took this step in his address to the United Nations in December, in which he announced to the world that the Soviet Union would reduce its army by two million men in the next two years.[41] No less than six tank divisions would be removed from Eastern European duty and additional forces from the China border, apparently as part of a prior agreement with the Chinese on a Gorbachev visit in the spring. The Soviet action would be followed in January by similar commitments to reduce forces on the part of all the other Warsaw Pact countries except Romania. Gorbachev spoke of having offered his initiative in the name of Universal Human Values, which now suggested a grand synthesis, he said, of the traditions of the French and Russian Revolutions. In service to the synthesis, international relations must be 'de-ideologized'.

Try as it might, it was not easy for the world to resist Gorbachev any longer. *Time* spoke of 'his vision, both compelling and audacious', and recognized his as 'the most commanding presence on the world stage'. A majority of those polled in the United States and West Germany (though not in France) gave a favourable opinion of him. He seemed to have broken through a logjam of scepticism and wariness. He was now over-whelmingly seen as a man who was radically different from the Soviet leaders of the past. It was more common to ask, not whether he was sincere, but whether he could succeed. Moreover, the original impression that he was attempting to rationalize the Communist system (an agenda that Gorbachev himself had not abandoned) now developed into the idea that he was going beyond all systems, discarding them as primitive prejudices of the Cold War. More and more, the belief grew that Gorbachev was not proceeding according to any orthodoxy but against all orthodoxies, not under any flag but against all flags, an impression that conveyed both his ruthless experimental spirit and also the fact that he was not at all sure what he was doing.

10 1989: The Year of Anger and Remembering

Most people deceive themselves with a pair of faiths: they believe in eternal memory (of people, things, deeds, nations) and in redressability (of deeds, mistakes, sins, wrongs). Both are false faiths. In reality the opposite is true: everything will be forgotten and nothing will be redressed.

Milan Kundera

In civil society each member is his own end, everything else is nothing to him ... the whole sphere of civil society is the territory of mediation where there is free play for every idiosyncrasy, every talent, every accident of birth and fortune, and where waves of every passion gush forth, regulated only by reason glinting through them.

G. W. F. Hegel

If you arrange it that a Pole can never become a Russian, I guarantee that Russia will never subdue Poland.

J.-J. Rousseau

Rudolf Spielmann begins his classic *The Art of Sacrifice in Chess* by explaining the difference between a sham sacrifice and a real sacrifice. Sacrifice, he says, is an immensely attractive option from the aesthetic viewpoint, in which one gives up some of one's pieces in order to open up chances for a more promising attack. 'A hallowed heroic concept!' says Spielmann, 'Advancing in a chivalrous mood, the individual immolates himself for a noble idea.'[1] Many seeming sacrifices, however, do not really deserve the name, for the calculating player chooses them only after looking deeply enough into the position to see that the material yielded will be regained subsequently, and usually with interest. Far different from this sham offering is the sacrifice that is made when one *cannot* see one's way to a sure advantage in the future. Real sacrifices have an element of risk. The likelihood of success is not calculated on the basis of the inherent facts of the situation, but involves instead a wager on the psychological effect of the act that itself creates a new position. One advances despite the inability to see over the horizon, assuming, or perhaps just hoping, that the boldness of the move will itself shape fresh opportunities. The attractions of the idea are obvious. 'The glowing power of the sacrifice is irresistible', Spielmann tells us, because 'enthusiasm for sacrifice lies in man's nature.'

Gorbachev's offer to remove substantial material from the European chessboard was in the nature of a Real Sacrifice. He had not thought things through; he was making moves that could not be thought through. They advanced from an extreme but logical tenet of New Thinking: security had not been achieved in the past by military measures, and must now be sought by 'political means'. Instead of plunging into labyrinthine negotiations of the type favoured by Gromyko, in which one painstakingly calculated the European security balance in terms of the different weapons in the different units, the force-to-space ratios, the tangled scenarios of possible threats, he now sought to alter the very political balance that had originally given rise to the military standoff. This suggested a willing sacrifice of this or that minor military advantage, in any case probably of little strategic worth in the larger perspective of Reasonable Sufficiency and minimum deterrence, in order to rout from Western minds the bugbear of a Soviet threat. It was exactly the approach once suggested by Sakharov to the threat posed by the American SDI, of which, by the way, the reader of the Soviet press now saw very little. The sacrifice was actually a kind of surrender, as was clearly recognized by Gorbachev's Western critics, who complained of his fiendishly clever negotiating strategy: asking the opponent what he wanted – and then giving it to him. Gorbachev felt himself succeeding in his larger quest for a new world harmony, and now he was moving toward the most momentous step of all. He was going to press a change of regimes on the bloc states, judging that he could afford to offer this to the West as a sacrifice, very like the 'surrender' of a lover, who gives up a part of him or herself to the other, in search of the happiness promised to both in the event of their reaching a meeting of minds and a communion of spirits.

THE ITALIANS ON MIDDLE EUROPE

Gorbachev did not come to this viewpoint entirely without assistance. The idea of an historic understanding between East and West had knocked around the European left for a good while. It had been extensively described in the writings of Western refractory Communists and other leftists, writings that had been ringingly denounced by Soviet authorities for their opportunism. That these notions had been attacked by the Brezhnev-era ideologists of course recommended them to Gorbachev all the more. He was yielding to the temptation to think that a contemporary revival of the broad left coalitions of the Popular Front era, in the thirties, and the immediate post-war period, might serve as a platform from which to lobby for an end to the Cold War. He had already suggested a broad outline for a grand reconciliation of the forces of the European left at his meeting with leaders of the European and other Communist parties in November 1987, when they were all in Moscow for his historical speech. At that time most felt that the main obstacle was the position taken by Suslov in 1980–81 in his last quarrel with the Italian Communists over a concept floated by PCI leader Enrico Berlinguer, the 'New Internationalism'. According to Berlinguer, Social Democratic and other left organizations, Greens, or feminists should henceforth be regarded by Communists as fraternal and co-equal with other Communist parties. Alliances should be formed with these 'as well as Christian, national liberation and many other democratic and peace-loving movements'.[2]

Berlinguer was pursuing a strategy first described in Palmiro Togliatti's Yalta Memorandum of 1964, a document that the Italian Communist leader had intended to present to Nikita Khrushchev on his trip to the Soviet Union. He was never able to do so. Togliatti died several days prior to the meeting, only a few weeks before Khrushchev's removal from the Soviet leadership. Togliatti's idea resurfaced at the end of the seventies in connection with Berlinguer's plans for a bloc of Italian Communists with the Christian Democrats, the *compromesso istorico*, and the concurrent vogue among the European Communist parties for independence from Moscow, a line they called 'Eurocommunism'. Suslov denounced New Internationalism and related tendencies as a danger to Communism. The International Department's Konstantin Zarodov, who championed the aggressive strategy of Alvaro Cunhal in Portugal, argued that New Internationalism 'panders to the anti-Communists by undermining the international Communist movement'.[3] In response the PCI defied Moscow by restoring its ties with the Chinese Communist Party in 1980 and unequivocally denouncing the suppression of the *Solidarity* movement by the Polish regime in 1981. Berlinguer announced shortly after that the PCI would in the future have 'no privileged relations' with Communist parties.

But *perestroika* made it possible to patch things up. In January 1989, V. K. Naumov, a lecturer at the Central Committee's Academy of Social Sciences, withdrew the anathema previously pronounced by Suslov and declared that Berlinguer and the New Internationalism had proven to be right after all, moreover that New Internationalism should now be regarded as a legitimate forerunner of New Thinking.[4]

Moscow made amends despite the fact that peace could no longer be got on the previous terms. The Italian Communists no longer believed in world Communist unity. The PCI's eighteenth congress in March 1989 made clear that it had moved on in the era of *perestroika*, and a new course had been chosen. The line of the Historic Compromise was considered to be dead. Achille Ochetto, the party's leader, spoke to the assembled delegates of a new recognition of democracy as one of Gorbachev's Universal Human Values, especially as this might refer to processes of renewal at work in Poland, Hungary, the USSR, and China. The PCI took the view that the international Communist movement had long since ceased to exist, so that, with Communist unity no longer a reasonable goal, the main task was to concert efforts with all the democratic tendencies of the European left broadly conceived. Gorbachev's remarks about a new role for the UN in his December speech were taken as evidence of his conversion to Berlinguer's ideas of the seventies on the importance of world government. The PCI's representatives spoke graciously of New Internationalism having been subsumed under New Thinking, but actually New Thinking was a surrender to New Internationalism. No one in the Italian party could doubt that Ochetto had got his way with Moscow. In a visit there prior to the congress, Ochetto and Gorbachev had met and conferred in an intimate atmosphere for five full hours.

The PCI had pioneered the rightward movement of the West European Communist parties for over a generation, from the days of Togliatti's Polycentrism to those of New Thinking, an effort that had nevertheless not proved sufficient to take them into power in their own country. They could cite the advice of one of their founders, Antonio Gramsci, about the necessity to find ways of exerting moral influence on their opponents and rivals, much as the British Fabians of the nineteenth century had thought that they could change the climate of opinion about socialism through the patient and gentle lobbying that they had called 'permeation'. It is highly doubtful that this approach has ever had an impact in the West. Now, however, the notions of the Italian Communists were winning over the Soviet bloc. The Italian Communists were pleased to note that New Internationalism seemed to shine forth from the speeches of the new Soviet ideology chief, Vadim Medvedev, and they were quick to agree with his suggestions about dialogue and cooperation between Communists and Social Democrats. The PCI was also interested in using its influence to extend *perestroika* to East Central Europe. The Italian daily *L'Unità* published Alexander Dubček's account of the events of 1968. At the end of 1988, Vasil Bilak, a Czechoslovakian Politburo secretary who had actively supported the Soviet invasion in 1968 and, alongside current party head Milos Jakes, had purged half a million party members, was summarily dismissed. Bilak, who was thought by some to have been the real force behind the regime of Gustav Husak, and who had earnestly urged Erich Honecker in November to resist pressures to imitate the Gorbachev reforms, was removed as a result of prodding by the Soviets, themselves encouraged by the Italian Communists.[5]

In their intervention in bloc affairs, the Italian Communists were holding out the hope of a revitalized and independent Central Europe.[6] They knew that there was fertile ground for this. In the summer of 1988 a Vienna conference brought together a gathering of European intellectuals on the theme of a Central European sensibility. Included were the Czech novelist Milan Kundera and the Hungarian writer Georgy Konrad, whose book, *Anti-Politics*, had earlier made a case for a broad and loosely organized bloc of disparate states in Central Europe. The conferees spoke of the cultural communion of the area, called to mind by mention of the names of Kafka, Schnitzler, Bartok, Mahler, Janáček, Kokoschka, Klimt.[7] It was suggested that there was such a thing as a Central European ethos, as distinguished from the Eastern European, with the latter's depths of passion and fury, its mercurial extremes, its fanatical logic. The Central European sensibility on the other hand was more ironic than logical, more cosmopolitan, more moderate and tolerant of the absurdities of things, and, while it was distinctive, was nevertheless more in tune with the West.

Since the fall of the Communist regimes in 1989 we have heard these themes cited as evidence of the inevitability of the revolutions that swept the bloc. Even prior to the winter of 1989, the feeling had taken hold that Vienna went naturally with Budapest and Prague, or even with Krakow and Trieste. Nevertheless, the *Mitteleuropa* idea was not at all a natural fit for the entire Soviet bloc, that is, for East Central Europe, except of course in the form envisaged by the Third Reich. An intellectual in Warsaw or Prague is less likely to have friends in Sofia or Bucharest than in Paris, London, Chicago or Berkeley. And yet there is a certain historical attraction to the idea of the unity of the former Habsburg Empire, especially in view of the long-standing ties developed through trade and tourism with Austria and North Italy, Slovenia and Croatia. It was tempting for the architects of New Thinking to seek to make this elusive Central European sensibility a solvent of the tensions of the Cold War. The subsequent Soviet initiatives toward changes in the political and military structures of the East European states cannot be properly understood without taking this temptation into account.

BACK TO 'LENINIST' SELF-DETERMINATION

The victory of Universal Human Values at the September–October plenum seemed to climax the tendencies toward revision of Soviet policy in the Third World. N. Simonia, of the Institute of World Economy and International Relations argued that changes in the Third World could not be seen as wins or losses for the socialist idea, as they 'do not alter the real balance of military and economic forces'.[8] A. Vasilyev, Director

of the Africa Institute, suggested that Reagan's Neo-globalism had demonstrated the impossibility of either superpower getting its way in the
Third World against the wishes of the other. Moreover, the spread of
nuclear and chemical weapons in the Third World gave regional conflicts
an 'ominous aspect'. His conclusion was that 'the age of wars of national
liberation is over'.[9] Vladimir Ilyich Maksimenko, of the Institute for Orien-
tal Studies, tied his own similar conclusion about the Third World to
those of the *glasnost* writers, especially Mikhail Gefter, about Soviet history: the theories of 'socialist orientation', 'non-capitalist development',
and 'national democracy' were simply survivals of the errors of the period
of War Communism during the Russian civil war, that is, before the Soviet
regime embarked on the 'many-faceted' experiment of NEP, which latter
should serve as a model for progress, in the Soviet Union and in the
Third World.[10]

There remained a good deal of old thinking among the habitués of the
Soviet research institutes, to be sure. Yet even the rearguard defence of
ideological prerogative in foreign policy showed in its own way how pervasive
the liquidationist mood had become. Vsevolod Ovchinnikov, a political
observer for *Pravda*, told a meeting of journalists that

> We must be more precise and responsible in our criticism so as not to
> put friends of the USSR under fire by our masochism . . . We are de-
> ideologizing our foreign policy, but we don't reject our views, our ideals,
> or our ideology.[11]

At a meeting with academics on 8 January, A. S. Ivanov, of *Molodaya
gvardiia*, while avoiding a direct assault on Gorbachev himself, criticized
Moscow News for its revisionism on foreign policy. The impact of the
criticism may be seen in Gorbachev's tepid defence of the new departures:

> The new political thinking, as is known, presupposes the de-ideologization
> of interstate (*mezhdugosudarstvennie*) relations. But this does not at all
> mean – as some may want to interpret it – the de-ideologization of
> international (*mezhdunarodnie*) relations.[12]

Gorbachev stressed that his foreign policy envisaged 'cooperating and
interacting' with the capitalist world 'while maintaining allegiance to one's
own social choice'. This would seem to answer and reassure the critics
who might doubt that Gorbachev was still a Communist and internationalist.
The formula he invoked was as old as the Soviet regime itself. Without
it, Lenin could scarcely have conducted a foreign policy that included
trade with England and the Rapallo pact with Weimar Germany. Nor
could Stalin have been part of the Grand Alliance against Hitler. Andropov

had denounced 'the carrying over of ideological contradictions to the sphere of interstate relations' and the 'transformation of the struggle of ideas into military confrontation'. Andropov was referring exclusively to the Reagan Doctrine and US 'Neo-globalism', and asserting that the Soviets positively excluded war as the extension of their ideological contest with the West.

It was not considered inconsistent to provide moral support for progressive forces in the Third World, even including the provision of some weapons – not in any case decisive in a conflict that depended ultimately on political forces and especially not a provocation to general war with the West. Prior to the Reagan Doctrine, the West was considered incapable of rendering historically effective support to the 'reactionary cause', in view of the existing balance of political forces. And even at this point, Andropov was objecting not so much to US aid to this or that guerrilla movement, as annoying as that might be, but to the apparent unwillingness of US policy to keep these actions separate from US–Soviet relations, thus destroying Detente.

Gorbachev restated the tested formula: it was a sin to project ideology into interstate relations. Yet he did not produce the same spin as Andropov. And in countering charges that the Communist cause in the Third World had been abandoned, Gorbachev cited only a continued fidelity to the socialist choice in the USSR. Socialist construction in the Soviet Union had been assumed to be its most important contribution to the international cause of socialism, at least since the days of 'Socialism in One Country' in the twenties. But this had never excluded a wide range of international activities, extending in recent decades to moral support for 'National Democracy' in Cuba, or for 'socialist orientation' in Angola. Now there was real reason to doubt Soviet steadfastness as a supporter of revolution, as the tense discussions between Gorbachev and Castro on the former's Cuba visit in the spring made evident. For his own part, Castro told all who were willing to listen that the socialist cause had indeed entered a period of grave danger.

Gorbachev treated such talk as panic-mongering by those who had no vision of transforming and renewing socialism and could only conceive of ruling in the old way. Nor did he lose faith in the policy of encouraging popular and national fronts in the Soviet constituent republics, even when the first bitter fruits of that policy became visible. Small handfuls of people had demonstrated in the Baltic republics on the anniversary of the Hitler–Stalin pact in August 1987; by 1988 there were 100 000 in Vilnius, 20 000 in Riga, 10 000 each in Kaunas and in Tallinn. Independent organizations sprang up like mushrooms, without really alarming the advocates of *perestroika*. Dmitri Kazutin visited Tallinn in February 1988, about eight months before Estonia declared itself sovereign within the Union (while at the same time claiming a right of veto over Union legislation). He judged that the administrative-command system had perhaps not taken into account the subtleties and complexities of the Baltic republics, but he was quick to point out that 90 per cent of Estonians knew Russian and should remind themselves that this provided them access to world culture.[13]

So, oddly enough, there was no unseemly panic over the eruption of spontaneous political activity in the constituent republics. In January the Moldavian and Kazakh fronts held large protests. Nagorno-Karabakh had to be put under Moscow's direct rule. In April the revolt of the Abkhazian Turks against Georgia had the peculiar effect of causing the latter to rise up against the Union. By May the different peoples' fronts in the Baltic republics met to coordinate their activity against Moscow, at about the time that the fighting between Armenia and Azerbaijan had broken into the open. In June Meskhetian Turks and Uzbeks were fighting in the Fergana valley, and in July Kazakhs in Novy Uzen were engaging in pitched battles with Armenians and other Caucasians. In September it was the Gagauz in Moldavia and the Galicians-Volhynians in the western Ukraine (who had been part of the Habsburg Empire until 1918).

As we now know, this was only the beginning. Many Soviet officials feared and deplored the eruption of nationalist sentiments on such a scale. Yet they also felt the temptation to think that spontaneous local initiative of this sort might be contained by *perestroika* and could not dismiss the thought that the protests might even contribute something to Soviet reform. For each time a disturbance broke out, there was an opportunity to rout the responsible local secretaries if they happened to be Gorbachev's opponents, as was the case, for example, with Semyon Grossu in Moldavia and Dzhumber Patiashvili in Georgia.

There was even a tendency, bizarre and alarming to those who still hoped to preserve a united Soviet Union, to view the situation in the Soviet republics and the bloc states as similar. At the time of the Nineteenth conference in July 1988, the Soviet Ambassador to Hungary, Boris Stukhalin, told a press conference in Budapest that the Estonian Peoples' Front and the Hungarian Patriotic Peoples' Front (the vehicle with which Communist oppositionist Imre Pozsgay had fought the Kadar regime in the name of radical reforms) were similar organizations, 'founded with the aim of supporting *perestroika*'.[14]

HUNGARY AND POLAND IN THE LEAD

Did the *perestroichiki* realize what they were saying? It is not an easy question to answer. Belief in the wonder-working powers of a presumed

'Leninist' nationality policy had spilled over into the realm of relations with the bloc states. For the latter, the men from the research institutes were hatching grand and visionary plans. At a conference on East Europe with American and Soviet academics, held in Alexandria, Virginia in the immediate wake of the Nineteenth conference in July 1988, Oleg Bogomolov, Director of the Institute of the Economics of the World Socialist System, outlined a significantly new conception of relations among the bloc states. In a paper that was drafted with prominent input from Dashichev, Bogomolov argued the case for considering the model imposed on Eastern Europe by Stalin as 'perverted', not in its socialism, but in its political forms. The interventions of Soviet forces designed to preserve these forms, in Hungary in 1956 and Czechoslovakia in 1968, had been the result of 'hegemonic aspirations' harboured by the Soviet leadership.[15] The states in the bloc must be permitted by the Soviets to go their own way, not just in respect to the Soviet Union, but to each other as well. The old theory of relations in the bloc being characterized by the absence of conflict was to be jettisoned.[16]

In fact, one could put it more strongly. In the phrase of Imre Pozsgay, 'the East European countries are individually embarking on the path to the Common House of Europe'. Central Europe was an idea rather than a programme, thought Pozsgay, but it was no coincidence that Hungary did 80 per cent of its foreign trade with the two Germanies, Czechoslovakia, Poland, Yugoslavia, and Austria. The Hungarian leaders wanted to end their relationship with the bloc countries through COMECON, an organization that party First Secretary Karoly Grosz called an 'overweight elephant'. He was referring to the fact that Hungary's trade with the Soviet Union was building up ever larger ruble surpluses, which were of no use in servicing huge Hungarian hard currency debts to Western countries, in per capita terms, the highest in the bloc. Moreover, Iran, Iraq, Libya, and Nigeria were not paying their debts to Hungary, and despite measures to realign trade by cutting down on exports to the Soviet Union, several Western countries, most notably Austria, were cutting their trade for fear that Hungary's credit was no longer good. Hungary had to make painful cuts in consumption to pay its debts, while it could only spend the East bloc trade surpluses on inferior East bloc goods.

Hungarian leaders could hardly blame this on the Soviet Union, which had given them raw materials, including oil and gas, at favourable prices for almost 20 years, essentially shielding them from the oil price 'shocks' in the West. This at a time when Hungary was building up both its debts and its reputation as a consumer paradise of the east. Hungarian leaders were not angry with the Soviet Union, but only desired to continue and increase their relationships with non-bloc countries. They thought that they saw in Gorbachev's reform struggle an opening that might permit them the changes they sought. It had been their reform ideas, under the rubric of the New Economic Mechanism of 1968–72, that had served as an inspiration for Gorbachev and Soviet economists such as Nikolai Shmelyov, who supported him. Rezso Nyers, the economist most closely associated with the NEM, had been in the shade since Kadar had turned from the reforms in 1972, but with the rise of Gorbachev and the struggle to retire Kadar in 1988, Nyers made his return. Together with Pozsgay, he sought to translate Soviet sloganeering about 'socialist pluralism' into a real multi-party political reform. Even Bogomolov did not advocate that for the USSR, but with the habit ingrained by their own pronouncements about the freedom of bloc countries to make their own decisions, the Soviet *perestroika* supporters resigned themselves to watching an 'out-of-town tryout' for the idea. Accordingly, they did not flinch when the Hungarian parliament passed a law in January 1989 permitting opposition groups to become parties.

A few months before, First Secretary Grosz had said of the post-Kadar regime's reform impulses: 'We are participants in an undertaking of a similar dimension and significance as in the period of the national democratic revolution between 1944 and 1948 . . .'[17] As he understood it, Hungary was returning to the policy of the immediate postwar years, a policy we have associated with the name of Varga, with coalition governments and NEP-like economic policies for the peoples' democracies.[18] As in the postwar period, the 'multiform' political and economic strategies of the Eastern bloc were to act as a solvent of potential tensions between forces defending different social systems. Grosz and the many other East European reform leaders who were eventually to be swept away by the anti-Communist revolutions of 1989 thus began with the idea that they were transcending the antagonism between east and west in a world which was itself dissolving the line of demarcation between capitalism and socialism.

Only a week after Hungary's decision for 'party pluralism', a central committee plenum of the Polish party decided for 'trade union pluralism', that is, the legalization of *Solidarity*. The decision was prompted by the seeming hopelessness of confronting *Solidarity* in an endless round of strikes and demonstrations, as had ensued in the spring and summer of 1988. General Jaruzelski had urged a move to break the impasse, but his opponents, including Prime Minister Mieczyslaw Rakowski, argued that if *Solidarity* were legalized the official trade unions would be destroyed and the regime itself would be threatened. Jaruzelski, finding himself in the minority in the debate, confronted his opponents with a threat to resign, whereupon Rakowski's opposition disappeared. It was agreed a week later to open round table talks with *Solidarity* in February.

'FINLANDIZATION' OF THE SOVIET BLOC

Bogomolov greeted the actions of the Hungarian and Polish parties by an extraordinary suggestion. He said that if Hungary should decide to become a neutral country 'like Sweden or Austria' it would not thereby threaten the security interests of the Soviet Union. Nor would this necessarily be a blow to socialism, for there had been exaggerations in the previous estimate of capitalism which, even in the West, was itself evolving toward 'elements of socialism'.[19] This was an advance on a formula Bogomolov had suggested only a few weeks earlier, according to which Hungary might have a market economy while remaining in the bloc. Bogomolov stressed that he was only speculating, and that Hungary would decide for herself, but in view of the relation between the USSR and the bloc countries in the past, his speculation really had the effect of advice to the Hungarians on how to proceed. To be sure, these were not the only expressions they heard on the subject; Gorbachev himself *declined* to state that the Brezhnev Doctrine was defunct, and even made numerous references to the common interests of the socialist countries. East European leaders had to conclude that there were differences in Moscow on these matters and that perhaps their own actions might also weigh in the balance. They did know that neutralism on the Austrian model might be interpreted in a new way in view of the statements made by the Austrian Chancellor Vranitzky after his October visit to the USSR. Vranitzky said that there would be no Soviet opposition to Austria joining the EC.

Bogomolov's ideas, like those of Grosz, had a prior basis in the Varga era, 1944–47, when Hungary's sovietization had been delayed and a broad multiparty coalition government permitted to wield power. While the Communists could easily have taken power at that time with the help of the Red Army, Soviet policy was to hold Hungary hostage in this intermediate position, lest the United States and England be unduly frightened, while the Communist parties advanced their cause elsewhere in Europe, namely in France, Italy, and Poland. In returning to these conceptions, Gorbachev and his supporters were trying to revive some of the features of Europe prior to the emergence of the two military blocs. Bogomolov nevertheless astonished Western and especially Eastern observers by putting the matter in terms of a 'Finlandization' of Hungary, which was only a step away from 'Finlandization' of the entire eastern bloc.

Influential members of President Bush's administration had anticipated for some time a retreat of Soviet power, but they thought in terms of a graceful Soviet decline, an 'Ottomanization', that would still leave the European balance roughly intact. In retrospect that actually seems to be the framework most suitable to American interests. But the Atlanticist

French President Mitterrand was not so sanguine, and expressed fears that the 'Yalta' division of Europe between the superpowers would be abruptly terminated. German President Kohl alone was delighted with the new Soviet speculations and claimed to see in them a splendid service to the grander cause of the 'Europeanization of *Ostpolitik*'.

While Poland could not be said to be less far along in the process, Soviet attention nevertheless focused on Hungary, for several reasons. We have already mentioned the vague hopes that had been raised about the *Mitteleuropa* idea as an alternative to the Europe of Yalta. There was also the fact that while General Jaruzelski was firmly in the Gorbachev camp, the Hungarian leadership was in the midst of a bitter struggle over how far to go with reform. Grosz, the nominal party leader, was not necessarily a favourite with Gorbachev and his group, and he himself was eager to defend socialist positions and to prevent the party from relinquishing its leading role. Grosz liked to quote George Lukacs: 'We are not taking off the hussar's pelisse – but we do want to button it differently.'

He might logically have fallen in with the opposition in the bloc to the Gorbachev reforms, but for the fact that that opposition was strongly anchored by Romania and Ceauşescu who, since the opening of the campaign to raze the Hungarian villages in Transylvania, had risen up menacingly as Hungary's primary adversary. Thus the national cause hemmed Grosz in and forced his reluctant agreement with the plans of Nyers and Pozsgay, who were far more radical than he. Ceauşescu had loomed up as Gorbachev's most entrenched opponent in the bloc since the tense meetings between the two men in May 1987. So the men of *perestroika* in the Soviet Union looked hopefully to Hungary to provide the next victory for reform, while the Hungarian Communist radicals looked to the Gorbachev supporters in the USSR for continued encouragement. Grosz referred to the Soviet reforms as the 'strong backdrop' for Hungary's, and Bogomolov called Hungary the 'proving ground' of Soviet reform.

In view of the complexity of the political struggle that was being waged in virtually all the East European countries, there would seem little reason to place confidence in the various interpretations of the events of 1989 that depend on a 'rational actor' explanation of Gorbachev's policies. According to this seemingly sensible *a priori* conjecture, the changes that led to the unification of Germany and the destruction of the bloc were not the result of Soviet bungling and political incompetence, but instead grew out of a conscious decision to jettison the increasingly onerous burden of an East European empire, together with the post-war balance of power presupposed by it.[20] This analysis has taken root in recent Western literature. Yet Gorbachev was the opposite of a rational actor. Neither he nor his cohort of allies foresaw the revolutionary events in the bloc. On the contrary, they spoke on numerous occasions of their horror at the thought of a united Germany.[21] The Soviet decisions were not agreed in advance but were contested at every point by fierce opposition in a political struggle that itself exerted profound influence on Gorbachev's thinking. Moreover he was not yielding to a new conception of international economic rationality, but pursuing a dream, a utopia perhaps, in which socialism and capitalism might find a grand historical synthesis. The dream itself took shape under severe external constraints, not least among which was the very sobriety and doggedness of the opposition of Ligachev and his other bloc opponents, which forced Gorbachev into ever bolder flights, but in these he was piloted by his own demon who no doubt whispered repeatedly in his ear about the advantages of daring and the salutary shock effect of Real Sacrifices.

THE PURGE OF THE DEAD SOULS

Gorbachev's Politburo reshuffle of September–October 1988 had removed the Gromyko centre with its moderating and conservative influence. Now Gorbachev and his opponents, the latter weakened but by no means silenced, stood face to face. Since Ligachev had been shunted off into agriculture, he had to carry on the fight in an arena of faltering performance that was painfully visible to all. Along with his fellow oppositionist Nikonov, with whom he had to share responsibilities in this field, he cited widespread dissatisfaction with food supplies. However, where Gorbachev usually spoke of the remedy lying in the increase of individual local initiative through leasing arrangements, Ligachev instead stressed that aid to the countryside should come from 'enlistment of the accumulated industrial and scientific potential of the cities', through the Agro-Industrial Complex, an institution that would, significantly, shortly be abolished. Vadim Medvedev spoke in favour of emulating Hungary and China in their leasing of kolkhoz land to the peasantry, thus urging that the soviets end a ban imposed originally in 1930. Most leasing at that time was on a short-term basis and got a reluctant response from the peasantry, who no doubt wondered how long this line would be in effect. Gorbachev wanted to provide further encouragement by offering 50-year leases. To this Ligachev answered: 'It was not for this that we established Soviet power.'

Criticism was not confined to the field of food supplies. As head of the Legal Commission, Chebrikov complained that the unofficial movements were proving to be a fertile field for various extremist elements who were fostering anti-party and anti-Soviet views. Crime was growing everywhere, he complained, and being abetted by the passivity of local cadres who feared that real resolution toward crime might be interpreted as interference with legal rights.[22]

This was not exactly open and explicit criticism of Gorbachev and his group, but rather the sounding of discordant notes – in favour of strengthening planning and central administration where Medvedev was speaking of the importance of transcending the ideological conceptions of the past and of accepting the advent of the market. Ligachev and his co-thinkers were not campaigning for the leadership themselves, but only trying to restrain Gorbachev from what they considered the worst excesses of people such as Shevardnadze and Yakovlev. *Ogonyok* editor Vitali Korotich, whose closest Politburo contacts were, he said, with Yakovlev, nevertheless reported no bluster or bullying from Gorbachev's chief opponent:

> Ligachev is such a soft man; he never cries out or shouts. It's never threatening. That tough way of speaking, which was quite normal for Khrushchev or sometimes Brezhnev, is not the way for him. He speaks this way: 'Why are you in such a hurry? Don't push progress faster than it's going. We have real problems.' He calls sometimes.[23]

Ligachev and Gorbachev certainly shared some of the same virtues and qualities that had caused Andropov to promote them. But now, if Gorbachev was going to continue his campaigns, Ligachev could no longer exert a moderating influence in the Politburo to bolster his admonitions against haste. He did, however, have some other forces that might work for restraint on Gorbachev. Among these was military opposition to the line of the UN speech of December 1988. Marshal Akhromeyev had resigned his post as Chief of Staff in the wake of the speech, not, he said, because of any disagreement with the policy of troop cuts. He became a Presidential advisor to Gorbachev. But it was known that the speech caused a clash between military and civilian experts, and Akhromeyev was widely thought to agree with Defence Minister Yazov, chief of staff Moiseyev, and General Lizichev of the MPA that the idea of Defensive Sufficiency should not be used to justify a proposed reorganization of the Soviet armed forces as a volunteer army.

There were constraints in the bloc as well: the GDR leadership under Honecker, with its ideologist Kurt Hager (whom some German Communists called a 'little Ligachev'); Jakes and Fojtik in Czechoslovakia; and Ceauşescu in Romania. Gorbachev was doing his best to move against these obstacles as he had against similar impediments in the Soviet Union. The planned troop reductions could easily be read as a removal of support for leaders who in their opposition to reform gave succour to his Soviet opponents. Gorbachev was also supporting *their* potential opponents. In March a daring letter urging reforms was sent to Ceauşescu by six elder statesmen of Romanian Communism, who included Silviu Brucan, former Ambassador to the United States. Brucan was also reported to be close to Gorbachev and had just spoken with him in November, shortly after the Politburo changes. It was thought that the letter could not have been composed and signed without the aid of either the *Securitate* or army intelligence.

The most powerful impact on the struggle in the Soviet leadership was the election to the new 2250 member Congress of Peoples' Deputies held in March. The results could only be interpreted as a crushing defeat for those perceived to be part of the party apparatus, that is, for 'conservatives' associated with the Ligachev opposition. Fully one-third of the seats had been reserved for the party and its allied organizations but, of the remaining seats, many were won by candidates who ran essentially against the apparatus. At the top of this list was Boris Yeltsin himself, who defeated the party's chosen opponent, Yevgeni Brakov, director of the auto plant that makes ZIL limousines. Yeltsin got over five million votes, slightly less than 90 per cent of the total cast. On the other side, the Leningrad party machine suffered a humiliating loss, with its top five candidates defeated. Leningrad regional party head Yuri Solovyov, who had railed against the unofficials for months, and who had been accused of rigging the elections to the 19th Party conference, was defeated, despite running unopposed. Voters turned out not to like this 'free ride', and simply crossed his name off the ballot. In retrospect it seems amazing that the apparatus should have tried to perpetrate this unopposed candidacy on the voters, and then given them a chance to veto it.

The apparatus itself could claim that, after all, 80 per cent of the winners were party members, and that Politburo Ligachevites Shcherbitsky and Vorotnikov survived. But 31 prominent regional leaders were defeated, including Leningrad city first secretary Anatoli Gerasimov, whose opponent got 74 per cent. And some of the 'victories' were not sweet. Algirdas Brazauskas, party chief in Lithuania, won in his district, but party candidates who showed less tolerance for Lithuanian nationalism, including the President and Premier, were defeated. By the end of the year, Brazauskas and the Lithuanian party had split with the CPSU. Both Gorbachev and the *Pravda* editorialists had to admit that there was such a thing as a 'Yeltsin syndrome', a tendency to vote against the apparat, which seemed to say that, from the party standpoint, the whole election idea had been a bad one. But their response was instead to call for serious soul-searching among the losers, and to ascribe their defeat to anti-conservative, rather than anti-party sentiment.

The defeated candidates had to look at things more realistically. In Leningrad, Yuri Solovyov immediately charged that the elections had permitted anti-party and anti-Soviet elements a field day, and that they had been provided with ammunition by the *glasnost* campaign. As a result, the youth had come to think of the CPSU as the 'party of crimes'.

It was time to add up the balance sheet of the last four years and admit that 'the minuses exceed the pluses'. Solovyov had no choice but to make a stand against the transformation of the party into a parliamentary machine, and to evoke old images of a fighting vanguard devoted to Leninist ideals. By July, however, Gorbachev decided that he had had enough of that. He went personally to Leningrad to oversee the removal of Solovyov. He was replaced by Boris Gidaspov, who would however soon be echoing the same complaints made by Solovyov.

Gorbachev's resolution in accepting the verdict of the voters struck some party members as another coup against the party. This was not wrong, if one concedes that the coup was delivered, not in the spring of 1989, but at the Nineteenth party conference when the decision was made to elect a Congress of Deputies. Ultimately such a body could only become a Long Parliament seeking the overthrow of Soviet absolutism. One can think of this in terms of Gorbachev the revolutionary systematically undermining the main pillars of the Soviet regime, in order to erect in its place a Western parliamentary democracy. But Gorbachev did not sanction the idea of a multi-party democracy at that time, nor did he announce a desire to force the party to reliquish its leading role.

Gorbachev saw a chance to enhance his own personal authority against the collective leadership under which he had lived his entire life in politics. If the party members were forced to run for office, Gorbachev and the popular masses together could transcend their conservative collective leadership. One might recall Stalin and his rebuff at the Central Committee plenum of October 1932, and how he later stirred the country into revolt against the narrow-minded local secretaries who had voted against him, accusing them of conspiring to subvert the construction of socialism. Of course, it will not do simply to compare Stalin with Gorbachev, the man who delivered through *glasnost* the most telling blows against the legend of Stalin. And Gorbachev was certainly not the founder of a new police state, nor a man who told fairy tales about conspiracies and enemies of the people, nor a jealous tyrant in the normal sense. Yet he was no different from Stalin in trying to make his leadership one with the people by attacking the intermediate layers of Soviet party bureaucracy, in order to raise himself to a personal status unlimited by any of the party's traditional restraints. Perhaps some may find these Tocquevillian comparisons out of place in discussing the habits and institutions of Soviet Russia. They may indeed be misplaced in one sense, for Gorbachev was no Stalin, if anything the very opposite. Perhaps: a Good Stalin.

His next coup was delivered at the Central Committee plenum at the end of April, when he forced the retirement of 74 members and 24 candidate members. In view of their having lost the positions that had entitled them to membership, these were deemed to be 'dead souls'. They included Gromyko, Solomentsev, and Aliev, as well as Marshal Kulikov and General Ogarkov. Former international committee head Ponomarev was cashiered along with former regional first secretaries such as Bagirov and Demirchyan. Their dismissal was clearly not intended as a follow-up to the defeat of party candidates in the March elections, but to Politburo changes of the September plenum. Solovyov was at that time still arguing heatedly, to little avail, against the parliamentarization of the party. Gorbachev had not only tamed the Central Committee, bringing it to at least the degree of compliance that characterized the Politburo but, with the retirements of Gromyko and Ogarkov, sent to the West a powerful message about the prevalence of the New Thinking over more entrenched foreign policy attitudes.

In fact, there is reason to think that the purge of the 'dead souls' was done largely for Western consumption, to convince observers that Defensive Sufficiency was indeed the prevailing national security orientation. Unfortunately this display had to be digested together with shocking news of the Soviet suppression of demonstrations in Tbilisi, in which gas, including apparently nerve gas, was used on the demonstrators. Gorbachev would, in characteristic fashion, be dodging responsibility for this decision for two years, but it must have weighed heavily on him in view of his cultivation of an entirely different image in the West. Still, Soviet studies claimed that he was indeed winning over American elites and even, it was claimed, causing them to lose interest in their own 'neoglobalism'.[24] A Times/CBS poll showed that those seeing a growing Soviet threat had been reduced from 64 per cent in 1983 to a figure of 26 per cent.[25] In West Germany, the victory was more sweeping. Whereas in 1980, 71 per cent of those polled had seen some Soviet threat, in January 1989, the figure was down to 11 per cent.

GORBACHEV HEALS THE SINO-SOVIET SPLIT

So Gorbachev had every reason to feel that he was subduing the refractory beasts of the old thinking and winning the world. In this buoyant mood he embarked on his visit to Beijing to end the rift with China, in an attempt at a *tour de force* like that of Nixon–Kissinger in 1971. One could say not only that reconciliation with China was at the very top of the Soviet foreign policy agenda, but that this had been the case since the end of Brezhnev's regime. Undoing the unnatural and anomalous bloc of China and the West against the Soviet Union was not only the first desideratum of New Thinking but also of Gromyko's visions of *recueillement*. Brezhnev had made a point in 1982 of granting that China was a socialist country with important interests that could be addressed frankly by the Soviets. And the issues were remarkably straightforward. At that time Deng Xiaoping and the Chinese leaders described an insidious Soviet attempt at encirclement of China on two axes, one cutting through Afghanistan to the Indian Ocean, and the other, through Cambodia to the straits of Malacca.[26]

In subsequent years the Soviet encirclement was described in terms of three obstacles to the improvement of relations: Vietnamese troops in Cambodia, Soviet intervention in Afghanistan, and the Soviet divisions and missiles in Mongolia and along the Sino-Soviet frontier. The old thinkers in the Soviet Union were making the argument that it was the United States who was doing the encircling, with the Third Fleet and bases from Okinawa to Diego Garcia, and with its assistance to the South Korean and Japanese armament programmes. Yet they could note with satisfaction that the 'strategic relationship' between the United States and China, which American officials had promoted in 1981–82 as a response to the Polish coup against *Solidarity*, had never gone very far, and was actually in a state of atrophy throughout the eighties.

Gorbachev's idea was simple: to remove the 'three obstacles'. In a speech at Vladivostok in September 1986, he spoke in Dobrynin's language of the 'growing inter-relatedness of universal human and national interests', and of the necessity for a security arrangement in the Asia-Pacific area comparable to the Helsinki pact in Europe. He reiterated the offer in September 1988 at Krasnoyarsk, as he had in the earlier speech, taking note of the ominous backdrop of Japanese rearmament and the inevitable reminders of 'the historical associations of the pre-war era'. Gorbachev plaintively asked the Japanese: 'Why discredit the unique experience of not being armed?'[27]

Some Bush administration officials were wary of Soviet moves in Asia, noting that, despite the removal of intermediate-range missiles from the China border, Soviet mobile missiles could still hit China.[28] One could also have cited the incomplete nature of the withdrawal of Vietnamese forces from Cambodia and the stalemate over the role of the Khmer Rouge in a new coalition government. The President, however, seemed to be as wary as Gorbachev about the Khmer Rouge, and despite being generally harder on the Soviets than on the Chinese, seemed also to endorse the idea of a Sino-Soviet thaw. While Gorbachev was in Beijing, Washington officials were complaining about continued Soviet aid to Nicaragua, and Presidential press secretary Marlin Fitzwater spoke derisively about Gorbachev, calling him a 'drugstore cowboy'. This prompted a *New York Times* editorial critical of Bush's caution in dealing with new Soviet realities. Yet in a speech at the Coast Guard Academy on 24 May, Bush argued that the United States was faced with the task of finding a new national security strategy, one that would go 'beyond containment'.

Shevardnadze had gone to Beijing to finalize the preparations in February, stressing that neither party wanted to restore the alliance they had had in the fifties. He did not discourage the idea that the Soviet Union might even have a good deal to learn from the Chinese, as various economists had been saying for some time. Ironically, China was at this time stepping back from the reforms initiated in the 1978–84 period, which had given rise to an increase in food production of 34 per cent. Private farmers had tended to grow specialty cash crops rather than rice and cereals, with the result that the state was importing millions of tons of rice and wheat. There was a spate of re-collectivization, and by the end of the year a reversion from the system of two-tier pricing to controlled prices.[29] Nevertheless, the reconciliation aimed at by Gorbachev promised, his supporters were convinced, not only the solution of a geopolitical problem but a blending of reform models.

If the Chinese seemed to be more advanced in the economic sphere, the Soviets appeared to most Chinese to be more advanced in the political. There were signs that the Chinese people wanted a *perestroika* of their own. Daily demonstrations of thousands were occurring at the end of April. By the time Gorbachev arrived at the beginning of May, there were close to a million marching in the capital and similar movements in other cities. Neither Gorbachev nor anyone else could determine precisely what this enormous outpouring was expressing politically. The demands heard most often were for more freedom of expression and measures against 'corruption'. The former was plain enough but the latter certainly was not. Did they mean the 'corruptions' introduced by the economic reforms? Was this a rebellion of collectivist impulses? What was the significance of the fact that one saw the red flag everywhere and heard the Internationale sung constantly? Was this a protest in favour of a kind of Communist legitimism, or a revolt against Communism itself, as the Chinese leaders finally concluded?

Gorbachev could not decide. He told Zhao Ziyang on 16 May, just four days before martial law was declared and Chinese troops carried out a massacre in Tiananmen Square, that 'we too have our hotheads'. But, after reading a letter addressed to him by a group of students, he said 'I value their position.' On 18 May, he was quoted as saying 'This shows what can happen if a government does not keep up with its people,' and, referring to his opponents at home, 'I wish they could see it.' The students, Gorbachev concluded, were part of 'a turning point in the development of world socialism'. On his return home, with a normalization in his pocket that ended some thirty-years of fierce enmity with China, Gorbachev spoke with great sensitivity of the problems faced by the Chinese people and the need to leave them to solve them on their own. This attitude was echoed in the Soviet press, understandable if one considers the magnitude of the foreign policy achievement that had been registered.

Gorbachev got from the experience powerful pointers of both a positive and negative kind. He could see that, in the eyes of the people of the world, he was the greatest revolutionary of his time. For four years the attention of the planet had been fixed on him. Now it was obvious that his mere presence was enough to cause enormous hopes to rise up from great masses of people the world over. This charismatic identification with his figure, he could only conclude, was building an international movement for the grand synthesis of socialism and democracy that was the foundation of his cause. Moreover, the act of repressing this movement, when the tanks rumbled into Tiananmen square on the night of 3–4 June, seemed to cast the Chinese leaders into a kind of dark hell of opprobrium before history. Gorbachev articulated the lesson in the language of the Russian Revolution, the language of his training, by speaking as Lenin had in 1917, of the party being 'behind the masses'. From this point on, he would constantly cite the relation between political and economic reform and the necessity 'not to lag behind'. By contrast, the East German paper *Neues Deutschland*, which had greeted with approval the letter of Mme Andreyeva in 1988, gave the actions of the Chinese leaders in Tiananmen Square a hearty endorsement.

REPENTANCE

One could not fail to notice that the movement for *glasnost* was lagging badly behind. Back in 1987, before Gorbachev had emerged as a democratic revolutionary, when the watchword was 'acceleration', *glasnost* had been conceived as a general movement for freer expression and better information, aiming to extend the de-Stalinization undertaken by Khrushchev in 1956. Gorbachev certainly did not imagine that the result of investigation into the 'white spots' in Soviet history as rewritten by Stalin would be an explosion of anti-Communism. The review of the history of the party was supposed to produce a Leninist critique of Stalinism, with a delineation of alternate paths of development and, by inference, historical backing for the new departures of *perestroika*. But it was always a highly controlled process, very closely tied to the prevalent need for ideological justification of a course of economic reform on the model of NEP, with the result that the only historical alternative that occurred to the publicists was that of Bukharin, and only that which he propounded from 1924 to 1928. The Bukharin who preached a 'theory of the offensive' in central Europe in 1920–21, who argued for 'red imperialism' at the Fourth Comintern Congress of 1922, who fought the 'right danger' as Stalin embarked on agricultural collectivization in 1929, who called for the blood of Zinoviev and Kamenev when they were arrested for Kirov's murder in 1935, this Bukharin was not of interest. Historian Roy Medvedev, who urged the Bukharin alternative, said frankly that the push for rehabilitations and the review of party history was only the 'logical result' of Gorbachev's economic policies.[30]

Gorbachev was only committed to the positions taken in the historical speech of 1987, which stressed, alongside the recognition of 'unforgivable' crimes, the heroic element in party history and Stalin's positive contributions. Despite his status as the father of *glasnost*, he seemed afraid of what it might produce. The party archives were not thrown open, nor was there any attempt at publication of the key documents of party history dealing with the rise of Stalin. Certain works of the various protagonists were published, but no systematic collection of documents. Historians do not work to order, but publicists are forced to do so, so it was the latter who advanced new opinions about Stalinism. 'There was no science', as Aleksandr Bovin put it, 'but only journalism.'[31] Historical lessons produced in this way are reduced to quarrels about protagonists, whose ideas themselves are caricatured on the basis of the worst of what their opponents said about them – thus the Bukharin who wants only the triumph of market forces, or the Trotsky who seeks only a return to the barracks communism of the civil war period.

Moreover, the debate on Stalinism was conducted in a very Stalinist manner, not a debate about history, but about history as described in the Stalinist textbooks. Instead of producing an appreciation for historical contingency and the context in which decisions were taken, the clash of cardboard figures against a backdrop of the monstrous crimes of the Stalin era produced in the intelligentsia an even greater animosity toward the regime and even a kind of hysterical response, visceral repulsion mixed with a passion for repentance. As one might verbalize this mood of repentance:

We are ourselves monsters to have been ruled by such monsters. Our history is not a legitimate tradition but a disease, one unknown to other civilized lands, who are fortunately ignorant of our utopias of social engineering. Before their stern gaze, we can only repent.

This was not what the promoters of *glasnost* had wanted. They had imagined a controlled rewriting of the party's history, according to the lights of historical commissions designated for the task. In this vein was the major work commissioned by the party, a full-scale biography of Stalin by Dmitri Volkogonov, a high official in the Army's commissar apparatus. The volume of 1200 pages, called *Triumph and Tragedy*, appeared in serialized form in 1988 and 1989. Volkogonov was obligated to package his work with the formulae of Gorbachev's 1987 speech, in which one found 'indisputable contributions' alongside 'unforgivable political mistakes', including the 'groundless repression of many thousands'.[32] He was able to see Stalin's correspondence, and the archives of Beria, Molotov, Zhdanov, and Malenkov. Volkogonov did his best to show that Trotsky was not an enemy of the Revolution, but only an opponent of Stalin. He noted that the Thirteenth Congress of 1924 had 'proved inconsistent' in failing to execute Lenin's Testament and remove Stalin from his post. He told the story of the Testament more honestly than Gorbachev. Yet he could not resist calling Trotsky the 'demon of the revolution', an appellation that Roy Medvedev thought more fitting for Stalin.[33]

Yegor Yakovlev of *Moscow News* wrote agonized essays about the struggle for the succession to Lenin in the twenties, which seemed to strike the most thoughtful publicists as the locus of the answer to the question of why the USSR ended up with Stalin. Yakovlev focused, as did others, on the period 1923–24, complaining of Lenin's comrades that 'They could not rally together to act on Lenin's demand and get rid of Stalin.'[34] And why not? The principal reason would seem to be unscrupulous use for factional purposes of the widespread fear of Trotsky. If this were the case, blame would rest not only with Zinoviev, Kamenev, and Bukharin, but with Lenin as well, for training them all in the art of restraining Trotsky. Economist V. Sirotkin, in a searching discussion of the dilemmas of NEP, cautioned against assuming the debate on economic policy to have been separate from 'a heated power struggle in the upper echelons of the party, where each of the opposing groups was drawing on the positive and negative aspects of NEP for arguments to use against its opponents'.[35] Albert Nenarokov in effect argued against the prevailing Bukharin cult when he suggested that Stalin accumulated power and the party became ossified precisely during the time when the Bukharin line was in effect, that is, 1925–26.

The historical debate might have been on a firmer footing if the participants had been able to show that Stalin's regime could not be finally associated with any particular line advocated by Bukharin, Trotsky, Zinoviev, Radek, or Preobrazhensky. The point is that Stalin sided with each of them at one time or another, usually with the others at his side, who in fact generally attacked more fiercely than Stalin. They all accepted NEP, even the 'left opposition' of Zinoviev that Trotsky joined in 1926. So the *glasnost* debate about the origins of Stalinism distorted things by claiming that the heart of the matter had been economic policy; if this discussion were to make any sense at all it would have had to address the question of the Stalinist regime, which was built unrelentingly whether the party pursued economic policies of the right or left. If the debate had proceeded along these lines, one would have ended with more complex judgements, and without devils. To be sure, there were those such as Nikolai Vasetsky and Grigori Pomerants who made special efforts to write dispassionately about Trotsky. *Moscow News* correspondent Mikhail Belyat remarked that 'the denunciations of Trotsky in our literature are about as convincing as the Immaculate Conception'.[36] But for some Bukharinists and most of the nationalists there was no incentive for such moderation.

Reading the documents of the *glasnost* debate on Stalinism leaves one with the impression that discussion about the twenties was within the compass of party-mindedness, while discussion about the thirties led inexorably to anti-Communism, not so much because of the economic or administrative issues but because of the massive scale of the atrocities in Stalin's purges. In a series of influential articles published at the end of 1988 and the beginning of 1989,[37] Aleksandr Tsipko managed a synthesis of two moods: the nostalgia among the Bukharinists for the NEP and the animosity of the nationalists for alien and cosmopolitan elements that had supposedly led the country astray. Tsipko wanted no complexity or moderation in judgements of the Old Guard of Bolshevism in the twenties, especially of its 'left wing'. For him these were wild men driven by insane visions of a utopian society where all normal life was excluded. They were as well heirs of a certain tradition among the pre-revolutionary nineteenth-century Russian intellectuals, one of infatuation with abstract schemes for the application of a science of society. To some extent, thought Tsipko, their ideas got a certain reception among the masses because of the 'Asiatic' nature of the latter, who lacked 'training in the sobriety and realism of capitalism'.[38]

So Tsipko returned to the famous warning of Dostoyevsky about the evils embodied in the infamous western principle 'two plus two equals four'. The tyranny of the intellectuals with their utopias based on mechanical materialism was sharply opposed to the organic development of pious Russian social life. He cited Bakunin's famous warnings about the potential tyranny of the managers and engineers in 'state socialism' – without, however, endorsing Bakunin's judgement of Marx, whom Tsipko at this point in his career, still refrained from attacking. Lenin emerged in Tsipko's rendering as the defender of normal trade and profit against the ideal of 'pure socialism' as embodied in War Communism. The theme was not just familiar by this time, but practically a ritual incantation heard from all the Bukharinists and promoters of the NEP model. Tsipko's twist, in some ways a typically Stalinist one, was to claim that the other former associates of Lenin, and particularly the 'left opposition' of 1926–27, had fought to scrap NEP for a return to War Communism. A number of the *glasnost* writers, along with those who have studied this question for decades in the West, recognize that no major Bolshevik figure suggested anything like what Stalin produced with agricultural collectivization, and that Trotsky, for one, was calling for a turn back from it by 1930. But Tsipko needed Trotsky, Zinoviev, and Kamenev to be the enemies of routine, people who hate to see the peasant or the worker well fed, who 'prefer the starvation of millions in the course of revolutionary activity'.[39] Some echo of this view may be found in Aleksandr Yakovlev's July speech on the French Revolution where this key icebreaker for Gorbachev said that Stalin had spilled so much blood because his predecessors had implanted in him 'a morbid faith in the possibility of forcing social and economic development'.[40]

Stalin, in Tsipko's rendering, was deeply affected by these conceptions of the 'un-trivial' destinies of the Russian people, his utopianism 'typical of the Marxists of his time'.[41] Lenin, of course, the font of all wisdom, is excepted. So Stalin represents a break with Lenin, but only in so far as he is in reality a disciple of Trotsky. Tsipko, even in arguing that democracy is incompatible with a socialist economy, and must be based on private property, still did not want to be a critic of Marx or Lenin.

Nor did he want to disparage the pre-revolutionary Russian 'patriarchal' culture, but rather to save it from identification with the conservatism of Mme Andreyeva. Indeed he seemed at this point in his intellectual odyssey to be engaged in a struggle with the apparatus for the soul of the Russian nationalist right. This may explain his insistence on the alien character and the absolute evil of Zinoviev, Kamenev, and Trotsky, all Jewish, and their malign influence over the Georgian Stalin. In this sense, Tsipko ran against a tendency in *glasnost* toward presenting the views of Stalin's victims as at least arguable. Tsipko was notable for being among the first to follow Selyunin and argue against socialism as such, but also for presenting the villains of the Stalinist past in almost the same way as Stalin had presented them.

Tsipko's writings showed a favourable attitude toward pre-revolutionary Russia's traditional political culture. But others were blaming that political culture for having set the stage for the scourge of Stalinism. Boris Kagarlitsky's book, *The Thinking Reed*, which could not be published in the USSR, nevertheless seems to have contributed to the designation of the CPSU regime as a 'partocracy', a phrase whose use so pained Gidaspov and the Leningrad authorities that they complained about it publicly. Kagarlitsky saw the central problem not in 'Stalinism' but in an 'Asiatic state' inherited from the tsars, along with a 'statocratic mode of production'. The Soviet *nomenklatura* was not a usurping ruling class, as Anatoli Butenko and some others had argued, but in effect a ruling 'estate'.

Can the Communism that issued from the Russian Revolution be viewed as a kind of *ancien régime*? A number of Soviet writers, some specialists in the history of the Third World countries, and some generalists such as

Burlatsky seemed to be going in this direction. Some found material in their studies of Marx's concept of the 'Asiatic mode of production' for the critique of the Soviet reality of their own day, as had other ex-Communist writers in the West. Serious scholarly discussion of the question can be left to the specialists in this field. But the ideological upshot of calling the Soviet *nomenklatura* a ruling 'estate', rather than a 'stratum' or even a 'class', is the implication that the Russian Revolution did not really accomplish for Russia what the French Revolution did for France in clearing away the impediments to a modern society, as the official Marxist-Leninist history had always taught.

In that case, to use the official language, the socialist countries were still historically ripe for 'bourgeois revolution'. In the year of the bicentennial of the French Revolution, of which Gorbachev was to take note on at least two ceremonial occasions, referring to the Russian and the French Revolutions as the foundations of the modern world, the idea of continued revolution against the existing order of things seemed to promise explosive consequences for the Communist regimes. Moreover, these regimes were passionately engrossed in what historian François Fejto called 'a battle for history'. The Baltic peoples and the Moldavians wanted to talk about 1939 and the Hitler–Stalin pact; the Poles wanted to talk about the Katyn forest massacre of 1942; the Czechs wanted to talk about the Soviet invasion of 1968; and the Hungarians about Imre Nagy and the revolt of 1956. A year of historical commemorations was developing into a year of angry upheavals.

NAGY REBURIED

The bloc regimes were following the pattern established in the Soviet Union, where *glasnost* had served to unhinge the popular support and self-confidence of Gorbachev's opponents by tarring them with the brush of Stalinism. Thus the Hungarian reformers coupled the announcement that they had established party pluralism with the news that they had reconsidered the Hungarian revolution of 1956. Specifically, they had decided that the official version that ascribed the revolt to a counter-revolutionary plot of the imperialist intelligence agencies was no loner tenable, and had arrived at a compromise formula designating the events as a 'Peoples' uprising' which, said Imre Pozsgay, 'as a formula, is fitting because it does not say revolution, it does not say counter-revolution . . .'[42] It was announced that the remains of Imre Nagy, executed in the aftermath of the Soviet invasion, would be exhumed in order to provide a proper burial. Nagy's reburial in May would provide an occasion for public education and ritualization of rebellion against the national disgrace suffered at the hands of the Soviets.

Nagy's reburial would also rally public opinion behind a rerun of the measures he had adopted in 1956. Nagy had introduced a series of policies analogous to those of Malenkov in the USSR. As Soviet writers such as Burlatsky were fond of pointing out, Malenkov himself represented reform attitudes not seen since the days when Bukharin himself had had influence. Pozsgay said that 'in 1953 he (Nagy) was the first statesman in Eastern Europe to take a truly determined step toward the dismemberment of the Stalinist structure, its destruction'.[43] Specifically this had meant to Nagy a return to the all-party coalition of 1945; Nagy therefore abolished the one-party system when the revolt brought him back to power in 1956 and in fact presided over a government in which non-Communists were the majority. The rehabilitation of Nagy did not sit well with Grosz, who was probably responsible for the process being drawn out over several months; Grosz in fact expressed continued confidence in the leading role of the Communist party. Despite being criticized by Yakovlev on the latter's visit in November 1988, Grosz still felt that the Soviet leadership could be convinced of the value of an intermediate position. But Pozsgay on the other hand was confident that reconciliation with the traditions of social democracy would win a new following for the Communists and even permit them, and him, to win a free election.[44]

The matter was settled by Grosz's visit to Moscow in March, where he was told in definitive terms that Gorbachev, at that time thoroughly immersed in preparations for the purge of the Dead Souls, would not interfere with even the most radical course of action in Budapest. Gorbachev said that the Soviet interventions in Hungary of 1956 and Czechoslovakia in 1968 had been very unfortunate, and that Soviet historians would soon provide documents on them. Hungarian officials left Moscow convinced that the Brezhnev Doctrine was dead. In the following month Soviet troops began to leave Hungary, and the Hungarian Politburo was reshuffled with the removal of hard-liner Janos Berecz. After that Grosz was powerless to slow the pace.

But a truly fateful event had already occurred in January, when Hungarian representatives convinced the Soviet foreign ministry to open the border with Hungary so that Soviet Hungarians living in Sub-Carpathian Ruthenia could visit their relatives on the other side. In the first two months there were over 250 000 such crossings with the checkpoints reporting little difficulty in handling thousands of cars a day. During Yakovlev's visit in November, Grosz had spoken of the Soviet 'safeguarding of the national profile' of Hungarians in the area as 'a factor of international importance'.[45] The contrast between Gorbachev and Ceauşescu on the treatment of their respective Hungarian minorities was stark; perhaps that was the reason that the Soviets were so willing to accommodate. The opening of this border in January prepared and encouraged

Hungary's similar action with regard to the Austrian border in May, with its fateful consequences for the East German regime.

Hungary had been designated by the Soviet reformers as an ice-breaker opening a path for the eastern bloc into Europe. This would have followed naturally from a supposition that the division of Europe arranged at Yalta was soon to end. But there is no evidence that Gorbachev and his supporters had made such a supposition, and much to indicate that they had something more modest in mind: a continuation of the political division of east and west at a greatly reduced level of military tension. Even so, they were encouraging an assertion of national interests in a part of Europe that Soviet literature had traditionally characterized as being without conflict. They were reviving old patterns in the international relations of east central Europe.

Before the First World War the states in question did not exist, as the entire area was divided among the Russian, German, and Austrian empires. In the Habsburg empire, as a result of the compromise of 1867, Hungary controlled a vast area peopled by Slovaks, Romanians, Serbs, and others, rather comparable in area to the lands she would gain in 1941–44 as a result of collaboration with Nazi Germany. The collapse of Hungary's empire occurred shortly after the defection of Russia from the allied war effort in 1917, so that both countries, for different reasons, were opponents of the post-war order primarily established and, as it turned out, solely defended, by the French. The Hungarians lost their control over the former subject peoples, now incorporated into Czechoslovakia, Romania, and Yugoslavia, soon to be arrayed in an anti-Hungarian 'Little Entente', and also had to endure these countries' control over substantial Hungarian minorities. Lenin saw possibilities in this situation, when in 1919 he learned that a Hungarian Soviet republic had been set up with the aim of defying the French. Lenin supposed that 'the land of the poets' would crack open capitalist Europe through a revolution in Vienna and, with any luck, something similar in Germany. That was not to be. Yet Revisionism (opposition to the Versailles treaty) continued to be the hallmark of Hungarian foreign policy under the subsequent regimes until its success in the victories of 1939–41, one of which was the acquisition of Sub-Carpathian Ruthenia from the partitioned Czechoslovakia.

The outlines of the Soviet bloc agreed to in Paris in 1946 re-established the states of the interwar period, leaving the Baltic states to the Soviet Union. Hungarians could see that the Treaty of Paris followed closely the lines of the Treaty of Trianon of 1920, in reducing Hungarian dominion over the surrounding peoples and in leaving sizable Hungarian minorities outside the state, as often put, 'ten million inside, four million outside'. In terms of territorial and minority questions, Hungary was by far the least satisfied country in the bloc. The attraction for the Soviet politicians and academics who encouraged Hungary in 1989 did not lie in revival of these potentially poisonous national quarrels. They were thinking in terms of a utopian image of 1944–47. Yet Gorbachev was embroiled in a passionate dispute over *perestroika* with some of the countries of the former Little Entente, with whom even the Hungary of Pozsgay and Grosz would still have a bone to pick. The front of anti-*perestroika* states was led in many ways by Romania's Ceauşescu, who was now razing Hungarian villages in Transylvania, and who had even suggested a revival of the Little Entente to Dubček in 1968.[46]

Hungary was therefore not simply an icebreaker into a new *Mitteleuropa*, but a weapon against the rejection front of East Germany, Czechoslovakia, Romania, and Bulgaria. Gorbachev showed no signs of realizing it, but he had in effect cast his lot in with a new Revisionism that would ease the military tensions of the Cold War at the cost of overturning the balance of power, not in east central Europe, but in Europe as a whole. Gorbachev was no 'rational actor' who had decided to permit the unification of Germany. He was trying to revive a central Europe of former Habsburg lands, while maintaining the division of Germany. However, as the Germans like to say, once you say A, you must say B.

LIQUIDATION OF THE COLD WAR BLOCS?

While the Hungarians were reviving the memory of 1956 and the Nagy Affair, other revolutionary commemorations were occurring all over the bloc. Jan Palach, the student who had immolated himself in response to the Soviet invasion in 1968, was remembered in six days of demonstrations in Prague in January. In March the Poles honoured the students who had risen up against the Gomulka government in March 1968, while the present Communist government restored Polish citizenship to General Anders, who had formed an army on Soviet soil in 1943 but took it to the West when the Soviets broke relations with the London Poles after the Katyn Forest massacre. The Soviets had up to this time regarded him as a traitor. Hungarians marched through Budapest to commemorate Hungary's 1848 revolution against Austrian rule; since the Hungarian rebellion was put down by Russian intervention in 1849, the manifestation was really against the Russians and the Brezhnev Doctrine

Soviet writers claimed to look to Hungary for the latest innovations but Poland was not behind in the process, and the role of *glasnost* was similar. A joint Soviet–Polish historical commission had for two years been mulling over the 'white spots' in the history of the two countries relations. These included Katyn, the destruction of the Polish Communist party in Stalin's purges, and the Nazi–Soviet pact of 1939.[47] The commis sion said in March that Katyn had indeed been the work of Stalin. At the end of May, *Pravda* reported that the commission had further concluded what had been known in the West for over 40 years: that Poland had indeed been divided in 1939 according to the provisions of secret protocols to the Hitler–Stalin pact. In July, Falin, who had been denying the existence of the protocols, confirmed them on German television. Gromyko died in that same month adamantly denying their existence even before all of his long-time colleagues and associates.[48] The bitterness in Poland at the admissions was probably outrun by the demoralization in the Soviet Union. On the 50th anniversary of the pact in August, *Moscow News* carried a piece by Dashichev linking Stalin's opposition to the German Social Democrats in 1930–32 to the later pact with Hitler: 'In both cases, German Nazism profited by Stalin's misjudgement'.[49] Aleksandr Bovin's article in the same number made a more conventional assessment, arguing that Moscow, 'seeing that the war could not be prevented, tried at least to steer clear of it for as long as possible'.[50] In December the Congress of Peoples' Deputies would declare that the pact 'violated the sovereignty and independence of other nations', but with regard to the Baltic republics and Moldavia, that this fact had no bearing on present borders. Yet the Lithuanian Communists referred to the issue that month when they broke with the CPSU.

The Polish Communists had hoped for some reward for their conscientiously facing up to the party's past. But none was to materialize. They negotiated a political reform providing for a new *sejm* with 65 per cent of the seats guaranteed them and their allies, the rest to be elected freely. They were hoping to marginalize *Solidarity* as a legal opposition. But *Solidarity* won almost all of the contested seats in the June elections. In the wake of its crushing defeat, the Communist party tried to cobble together a coalition government. Jaruzelski was elected President in July, but he failed to gain the assent of *Solidarity* to a Communist Prime Minister or a coalition in which *Solidarity* would be a minority.[51] Lech Walesa saw through this immediately: 'if we are puppets with the worst portfolios, we will achieve nothing except losing public confidence and being swept away'.[52]

Moscow was seconding Jaruzelski in denying *Solidarity* a cabinet majority, at least until Gorbachev, in a telephone conversation with newly appointed party head Rakowski at the end of August, changed position and gave his imprimatur to a *Solidarity* government. This array of political forces bore an outward resemblance to the governments of the immediate postwar period, headed by non-Communists but dominated by Communists by virtue of their control over key ministries such as defence and the interior. This was the case with the first *Solidarity* government headed by Tadeusz Mazowiecki, except for the fact that almost all its economic staff was committed to the radical free-market reforms later to be known as 'shock therapy'. If this was an attempt to return the models of 1944–47, the historical film was being run in reverse, and more rapidly. When the negotiations had begun, the social-democrats among the *Solidarity* people had appeared to be in charge. Perhaps this misled Gorbachev, or perhaps it was just a case of his not getting up early enough to keep up with Lech Walesa.

Through the spring and summer, Soviet diplomats had still tried to maintain that they were parcelling out changes in Eastern Europe to the West in return for European arms reductions.[53] Western experts welcomed the Soviet moves in the east, with both Henry Kissinger and Zbigniew Brzezinski agreeing that Moscow must be reassured against US interference as the Soviets reduced their grip. At the same time, however there was no general agreement about the efficacy of modernizing NATO's short-range missiles, with Thatcher, who favoured modernization, pressing Kohl, who wanted sweeping reductions. In the middle of May Shevardnadze even attempted, without success, to link the matter to implementation of the INF agreement of 1987, threatening to hold up the dismantling of SS-23s aimed at West Germany. This was brushed aside by Bush at the end of May when he announced that US forces would be reduced by two divisions if the Soviets were to reduce to 'conventional parity', and only then would there be a 'partial' reduction of the short-range missiles. This ensured that they would be part of the NATO arsenal in the future, a solution that seemed to satisfy the Germans.

The Gorbachev strategy was not unhinging Western support for the alliance. While many in the West were deeply impressed with the changes brought about by the New Thinking, others could point to adverse developments like the sale of Soviet bombers to Libya that were capable of attacking Israel. On his visit to Syria in April, Defence Minister Yazov told his hosts that four countries remained central to Soviet foreign policy: Syria, Cuba, Angola, and Vietnam. The tough posture was not very convincing. There was no mention of Nicaragua, Ethiopia, Laos, or Kampuchea. An impression of serious retrenchment was reinforced by data from intelligence sources to the effect that the Soviets had reduced ship days and distant deployments, reduced their use of Cam Ranh Bay, and conducted fewer joint operations with Cuba.[54] Soviet shipbuilding, with its long lead-times, was still on the increase. Some administration officials were wary; these included Defense Secretary Cheney, and Undersecretaries of State Paul Wolfowitz, and Lawrence Eagleburger, CIA's Robert Gates and Vice President Quayle. But other observers were equally convinced that the Soviets were winding down the navalism of Admiral Gorshkov's time, and trying to set the example for large naval arms control agreements with the United States. It seems that that was actually the Soviet *hard-line* position. When Akhromeyev came to Washington in July, he made a plea for naval cuts as a kind of even trade-off for the Warsaw Pact cuts already in train. Aleksandr Bovin undermined him by saying of Akhromeyev's idea that 'it is hard to see how NATO's navy can stop the Pact's tanks unless they cross the Atlantic'.[55]

CROMWELL, BONAPARTE, AND GORBACHEV

While matters like these were at the heart of the Gorbachev programme of reforms as originally conceived, Gorbachev by this time had more to worry about, in particular, the alarming slide downward of the Soviet economy. When Nyers paid him a visit in July, he found him preoccupied with internal affairs.[56] The party had taken a drubbing at the first meeting of the Congress of Peoples' Deputies in May. Aside from Gorbachev and Ryzhkov, no Politburo officials had spoken, and party people in the assembly had appeared practically neutral in the face of attacks on Ligachev. Instead of defending out of party discipline, they were somewhat at a loss, perhaps assuming that these attacks served Gorbachev's cause. The party was unequipped and ill at ease in its new parliamentary role.

Gorbachev had to admit to the Congress that economic and social problems had worsened under *perestroika*, but he only urged pressing on. He was elected to the post of Chairman of the Supreme Soviet, in effect the President of the USSR.[57] Refractory delegates, led by Yeltsin, Afanasyev, Sakharov, and Popov, responded by forming an 'Inter-regional Group' that was unmistakably a parliamentary opposition. Moreover, ethnic violence intensified as Abkhazians and Chechens fought the Georgian authorities, clashes occurred between Uzbeks and Kazakhs, and pogroms broke out in Fergana *oblast* by Uzbeks against Meskhetian Turks. In July, a massive strike of 300 000 coal miners spread like wildfire from Kuzbass to other coalfields all over the Union. Gorbachev said that he was 'inspired' by the strikers and praised their 'taking matters into their own hands'. He pronounced them the spearhead of *perestroika* and promised them big wage increases and legislation on the right to strike. This was really no solution; inflationary price increases for coal followed and the condition of the miners was not noticeably improved. And the miners were not to be trifled with: the next time they came out on strike they would be demanding the abolition of article six of the constitution, which acknowledges the 'leading role' of the party.

As a result of the elections and the case of Solovyov in Leningrad, many leading oppositionists who had criticized the reforms changed the tenor of their critique. Where previously they had spoken of the too rapid pace of change and of certain excesses (this was even the case with the letter of Mme Andreyeva), they now said openly that the party was being led over the side of a cliff. At a special Central Committee conference in July, it was said quite straightforwardly that taking the party out of the economy and turning the latter over to the soviets had been the reason for the shortages and dislocations. Zaikov, who seemed to some to be acting as a second secretary in the wake of the demotion of Ligachev in September 1988, gave credence to suppositions about his own ambitions by attacking the new leadership of the Central Committee's Ideological Department.[58] L. F. Bobykin, Sverdlovsk first secretary, complained in the same vein about the weakening of the Secretariat: 'Obviously we need a central committee second secretary, whatever you call it officially'.[59]

Ligachev himself gave a measured speech, a careful speech, more coherent than those that Gorbachev gave at this meeting, in which he implicitly invoked the heritage of Andropov and even Brezhnev. He spoke approvingly of the transition 'from confrontation to cooperation' in superpower relations, making clear that the achievement of strategic parity had made it all possible. He seemed to echo Selyunin's 1987 ideas about conversion of military to civilian economic priorities. He agreed with the others that the turn to 'soviet power' in the economy had been ruinous for the party. He also viewed with alarm calls for a multi-party system, a development that he judged potentially to be 'simply fatal'. Ryzhkov, cleaning up, foresaw calls for the repeal of article 6 and concluded, with obvious reference to the rise to power of a new President Gorbachev, that a rearming of the party must begin by helping to ensure that Gorbachev 'as general secretary of the central committee, devotes more time to his party duties'.

Gorbachev was oblivious, or perhaps simply contemptuous. He answered these worried queries and plaints by reminding the comrades that they were in the 'revolutionary phase' of *perestroika*. Anticipating the crisis at the January 1990 plenum over abandoning article 6, Gorbachev insisted that 'It is impossible to decree the party's authority.' Moreover, another drastic renewal of cadres was needed, and this must occur 'everywhere, at every level'. It was January 1987 all over again, as Gorbachev, mincing no words, told the comrades explicitly. Another 'escape forward'.

The Soviet Union has always had a unique political culture, a 'movement-regime' as political scientist Robert C. Tucker has called it. The sense of shaking things up from above, and especially shaking up the cadres from above has been in evidence in periods when the social and economic policy was moderate almost as much as when it was most extreme. The moderation in cadre policy that had characterized the Suslov era, the ideal of Ligachev and the other opponents of Gorbachev, is somewhat anomalous in comparison with the rest of Soviet history. Now the heirs of Suslov, as opponents of Gorbachev, were floundering for the lack of a way forward. They had never developed the perspective of seeking an alternative to Gorbachev, but only that of limiting him as Suslov had limited Brezhnev. But now they were up against something more formidable than that. Gorbachev was not satisfied with the party that had brought him to power, and clearly found it an obstacle in his search for superiority over rivals such as Yeltsin in the democracy at large, a democracy which now had a parliamentary instrument. Gorbachev wanted, or so he said, to reshape the party according to the specifications of the democracy. In this rendering the party would have become a more effective vehicle for his own personal power. Could he be believed when he promised the party that its reformation would equip it to lead in a democracy?

Gorbachev supporters were, as usual, willing to help him to build his own dominance. Andranik Migranyan provided a theory for the personalization of power under an authoritarian leader. He cited as a model the English revolution of the seventeenth century, by means of which civil society had secured its rights and representative institutions as an indispensable prelude to the later development of democracy itself. It had been normal in history, he said, for civil society to record its initial triumph under a regime of dictatorship. Migranyan did not make the point elegantly. The government of William and Mary after 1689 was far from a dictatorship, even if it was not a democracy. The military dictatorship that did develop in the English revolution was the ephemeral one of Oliver Cromwell in the 1650s. Civil society was already robust and demanding rights. The comparison with the Soviet Union, in which there was no propertied civil society, was awkward. In the regime of *perestroika* as depicted by Migranyan, however, Cromwell-Gorbachev was faltering, his popularity declining for the absence of economic successes. Moreover, the voters were anti-apparat, as the last elections had demonstrated, and Gorbachev was a man of the apparat. Thus he was a revolutionary and a conservative at the same time, 'both Luther and the Roman Pope'.[60] He was, as well, faced with a powerful challenge from the 'neobolshevism' of Yeltsin, whose appeal stemmed from his promise to redistribute the ill-gotten gains of *nomenklatura* privilege. On the other side the *nomenklatura* might try to stave this off by a coup that ended the reforms altogether. To defeat this dual challenge, Gorbachev must uncouple his own 'personal power' from that of the power structures of party and soviets, thus the better to enhance that power in the interest of a transition to democracy. Gorbachev's struggle to rise above the collective leadership had acquired a new champion and a 'democratic' rationale.

Migranyan was not alone. Ambartsumov defended Gorbachev's wearing two hats, those of President and of General Secretary, in the name of stability. Len Karpinsky spoke explicitly of Gorbachev's 'authoritarianism'

as necessary.[61] Gorbachev himself seemed to find the analysis most con
genial, to judge not only by his actions, but also the language that en
tered his political utterances. From this point on, he formed the habit o
referring to Yeltsin and other critics among the democrats as 'neo
Bolsheviks'. Migranyan's description of his lonely battle fitted perfectl
his frequently professed idea of a 'struggle on two fronts'.

Western newspaper reports of these developments were wrong to as
sume that Gorbachev was being attacked, or at least criticized, by thos
who marked his rise above the collective leadership and above the part
as such. To be sure, there had been such criticisms at the parliament'
first meeting. In the debate on the vote for a President, Ales Adamovic
had worried about a 'new Stalin', and L. J. Sukhov, a Kharkov truc
driver, had cited Napoleon's guided transition from republic to empire
as well as 'the power of his wife's influence'.[62] Yet the liberation o
Gorbachev from the constraints of the party was in general seen by som
of his most prominent supporters as a natural result of his struggle agains
Ligachev and company, and a useful weapon in the fight for democracy

This was a landmark in what has been called the 'paradigm ecstasy' o
the liberal writers. Migranyan's theory had capped two years of discus
sion in the press of the idea of 'civil society', an eighteenth-century ter
which, in the Marxist canon, refers to the whole of social relations in ;
market economy. Hegel had used the term *bürgerliche Gesellschaft*, whic
can also be translated as 'bourgeois society', to describe everything be
tween the family and the state unbound by public interest, an arena 'wher
waves of every passion gush forth'.[63] In the context of urbanized Sovie
society, even with its overpowering statism, it was nevertheless possibl
to think of an emergent civil society by reference to an aggregate o
informal networks and institutions acting independently of the state.[6
Once it was permitted to use the term in this way and even to speak a
Gorbachev did, of 'combining personal interests with socialism',[65] the wa
was open to 'infuse the term with new content', as had been the rul
with *glasnost*. The transition to democracy could be discussed in terms o
the classical experiences of England, Holland, and France in the estab
lishment, not of democracy, but the ascendancy of the rights of civil so
ciety. The term 'bourgeois revolution' was no longer used, but if it ha
been, then Gorbachev would have been the latest in the line of 'bour
geois' revolutionaries, or, as Migranyan saw it, the great despots of th
European revolutions. It was only a short step from there, quickly taken
to invoke the model of General Pinochet in Chile, whose dictatorshi
had provided a transition to a market economy.

While Gorbachev was flirting with these pretensions, his opponents wer
not asleep. They recognized the threat to what they perceived as tradi
tional party prerogative. Actually Gorbachev's personalization of powe
was very much in keeping with Soviet experience, and it is entirely pos

,ible to view the Suslov reign of stabilization and entitlement in the six-ies and seventies as a mere interlude. The latter remained the ideal of Gorbachev's opponents, but now they had to fight on the ground of the democracy. Instead of complaining about the 'unofficials', they had to build unofficial organizations themselves.

Their counterattack began in Leningrad, at the initiative of Solovyov, after his defeat in the spring elections. A club bearing the name 'For Leninism and Communist Landmarks in *Perestroika*' was formed as an alternative to the Leningrad Popular Front, and it published a paper, *Smena* (Change). It criticized the Popular Front for having abandoned socialism in its infatuation with the cooperatives, for attacking the sys-tem of planning in the name of a prospective market economy, and for having reduced the representation in its ranks of workers from the fac-tory bench.[66] In September, a Congress of the United Front of Workers of Russia was held in Sverdlovsk with 110 delegates from 29 cities, as well as Russian groups from Moldavia, Latvia, Estonia, and Tadzhikistan.

Its political centre was, in keeping with the history of struggles be-tween right and left in Soviet history, in Leningrad. Boris Gidaspov, whom Gorbachev had chosen to replace Solovyov when the latter was removed in July, became its most prominent leader. The organization presented itself as a vehicle for the recrudescence of a Russian nationalist presence in *perestroika*. It followed up on the calls that had been issued at the special July Central Committee meeting, along with criticisms of Gorbachev, for a Russian Bureau of the Central Committee. By December, such a bureau had been established in the party, a step in the direction of the formation of a Russian Communist Party in the following months. This organization would be the centre of efforts to restrain and control Gorbachev. Thus the notes sounded by Nina Andreyeva's letter in 1988 had become a permanent motif of opposition to Gorbachev in the name of Russian national Communism. Along with support in the Union re-publics, the oppositionists had co-thinkers in the bloc among the senior Politburo ideologues from the GDR, Czechoslovakia, and Romania, who had convened in Karlovy Vary in July to concert their propaganda cam-paigns against Gorbachev's excesses. The next tumultous chapters of the struggle would unfold in their theatre of action.

But first the sword of Gorbachev descended with a crushing stroke. At a two-day closed session of the Central Committee at the end of Sep-tember, Gorbachev decimated his opponents. Shcherbitsky, Chebrikov, and Nikonov were all removed from the Politburo. It will be recalled that Gorbachev had been fighting to remove Shcherbitsky since the be-ginning of 1987. His success would be followed in the next weeks by the removal of Viktor Afanasyev as editor of *Pravda*. Afanasyev, who, after the time of the Berkhin Affair in 1987, had been solidly behind the Ligachev forces, was replaced by Gorbachev's friend Ivan Frolov, up to this time editor of *Kommunist*. Ligachev, realizing that his faction had been torn up by the roots, lashed out desperately. 'A bitter struggle' was being waged, he claimed, a 'class struggle' against intellectuals and parvenus, who wanted 'capitalism and bourgeois democracy, the introduction of private property and a multiparty system'.[67] But he neglected to say who was leading these forces. And he was of course careful not to criticize Gorbachev. He had only ten days earlier been exonerated of charges of corruption made against him back in May by the energetic, not to say unscrupulous, investigating team of Telman Gdlyan and Nikolai Ivanov from the State Procurator's office.[68] The pressure had helped to keep him relatively quiet over the summer, and he was no doubt psychologically unprepared, after having felt it suddenly eased, to attack with much energy. Gorbachev was dispensing just the right amount of dosage.

This may also have been the case with his handling of the Yeltsin problem. On 28 September, Yeltsin showed up in a police station drenched to the skin, claiming that several men had thrown him into the Moscow river. When Yeltsin got into some dry clothes, he retained his composure and claimed it had all been a 'joke'. Asked about the bizarre incident, Gorbachev replied menacingly: 'Maybe he made a joke and maybe he didn't.'

Gorbachev in fact claimed that, at this moment of his greatest personal triumph, when his Politburo opposition had been dispersed and intimidated, he was besieged on the left and on the right, by forces that were 'both dangerous and connected'. And he acted as if he thought this to be the case. He came to the aid of the trade union press, accused in articles published in *Moscow News* and *Izvestiia* of unjust attacks against private traders and cooperative ventures. Was this a harbinger of a future turn to the 'right', that is, to the position occupied by Ligachev? At a meeting of editors, Gorbachev lashed out at one of the staunch supporters of reform, *Argumenty i fakty*'s editor Vladislav Starkov, who had recently published a poll showing Gorbachev's popularity to be inferior to that of, among others, Andrei Sakharov. Gorbachev tolerated almost any position on issues, but attacks on his personal power affected him in his viscera and caused him to bite back. 'Recently certain political figures have arisen', said Gorbachev with contempt, 'who, in their chase after popularity, are ready to speak out against their own mother.' Ligachev was said to have called Starkov at this very time to express his support.

THE SINATRA DOCTRINE

So the Politburo opposition to Gorbachev was routed. There remained however, the surviving nests of unbelievers in the Soviet bloc. None of them could have been very confident in view of the repeated hints by

oviet officials and by Gorbachev himself that the Brezhnev Doctrine
vas no longer functional. These hints were contradicted by various de-
ials, but they seemed to confirm to the anti-reform Communists of the
Rejection Front that a struggle was going on in Moscow on the subject.
At any rate no one could say: 'yes, we hope to be saved by Soviet tanks
n the event things get out of control'. Indeed, the foreign ministers of
he Warsaw Pact renounced the 'so-called Brezhnev Doctrine' in a War-
aw meeting on 27 October. Soviet foreign ministry spokesman Gennadi
Gerasimov remarked two days earlier that they had adopted the 'Sinatra
Doctrine' and that henceforth the bloc countries could 'do it their way'.[69]
Gyula Thurmer, a foreign policy adviser to Hungarian General Secretary
Grosz, invoked a 'Gorbachev Principle', implying continuity in foreign
olicy but the establishment of coalition governments with Communist
ontrol over departments of home affairs and defence.[70] The Hungarian
eadership was still hoping to rerun the era of New Democracy of 1944–47.

The Hungarian leaders were providing the model. In the beginning of
October they changed the party's name to the Hungarian Socialist Party,
dopted a set of reform principles, and chose Imre Pozsgay as their can-
idate in a projected popular Presidential election to be held the follow-
ng spring. This would never materialize – a petition campaign led to a
eferendum that decided to elect the President from the new parliament.
ozsgay was reportedly thinking about a prime minister from the Hun-
arian Democratic Forum, a Christian Democratic formation, in a future
overnment. Geza Jeszensky, dean of the school of Political and Social
ciences at Karl Marx University, who would be foreign minister in an
IDF government within a year, said confidently of the new arrangements
hat 'we are all liberals now'. This reminded me of a conversation I had
ad with Jeszensky a few years prior to this, in which he estimated that
here is not one Marxist in Hungary'. It was now evident that he had
ot been exaggerating much. On 23 October the appellation 'People's
Republic' was dropped and the nation was renamed simply the Republic
f Hungary.

The Hungarian ex-Communists also actively intervened in the leader-
hip struggles in the countries of the Rejection Front to spread the revo-
tion. The really decisive action was the opening up of the border with
Austria in May, which had been prepared by the opening of the check-
oints with Soviet Sub-Carpathian Ruthenia in January. East Germans
ere crossing over into the West through Hungary at a furious rate. From
ugust to October, some thirty thousand made this exodus. The GDR
overnment warned menacingly that it was taking names but this had
ttle effect. Refugees were crowding into the West German embassy in
udapest. And the West German government was by no means ignoring
hem. In fact it agreed with the Hungarians to fly over a hundred to

Vienna, thence to take them by bus to West Germany. The GDR, whos
actions had been somewhat halting because of Honecker's illness (a
operation for gallstones) that kept him inactive until September, made
series of threatening statements to the Hungarians at the beginning c
September, reminding them of an agreement made in 1967 to help eac
other in matters of this kind, but the Hungarians answered on 10 Sep
tember by suspending the agreement!

The GDR denounced this as a violation of legal treaties and an inter
vention in East Germany's internal affairs. Gerasimov, he of the 'Sinatr
Doctrine', blamed the West Germans for trying to 'de-stabilize' the GDR
but a *Washington Post* report of 16 September maintained that Miklo
Nemeth, the Hungarian Prime Minister at the time, had told Germa
Chancellor Kohl that 'we have taken a sovereign decision', adding tha
'the Soviets have not said no'. On 13 September two West German bank
announced a loan of 500 million marks to finance Hungary's export in
dustry. Ligachev came to East Berlin on 12–15 September, only days a
ter his exoneration on corruption charges, promising that Soviet suppo
for the GDR would be constant. But Gorbachev himself expressed adm
ration for the humanitarianism of the East German authorities in pe
mitting the refugees to go. A week after Ligachev's return to Moscow
his faction was crushed by the Central Committee plenum. That sent
chilling message to the hard-liners in East Berlin. Perhaps this helps ex
plain why Kurt Hager, 'the German Ligachev', was actually to lead th
Politburo forces *opposed* to suppressing the exodus of refugees.

The issue came to a head with Gorbachev's visit to Berlin for the 40t
anniversary of the GDR on 6–7 October. At his public appearances thou
sands chanted 'Gorby! Gorby!' and 'We want to stay!' He refrained fror
openly attacking the Honecker leadership or its programmes, but he ap
proached Egon Krenz, Hans Modrow, the Mayor of Dresden, and Marcu
Wolf of the GDR security forces, and he suggested at one meeting tha
those who resisted masses of people 'put themselves in danger'. Neve
theless, no sooner had he left for Moscow than GDR police started t
use force against the massive demonstrations in East Berlin, Leipzig, an
Dresden. Honecker, who had closed the Czech border on the occasio
of Gorbachev's visit, was now resolved on a 'Chinese solution' and ha
given the appropriate orders.[71] Hager, supported by a majority in th
Politburo and most of the regional leaders, led the opposition in mee
ings on the 11th and 12th that decided to force Honecker to resign. Th
new leadership was led by Krenz (who had supported the Chinese ac
tions of June no less than Honecker). He promised 'continuity and re
newal' and resolved to protect the political lead of the party.

This naiveté was based on the fact that the opposition appeared to th
GDR leaders to be centred in the intelligentsia of Berlin and other bi

cities and led by *Neues Forum*, whose leaders cautioned against continued flight to the West, and did not advocate unification of Germany or capitalism. They only wanted *perestroika* and democratic socialism for the GDR. That was all the Krenz–Modrow leadership wanted. When the Berlin Wall was first creased on 5 November, Politburo member Gunter Schabowski said firmly that it would remain. Then a day before it came down, he proposed 18 breaches in it and 30 days' travel to the West for all, a policy to be put in place by Christmas. At this point even Chancellor Kohl was stating that 'Our interest must be that our compatriots stay in their homeland.' There did not at first appear to be any sense of fatal inevitability, but after the first few days of witnessing the crowds pouring through and the demonstrations continuing, Hans Modrow said 'I can see the end coming.'

Today it all seems obvious. Having yielded to the spontaneity of the democracy, the East German leaders had signed the death warrant for their party and their state. But they themselves did not realize it at the time. Modrow, who spoke of seeing the end coming, also opined three weeks later, when he became prime minister, that 'German unification is not on the agenda'. Opening the wall might not have decreed that they be swallowed up in the ensuing Round Table measures to provide for free elections. Gerasimov himself said that even the most drastic domestic changes would not alter the fact that East Germany remained a 'strategic ally' of the USSR. 'Poland is a good member of the Warsaw Pact', he said, 'and in Poland you have a coalition.' That was true enough. The Polish Solidarity government led by Mazowiecki had, to be sure, a disciple of Milton Friedman as Finance Minister in Leszek Balcerowicz, but it also had Communists in charge of the interior ministry, national defence and foreign economic affairs.[72] The party was trying to move with the times. On the 50th anniversary of the Soviet invasion of Poland, the party daily *Tribuna Ludu* had published an editorial critical of the Soviet Union of the Stalin era and its 'crimes involving millions of Poles'.

But how could the bloc Communists retain any control over the reform process when Moscow seemed to demand no less than the revolutionary overthrow of the leaderships in place? In Bulgaria, Todor Zhivkov was told by the Soviet Ambassador that if he did not step down he risked the same treatment as had been meted out to Honecker. Gorbachev himself conferred with Zhivkov's opponent, Petar Mladenov, only days before the Bulgarian Politburo voted 5–4, on the day the Wall fell, to oust Zhivkov. The situation was similar with the Czech leadership under hard-liner Jakes, which had to deal with Soviet and Hungarian television rehearsing the events of 1968, and broadcasting interviews with Dubček. There were those in the Soviet Politburo who felt differently, as then Foreign Minister Shevardnadze has since revealed, telling of 'pressures' to use force in defence of the Communist parties of the bloc, especially that of East Germany: 'we were told to act according to the scenarios of 1953, 1956, and 1968'.[73] But they were overcome. The political differences in the Politburos of the bloc countries mirrored the differences in the Soviet political and military leadership. And Gorbachev acted on the idea that, in order to win anywhere, he had to win everywhere.

The last domino was Ceauşescu's Romania, by consensus the most grinding internal regime and the one on the worst terms with Gorbachev since his tense visit of 1987. Brucan's Open Letter to Ceauşescu in March 1989 was evidence of an ongoing clandestine opposition with support in the armed forces or secret police, and no doubt, the encouragement of Moscow. The actual overthrow of Ceauşescu was triggered by the demonstrations that broke out when Laszlo Tökès, a Hungarian Calvinist pastor in the Transylvanian town of Timişoara, resisted a change of assignment and openly challenged the regime. Again Hungary intervened. Hungarian television broadcast exaggerated accounts of casualties in the repression of the demonstrations by the Romanian police, including gruesome reports of a massacre of children on a candlelight vigil in support of Tökès. Later more accurate reports from Western sources made clear that the 'massacre' had been a fabrication. When Ceauşescu addressed a crowd in Palace Square in Bucharest on 21 December, the crowd jeered back at him. That began the insurrection. Curiously, at no point was it opposed by the army, but only by sections of the *Securitate*, Ceauşescu's elite guard. Even the helicopter pilot who spirited Ceauşescu away from Bucharest in an attempt to escape apparently aided in the capture of Ceauşescu by army officers who shot him after a hasty trial three days later. Many mysteries continue to surround the events of 21–25 December, with persistent rumours of the anti-Ceauşescu activity of Hungarian army and intelligence. As in the case of the border barriers and the East German refugees, official Hungarian forces had exerted important influence in the overthrow of the Ceauşescu dictatorship.[74]

The revolutions of 1989 capped a year of smashing victories over Gorbachev's Soviet and bloc opponents. The obstacles to *perestroika*, as defined by Gorbachev, had been swept away. He made clear that he viewed the dynamic changes in the bloc as part of the process that promoted security, trust and cooperation in the House of Europe, provided, of course, that they enjoyed the necessary stability. The new Europe could not be viewed with the blinkers of the Cold War, but only with the broad perspectives of the New Thinking. Western countries, he noted, were generally refraining from taking advantage of the new situation, something that showed that they understood 'the inviolability of the postwar borders' (that is, the existence of two German states) to be the most important condition for the taming of east–west tensions and the creation of 'a new world order'.[75] The security of Europe, never attained in the past by military means, was now to be constructed anew by political virtuosity. And Gorbachev, finally free of constraints, considered himself its historic architect. His bond with the masses of the socialist countries had proven superior to the petty complaints of his detractors. Gorbachev thought that this was because he understood the masses better than had the defenders of the old ways. Yet he would soon learn that he had not understood them at all.

11 From the Wall to Stavropol: Gorbachev's German Policy

> Europe is getting ready to conclude a new treaty of Versailles. There are many indications that while one great European power is losing territory and zones of influence, another European power is turning into a great power... The first power is the Soviet Union, the second is Germany. The great power that won the Second World War is losing it after 45 years of peace... some irreversible changes are bound to take place in the world. They would consist mainly in drawing a line between the spheres of influence of Germany and Russia.
>
> Laszlo Lengyel

Gorbachev could now claim a stunning victory. His opponents were scattered and defeated. A roster of the Kadars, the Honeckers, and the Ceauşescus was now duly entered into the register of discards of history. Ligachev was permitted to hold on only because the leader was reluctant to retire him and awaited the proper moment to deal the fatal blow. Gorbachev had got everything he wanted. But what had he got? He had to pause to remind himself of the initial aim in search of which he had set out on this vast journey. It was to achieve for the Soviet state world levels of technical proficiency, was it not? Yes, and to cleanse and renew the Communist idea. Or perhaps to save the world from the threat of nuclear war by the complete elimination of nuclear weapons by the year 2000? Was it to liquidate the tensions of the Cold War by promotion of the concept of the House of Europe? Or perhaps all of these. Well, had these great ideas not won a great victory? At any rate, their triumph was ensured, was it not, by the fact of their opponents and the other sceptics having been so convincingly vanquished?

'LET HISTORY DECIDE'

Hans Modrow later reported that in their meeting in early December, some three weeks after the Wall had fallen, Gorbachev was full of illusions about the unification of Berlin and the defeat of the Honecker leadership having contributed to the strengthening of socialism in the

GDR, whereas he, Modrow, realized that the system would have to be liquidated.[1] Modrow may have been right about Gorbachev, but had he himself foreseen everything? At the time, Modrow had said nothing of the kind. He rejected 'speculations about re-unification', assuming that the East German Communists would prevail in the elections that were eventually held on 18 March, or at least that they, together with other forces of the left, would have a bloc capable of controlling events.

But the flight of refugees and the demonstrations against the Communists did not let up with the fall of the Wall. By February Modrow could see that none of the major parties wanted to delay or even to haggle over the terms of unification. So he yielded to the plan of West German Chancellor Kohl, for a three-step transition to German unity: for 'contractual community', confederation, and then federation.[2] Modrow recast this as the Modrow Plan, presumably having for its aim a neutral unified Germany. In that form it was accepted by Shevardnadze (who had said on 19 December that 'we will protect the GDR from any offense'). By the beginning of February, Shevardnadze had convinced himself that 'there is logic in Modrow's concept'. Even at this point Modrow and Shevardnadze were clinging to the idea of a round of hard bargaining that would essentially reshape Europe. Kohl, however, unhesitatingly expressed the view that the future could be discussed, not with Communist politicians, but only with an elected government – and that in any case, any German exit from NATO was entirely out of the question.

Kohl and the FRG were acting in the manner of a great power. Gorbachev was acting like a statesman whose country was about to be partitioned. And that was what he was. What authority could he bring to international negotiations? According to French President François Mitterrand, Gorbachev had once assured him that the two Germanies would never be made one on his watch: 'The very day that German unification is announced, a two-line communiqué will report that a marshal has taken my place.' In Italy in December, however, Gorbachev was asked about East Germany, and he answered: 'Let history itself decide.'[3] That really means: let someone else decide. He may still have thought that history was on the side of a general dissolution of NATO and the Warsaw Pact and a corresponding apotheosis of the Helsinki process, perhaps in a 'Helsinki 2'. A new security arrangement could be grouped around the two German states, the only obstacle consisting in the possibility that the United States might be tempted to speculate in the East European changes, seeking the dissolution of the Bloc and the liberation of the Baltic republics. Gorbachev thus tried to get President Bush's assent to restraint at their Malta meeting in December.

But it was not Bush who could provide him with the necessary assurances. The opening of the borders between the Germanies meant that

West German political parties, led in the event by the Social Democrats, were given a free field for political organizing in the East. And Modrow himself put West German financial aid as the first object of his own policy. For his part, Kohl tied all aid to the prior adoption of a monetary union, which more or less excluded any hard bargaining over reunification. Gorbachev had warned consistently for several years that tampering with the status of the two Germanies was the primary threat to the stability of Europe, but he foreclosed even on any slight chance of multilateral diplomatic pressure when he assented to the '2 plus 4' agreement of the foreign ministers in their Ottowa meeting. This idea was apparently the brainchild of Secretary Baker's assistants Robert Zoellick and Dennis Ross.[4] The actual unification would be decided by the two German states, while the Big Four (Britain, France, the United States and the USSR) would agree on its foreign policy extensions. That tossed everything to Kohl. For the Four to agree that things should be decided by the Two (the GDR and the FRG) meant that things would in the end be decided by one – the FRG.

Not that Gorbachev showed any indication of knowing what he was doing. Kohl and others even thought they heard in some of his utterances a hint of accepting an eventual German unification in NATO. This produced a shock in Moscow. The International Department's Valentin Falin intervened in an interview with *Der Spiegel* in mid-February, saying that 'If the western alliance sticks to its demand for a NATO membership for all of Germany, then there won't be any unified Germany.' It was obvious that there was no one Soviet policy. Gorbachev shifted gears and went along with the Falin line in an interview on West German television on 8 March, saying of a new Germany in NATO that 'this is absolutely ruled out'. But a sometime advisor on German affairs, Nikolai Portugalov, later explained that Gorbachev's statement 'was just a starting point in negotiations'. The redoubtable Vyacheslav Dashichev even suggested that 'you have to bind Germany to the framework of NATO'. The extraordinary disarray and ineptitude among the Soviet officials and experts demonstrates beyond doubt that the historic changes in East Germany could not possibly have been thought out or coordinated in advance.

The Soviet hard-liners thought Gorbachev's policy catastrophic. Aleksandr Prokhanov argued that the 'sentimental theory of the Common House of Europe' had led to the destruction of the Communist parties of the bloc. This was a reflection, he said, of the general drift of the Gorbachev policies since 1987. Ideology had been cast off, leaving behind 'an exploding galaxy with a black hole at the center'.[5] The attacks on the 'command-and-administer methods' in the economy had meant the destruction of the economy itself, and attacks on discipline in the army and the party were producing a similar result. The reforms had given the country nothing but chaos, and now bore the threat of genuine catastrophe. Prokhanov reminded his readers of the possibilities in the event that a civil war, already developing, were to break out in earnest, and among these possibilities were 'unintended ballistic missile launches'.

Ligachev himself warned that the reforms were unravelling everywhere. 'The first two years saw a certain revival', he said, 'but now *perestroika* has gone into decline'.[6] The cooperatives were getting rich as the country sank into near-famine conditions. Localism was so rampant that the essential food supplies, especially meat, were not being delivered to all-Union stocks. Ligachev railed against the campaign of insinuations about him in the press, objecting especially to intimations of 'a special Ligachev role' in the decision to suppress the Tbilisi demonstrations of May 1989, a decision which was taken, he said, 'by the whole Politburo'. He reminded his party comrades of the reduction of his influence in the last year. One could not blame the opposition, he said, for the unprecedented bungling of the Gorbachev–Shevardnadze foreign policy, which had inadvertently revived the German problem with a vengeance. Giving up the GDR to German unification on Western terms would amount to nothing less than 'a new Munich'.

Since Europe was in the process of being delivered from the fate assigned it at Yalta, there was an understandable tendency for Ligachev to invoke analogies from the pre-war period. But the bargain that was shaping up between Gorbachev and Kohl could have called to mind another project of the thirties. In 1934–35, French statesmen tried to pressure Hitler into accepting what was then called an 'eastern Locarno', reproducing in the east what the 1925 treaty of Locarno had given the West: a guarantee of the borders between Germany, on the one side, and France, Belgium, and Holland, on the other. For this guarantee to have obtained in the east, the Soviet Union and Germany would have had to guarantee Poland and Czechoslovakia, that is, to proclaim their own self-containment. The Baltic states formed an entente in 1934, against the possibility, among other threats, of an Eastern Locarno that might result in a Soviet–German deal assigning them to a Soviet sphere of influence. The Scandinavian states, led by Finland, briefly showed signs of being willing to support a bloc with the Baltic states, then backed away. By 1939, the hope for a bloc was abandoned when Finland, Norway, and Sweden refused the Soviet suggestion of a Soviet–British–French guarantee. They stood clear of a Russo-German deal assigning the Baltic states to Soviet control. Before November 1989, it looked as if West Germany would want to come to terms with the USSR on unification and would as compensation agree to look with sympathy on Soviet efforts to suppress the Lithuanian independence movement. The Scandinavian states, in this same cautious mood, were careful not to encourage the Soviet Baltic republics to carry their struggle against Moscow to extremes. Their counsels of moderation exerted an influence in suppressing, at least for the moment, the perspective of national independence.

However, after the revolutions of 1989, the Scandinavians changed their line and the Nordic Council became an enthusiastic supporter of the breakup of the Soviet Union and the full national sovereignty of the Baltic states. The rise of hopes in the Baltic republics ran parallel to the explosion of discontent in the bloc, but when the bloc regimes gave way, the FRG and the other Western powers seemed to draw back cautiously from encouraging the Balts. The reasons were obvious enough. It was thought that Moscow saw the bloc states and the Soviet republics quite differently, and would not hesitate to use force to keep the republics in line. As a case in point they could cite the violent Soviet suppression of the Tbilisi demonstrations in April 1989. The FRG seemed to be coming to terms with the USSR on a division of spheres of influence that certain of the diplomats of the thirties, who had sought to trade German control of the Polish corridor for Soviet control of the Baltics, would have recognized.

Yet the 'children of the Arbat', as the hard-liners called the supporters of Gorbachev, refused to see the looming German unity. They quoted with approval the famous remark of French writer François Mauriac: 'I love Germany so much I am glad there are two of them.' Evgeni Ambartsumov looked forward to a 'parallel and symmetrical merger of blocs in the Common House of Europe'.[7] Egon Krenz promised that the GDR would remain socialist. Even Jens Reich, one of the founders of *Neues Forum*, told *Moscow News* that socialism would be revitalized in an independent GDR undergoing *perestroika*.[8] Hopes persisted right up to the time of the East German elections in March, largely because the polls showed an impressive lead for the Social Democrats. It was thought that even in the worst possible case, that is, a ringing defeat for the Party of Democratic Socialism (as the East German Communist party had been renamed), the resultant Social Democratic victory would nevertheless be enough to overturn the old balance of political forces even in a united Germany.

THE BALTICS AND ARTICLE SIX

Gorbachev and his supporters seemed to think that the internal fissures on the republic level in the Soviet Union were as menacing as anything that could happen in the bloc. In December the Lithuanian Communists had announced their break with the CPSU. This was a shock for those Communists who remembered how hard Lenin had fought against party federalism. Gorbachev was immediately dispatched by the Central

Committee to Lithuania to work another of his miracles. It was a tense trip that climaxed in a dogged discussion with Brazauskas and the other rebels that lasted over four hours. Gorbachev warned them of being cut off from the centre: 'comrade Brazauskas will be reading the papers to learn what we have decided about the fate of Lithuania'.[9]

But he did not succeed in changing their minds. They may have sensed what others had already sensed about Gorbachev: that no matter how fiercely he proclaims it, he seldom sticks to a position, and that patience and persistence will usually wring any concession from him. In this case his action seemed for the moment to be in solidarity with the party hards, and if Lithuania is taken as a test of Gorbachev's attitude toward his struggle with them, he seemed to be making a turn. Yet, in the same meeting, he said that the CPSU itself was taking on a new role, that he was 'ridding the party of the complex of infallibility . . . and of political monopolism'.[10]

That seemed to say that Gorbachev was going to follow the lead of the Hungarians, the East Germans and the Czechs who had already renounced their former pretension to a 'leading role'. A second wave of miners' strikes in the summer of 1989 had demanded that the party abolish article 6 of the Soviet Constitution which proclaims the CPSU's leading role. The central committee of the Estonian party had voted to annul the article. There had been demonstrations in Armenia against it, and a crowd of 20 000 in Leningrad, rallying against Gidaspov, carried placards saying 'Down with Gidaspov! Down with article 6!' Gorbachev's remarks in Lithuania cannot have been a slip; he must have been prepared to move against article 6 for some weeks, or perhaps months, before he brought the matter before the January 1990 Central Committee plenum. And at this gathering he again amazed the world with his ability to dominate party meetings, in this case getting the Central Committee to vote for what amounted to its own political self-abnegation. Some party people saw this as just another bandwagon to jump on with aplomb. Gerasimov said straightforwardly of dropping article 6: 'If you can't beat 'em, join 'em.' But others needed some rationale. Gorbachev told them that renouncing the legal claim to vanguard status would be the prelude to even greater influence in fact. No doubt the dumbfounded delegates reasoned that they did indeed have control of the armed forces, the intelligence agencies, and a great deal more, and that perhaps it was harmless to let the leader have his way. They must also have been seduced by the luminous prospect that Gorbachev set before them of winning genuine popular acclaim and legitimacy for their leading role by mastering the arts of democracy.

For his own part, Gorbachev was building executive institutions capable of freeing him almost entirely from party control. In February and

March 1990 new powers were assigned to the executive presidency by the Supreme Soviet and the Congress of Deputies. Among these was the right to declare martial law and civil emergencies. The President was to have a Presidential Council providing him with his own body of advisors. It was all 'law-based' and constitutional, to be sure, but in effect Gorbachev now had an apparatus of personal authority outside the purview of the party. In fact the Presidential Council was to conduct the business previously conducted by the Politburo and the Secretariat.[11] It was the second wave of blows against these party bodies; the first had been the formation of Central Committee commissions by the September–October plenum of 1988. After their formation, the Secretariat 'did not meet for a long time', as Ligachev was to put it at the 28th Congress.[12] But the exact role and function of the commissions was never made clear, and they apparently tried to do their work in such a way as not to interfere with the work of the Supreme Soviet, the Congress of Deputies, and other state bodies. Now Gorbachev was trying to subvert the entire Central Committee with his Presidential Council.

This body was constructed on non-party lines. Primakov stated that it 'must not be shaped on the principle of party affiliation'. Stanislav Shatalin said that 'it will soon turn into a coalition Council'.[13] That would seem to suggest that its members thought it a prelude to a Round Table like those in Poland or Hungary. At the same time it appeared to subsume the functions of the old Brezhnev-era Defence Council, which had previously been under the Ministry of Defence. Now that the Defence Council was attached to the President's office, the latter took on the visage, perhaps consciously, of a National Security Council on the American model.[14]

Could it be that the party had lost control over the Armed forces and the KGB? It was not an easy question to answer. One could at least say that the bureaucratic routine of party supervision over the army and the security organs had been disturbed repeatedly by the reorganizations of the last two years. The Administrative Organs Department of the Central Committee, traditionally the overseer of the security organs, had been subsumed under the State and Law Department by the plenum of September 1988. The status of the latter was unclear after the 28th Congress, but it was the last party institution known to exercise this function. The closest thing to it was the General Department headed by Valeri Boldin, who had a seat on the Presidential Council. Boldin's work for the President seemed to be that of a chief of staff, and he made pronouncements on military-organizational matters.[15] It may be that his presence in that position was designed to encourage the party to comfort itself that it was maintaining control behind the scenes.

Yet no one in the party could doubt that the traditional authority that it had wielded was crumbling fast under Gorbachev's imperium. More-over, the path seemed to lead straight to the same fate that had befallen the Communist parties of the bloc. It was said that Gorbachev had promised the Estonian deputies negotiations on independence in exchange for their votes against article 6. Brazauskas, on his election to the Presidency of Lithuania, said that Gorbachev's opposition to Lithuanian independence was 'just for public consumption'. Vadim Medvedev went to Vilnius to defend the integrity of the party, but at the same time Maslyukov said that it was all right for the Lithuanian Communists to split. Then on 11 March came the news that Lithuania had formally declared its indepen-dence. Estonia, Latvia, and Georgia seemed destined to follow suit. The partition of the Soviet Union appeared to be in process. Small wonder that the party loyalists were distraught at the prospects of continued Gorbachev reforms.

'EXPELLED WITHOUT A FIGHT FROM LANDS OUR FATHERS FREED'

The news from the former bloc was not encouraging. The Soviet party press had not yet written off the GDR, insisting that the German work-ing class was going to have its say in the coming elections. A big vote for the ex-Communists or for the Social Democrats was thought to be suffi-cient to ensure that the GDR would not simply be turned over to the West German state. The returns on 18 March dashed these hopes. The PDS got only 16 per cent, and the Social Democrats only 21.8 per cent, while the Christian Democrat-dominated Alliance for Germany got 40 per cent. With the Christian Social Union's 6.3 per cent and the Free Democrats' 5 per cent, a coalition much like that of the ruling FRG government was more or less inevitable. The Christian Democrats had collected a whopping 58 per cent of the industrial workers' vote, but only 32 per cent of the vote of university graduates. The PDS was hu-miliated in industrial Saxony, the traditional stronghold of the left since the twenties, while it got 30.2 per cent in Berlin.[16] The appeal of the left was reasonably successful with the intelligentsia, but the working class had thrown everything behind the hope of rapid unification offered by the CDU.

The Hungarian and Czechoslovakian elections returned similar results for the ex-Communist parties who, while they were not ruined for all time, still failed to improve on what their counterparts normally got, for example, in France. Communism in the ex-bloc turned out to have real electoral strength only in Bulgaria, Romania, and, as would be shown in 1991, Albania – and in those countries the vote came mainly from outside the capitals. There seemed to be a curious division of sentiment between the areas of traditional Latin Christendom and those in which Orthodoxy or Islam had been the old confession.

The idea that Soviet reforms would open a broad field for the fusion of Soviet Communism with Western social democracy was punctured painfully. Increasingly, the response of the party hard-liners took on the language of Russian national Communism. We have already considered this ruinous ideological departure in regard to the 1988 letter of Mme Andreyeva. Demands for the establishment of a Central Committee Russian bureau had been raised in the spring of 1989. At the July plenum that saw calls for the revival of the Secretariat and reinstatement of the post of second secretary, Vorotnikov had argued that a specifically Russian party body was necessary to deal with the exploding nationality problems. At the December plenum a Russian Bureau with 16 members was established, with Gorbachev in the chair. Among the 16 were Politburo members Vorotnikov and A. V. Vlasov, Central Committee Secretaries Iu. I. Manayenkov and G. I. Usmanov, Prokofiev and Gidaspov for the two capitals, first secretaries for a number of other big towns, and Chikin of *Sovetskaya Rossiia*. It appeared to be a potentially powerful formation, but Gorbachev, sensitive to charges that he was undermining CPSU authority for the sake of political support from fractious Baltic comrades, did his best to deny it any real authority. During the period when Gorbachev was preparing his campaign against article 6, Gidaspov and others agitated for the holding of a Russian party congress, and argued against Gorbachev's subsequent suggestion that the congress be held *after* the coming CPSU congress in July. Gidaspov got his way. The path now lay open for a showdown between Gorbachev and his critics.

At the time of the January plenum, Aleksandr Tsipko had charged that 'several dozen dogmatically-minded members of the central committee' were holding *perestroika* hostage.[17] By April, the dozens had managed to convene a meeting of 600 in Moscow, largely with the preparations provided by the Leningrad regional and city committees. Charging that Gorbachev's 1988 economic reform had given the real power in the economy not to the soviets, but to the 'shadow economy', one delegate said that 'it is too late to complain about some "creeping counter-revolution", because it is already standing proud'. There were numerous attacks on the democratic press ('that pro-Democratic Union paper *Ogonyok*') and demands for a founding congress of the new party. Some delegates expressed anger at the situation in the former eastern bloc, and at their having been 'expelled without a fight from countries our fathers freed'. They resolved to form a new party in June. The rise of their critique was parallel to an apparent general swing in the CPSU, which announced the same month that it would purge its ranks by the removal of all who were not 'convinced Communists'.

This was enough for Ilya Chubais, I. Yakovenko, V. Lysenko and confrères to conclude that the CPSU had been captured by a 'conservative coup' led by Ligachev. Accordingly they put together the 'Democratic Platform', a caucus that had grown out of Moscow meetings held at the time of the January plenum. Pushed along by Vyacheslav Shostakovsky of the Moscow Higher Party school, the faction claimed the adherence of 100 000 Communists, with 40 per cent represented by 'learning establishments, design bureaus, and research-production associations', and some 20 per cent by full-time party functionaries. Over 60 per cent, it was said, had been party members for at least ten years.[18] The platform itself demanded the abandonment of Democratic Centralism and of the ideology of Marxism-Leninism, and the adoption of the outlook of the international social democratic movement. But the faction was divided between those like Chubais, who wanted an immediate split with the CPSU, and a majority who, like Yakovenko and Lysenko, wanted to issue an ultimatum to the coming CPSU congress, with the intention of splitting only if it were refused. To round things out, there developed in addition a 'Marxist Platform' with a programme of inner-party democracy and rededication to Marxism in a critical spirit.

Some 2800 co-thinkers of Ligachev and Gidaspov gathered in June for the founding congress of the Russian Communist Party. They reaffirmed faith in Marxism-Leninism as defined in the 1987 historical speech of Gorbachev, and rejected a market economy. They elected as First Secretary Ivan Polozkov, formerly First Secretary from Krasnodar. In his former capacity, Polozkov had actually suspended the law on cooperatives in Krasnodar, saying that 'one must react not according to the law, but as is necessary'. Polozkov and the Russian Communists were alarmed at Gorbachev's attempts to reduce the party in favour of the new state intitutions. They saw additional menace in Boris Yeltsin's having been elected chair of the federation Supreme Soviet.

The unity of their ranks was not so solid as their unanimous votes at the congress seemed to imply. There was tension between the Leningrad and Moscow party organizations. Yuri Prokofiev, who had replaced Zaikov as head of the Moscow party back in November, was not a particular enthusiast of Polozkov's leadership. He did echo many of Polozkov's complaints, to be sure. When the Russian Bureau had been formed in December, he had complained that a political vacuum was being opened, to be filled by 'bourgeois democratic, liberal, social democratic, Euro-communist, anarcho-syndicalist, neo-Stalinist, monarchist, and national socialist' viewpoints. In his opinion, various party collectives, such as the journal *Kommunist* and Bogomolov's Institute of the Economy of World Socialism, were *in* the CPSU without wanting to be *of* the RCP. Prokofiev was at this time mistakenly regarded in the West as a supporter of

Gorbachev and reform. In fact he did remain on record as supporting the leader, but he was no democrat. He spoke enigmatically about dictatorship as the *sine qua non* of 'either a return to capitalism or the past totalitarian state'.[19] He regarded Polozkov on the one side and Yeltsin on the other as being 'authoritarian enough', and expressed doubt as to whether Gorbachev would be able to do his 'balancing act' forever. How to judge the loyalties of this man at the head of the Moscow party apparatus? He appeared to be one who, at least outwardly, was comfortable with any option, as long as it was dictatorial.

GLASNOST AGAINST THE RUSSIAN REVOLUTION

At previous crucial times such as this, when the party stood before a crossroads, great attention concentrated on the material appearing in the *glasnost* press on party history, as the party faithful searched for an ideological sign of which way the line was turning. This had been the case at the time of the 70th anniversary of the revolution in 1987, and again at the 19th Party Congress in 1988. The reason for that was that the country was then still under the sway of the party and of Marxism-Leninism, and any historical revelation or admission by the party could be a signal of a general change of social and political orientation. Articles in the press dealt with questions regarded as crucial for the reform and reshaping of Soviet Communism within the framework of the traditions of the revolution in 1917. Yuri Afanasyev had set the tone by calling for special re-examination of the twenties and fifties, to determine what party weaknesses had permitted first Stalin and then Suslov to get their way over the 'healthy forces'. As we have already indicated, not much came of that beyond the promulgation of a virtual cult of Bukharin, as a justification of the revival of ideas associated with the NEP of the twenties.

The 'repentance' school weakened the search for a renewal of the Communist ideology of Lenin's era, in favour of a more passionate and bitter examination of the Stalinist purges of the thirties, with the result that the discussion, which was within the boundaries of Marxism in 1987–88, brought forth anti-Communist perspectives in 1988–89. We have already considered the impact of the ideas of Selyunin, Gefter, Tsipko and others in this regard. The events of 1989 rendered the discussion of party alternatives of the twenties and thirties entirely irrelevant. Now the discussion was more likely to be interested in the events of the year 1917 itself. Instead of debating the possibility of a Bukharin alternative, the questions more likely to be raised concerned Lenin's break with the Mensheviks and other men of the European social democratic tradition, or the dilemmas of the period of Kerensky's rule under the Provisional Government. It was as if the writers and their readers were searching for a reason to say that the turn taken by Russian history in 1917 need not have been taken and, if not for that, Russia might still be part of the community of civilized countries.

Ironically, this shift in the perspectives of the discussion promoted the closest thing to a dispassionate view of the historical opponents of Stalin that had yet come from *glasnost*. The demonization of the Jewish Bolshevik leaders, Trotsky, Zinoviev, Kamenev, and Radek, seemed to lessen, except in the nationalist press, and those historical figures who had been associated, not always accurately, with the extreme left positions at least got their due as revolutionaries. It became normal to speak of the actions of the Bolsheviks in 1917 as being 'led by Lenin and Trotsky'. Photos of the two from the time of the revolution and civil war appeared normally, without cropping, in the press. Whereas the standard fare in 1987–88 contained the assertion that Stalin 'realized in his own political projects the central ideas of Trotskyism',[20] now that Lenin himself was coming under critical review, it was just as confidently asserted for the first time that Lenin and Trotsky were the most intimate co-thinkers. There are of course varying elements of truth in both these views, which could be debated at length, something space does not permit here. The striking phenomenon from the standpoint of our investigation of the struggle for power, however, is that whatever was admitted about Trotsky's historical acts, he remained in the status of devil. It was likely that Trotsky and Lenin were being shown together in order to make of Lenin a Trotskyite. The point seemed to be that the *glasnost* papers were evolving into organs of an opposition party or parties, and that they were cognizant of the need to avoid offending nationalist sentiment by too friendly an attitude to this lodestone of Great Russian hatreds.

Aleksandr Tsipko, who had led the chorus against 'left-radicalism' in the party's past, now criticized the present party platform for trying to steer too cautiously between the other two platforms. Arguing for the abandonment of Leninism and Marxism, he cited his long-time support for the idea of universal human values.[21] This was not false. Tsipko's calling pre-Gorbachev had been that of a party ideologist and propagandist, a purveyor of what might be called Stalinist devotional literature. In this role he had pronounced on the compatibility of Communism and universal values, writing in support of Brezhnev as of Gorbachev. Along with many others, he had also in the seventies already given voice to the central idea in the economic reform of 1988:

> Only when the direct producer takes a very direct part in the organization and management of production can we count on the awakening in him of the sentiments of a master of production, and thereby the establishment of a new direct bond between him and the nationalized means of production.[22]

Now he had come to the view that the sentiments producing this direct bond were best found by private property. This was a substantial conversion. Along with many others, Tsipko did not want to lag behind. He was deeply impressed by the rejection of social democracy in the elections in the former bloc countries, and he concluded that it had 'a certain logic', in that, starting from scratch with the market economy, there was no point in supporting social democrats whose ideology was one of limiting its effects. He wanted the CPSU itself to follow the trend and draw the same lesson, rejecting social democracy in favour of a 'centrist, general democratic ideology . . . capable of standing up to right-wing and left-wing populism'.[23] This sounds rather like the philosophy of what would be called in the United States a Neo-Liberal.

Tsipko claimed to find the inspiration for his current thoughts in a rereading of the *Vekhi* symposium of 1909, a collection of essays by a group of the leading former left intellectuals of that period, who rethought their commitment to revolution against the tsarist order in the wake of the repressions that followed the revolution of 1905. They had concluded that the people could never be at one with the intelligentsia, and that, in the event of another revolution, the 'Acheron' of the masses would surely tear down the traditional edifice of Russian culture and civilization.[24] Lenin had called *Vekhi* 'an encyclopedia of renegacy'. Its contributors had agreed that the most important need for their country was not democracy, but the rule of law. At least one, S. L. Frank, in attacking the mystique of revolution, lamented that the spirit of populism had swallowed up Legal Marxism, which stood at that time for progress toward a capitalist system on the Western model. And another, M. Gershenzon, had remarked that the intelligentsia should thank the Tsar for his prisons which alone protected them from the wrath of the people.[25]

Tsipko's enthusiasm for *Vekhi* does not need interpreting. In praising it, he only meant to say that the events of 1917 had been a terrible tragedy. The writers of the 1909 essays had advocated cooperation with the Tsarist bureaucracy as preferable to its overthrow. This matches with Tsipko's fears of populism. In view of the position of the CPSU as presiding over an *ancien régime*, should he not therefore have advocated a similar cooperative attitude toward it? In fact, some of his disavowals of the more extreme statements of Yeltsin in 1991 would show exactly that.

Gorbachev was using more and more of Tsipko's language for his own purposes, as for example, in saying that the socialist idea itself was not in crisis, but only 'utopian, left-radical military Communism using socialist terminology'. Perhaps this was a fairly accurate reading of the moods of the Soviet people, who seemed to want some sort of system describing itself as socialist. A poll taken in October 1990 would show 36 per cent in support of the idea of 'democratic socialism', and 30 per cent in favour of 'Swedish socialism', with only 6 per cent in favour of 'capitalism'. Among the Scandinavian-minded citizens of the Baltic republics, 'democratic socialism' garnered only 3 per cent, while fully 52 per cent professed to approve 'Swedish socialism'.[26] The latter term seemed to be synonymous with 'market economy', which received similar amounts of support from the Soviet public. So the Balts and, to a lesser degree, the other Soviets apparently wanted a 'market economy' run in the manner of the Swedish Socialists. Arguing from this, one could assert that there was still some faith in the word 'socialism', however defined. But there was very little faith in the existing Soviet economy: an all-Union majority expressed doubt in a July poll that the government was capable of freeing the country from its economic crisis.[27]

This could be a surprise in view of the fact that the economic decline that had begun in mid-1988 had by this time reached the point of a *net decrease* in the national income.[28] Buffeted by this and the depressing news from Eastern Europe, Gorbachev's popularity was now also in a dramatic downward slide. A running poll asking respondents whether they approved of Gorbachev's policies showed his support fade from 52 per cent in December 1989 to 39 per cent in May 1990.[29] In Red Square on May Day, Gorbachev had to face something none of his predecessors had ever seen; after the official parade organized by the trade unions, itself a far cry from the once traditional military displays, a second opposition parade of various groups ranging in political complexion from Christian Democrats to anarchists went past Lenin's tomb carrying placards openly denouncing Communism and Gorbachev.

NEGOTIATIONS

In the past he could usually pump up his popularity with a US–Soviet summit, but his June visit to the United States did not do the trick. He now appeared in the West in a more pathetic persona. The USSR had not been paying its bills for several months and its trade with the outside world was down. Some in the West offered the ingenious and encouraging suggestion that Gorbachev's apparent weakness gave him a certain strength, in view of undeniable Western fears of chaos should he fail. But the election of Yeltsin to Chair of the Supreme Soviet blunted that argument by providing an alternative to Gorbachev who might be even more pliant than he.

In the event, Gorbachev could not achieve much. He did not want to make any new concessions on Germany with the 28th Congress (and, as he anticipated, a concerted move to oust him as General Secretary) approaching. It was Falin who weakened in his opposition to German unity.

In April, while advocating an all-European security system in the future, he nevertheless allowed that in the meantime Germany could be a member of both pacts.[30] In May, Shevardnadze said that Germany could *not* be in NATO, except in the case that the Warsaw Pact countries could join too! But, again, Major General Geli Batenin, an advisor to the Central Committee, suggested on the contrary that inclusion of Germany in NATO might help 'stability'. Discipline in foreign policy was a thing of the past.

And Gorbachev's once consummate summit diplomacy was looking threadbare. When he went to Washington to meet with Bush in June, he at first complained that the American President held a 'rather rigid' position on Germany's status in NATO, but after his first meetings with the American President he claimed that 'something emerged'. This was nonsense. Actually Bush yielded nothing. He allowed that Soviet troops could stay on in Germany 'for a few years' with West Germany contributing to their support. There would be no NATO troops in the former GDR during that period. NATO would in the future 'revamp its force structure'. But the new Germany would in any case be in NATO, or rather: the new Germany would be permitted to decide, which was the same thing. Again Gorbachev's summit entourage expressed more than one view on the matter: the International Department's Vadim Zagladin said that Germany could not join NATO 'as it exists today', but Andrei Kortunov, of the United States and Canada Institute, allowed that Germany could be in NATO if there were no NATO troops in the east. That was Bush's position. Gorbachev was not conducting any diplomatic *tour de force*. In fact the American Congress was a step ahead, actually pressing for linkage of future aid or Most Favoured Nation trading status to Soviet acquiescence in the secession of the Baltic republics.

THE MASTER OF CHAOS

The international *pourparlers* about the liquidation of the Soviet bloc and the partition of the Soviet Union cannot have looked very impressive to Gorbachev's opponents at home, preparing for their showdown at the 28th Congress of the CPSU. But what kind of showdown could they actually have? If they were to defeat Gorbachev as General Secretary, he would no doubt leave the party to form another. Whom did they have to compete with him and Yeltsin for public favour? Were they prepared under these circumstances, to dispense with the new democratic institutions and arrest their opponents? The idea must have occurred, but it must also have terrified them, hence the extraordinary timidity with which they confronted Gorbachev, the presumed serpent of their agonies. Gidaspov, Polozkov, and others who had been breathing fire at the RCP gatherings began to sound like more moderate fellows. Instead of assaulting the leader, they cast their bolts at his supporters, Yakovlev, Shevardnadze, and Medvedev. In fact, the critics, many of whom said frankly that a return to the Andropov line was needed, did not actually seek his fall but only the removal of his supporters so that they could 'turn back with Gorbachev'.[31]

On the first day of the congress, Polozkov spoke of the coming of a new political phase, in view of the general recognition throughout the party of the need for 'consolidation' on the basis of 'the socialist choice, Communist perspectives, and democratic centralism'.[32] Gorbachev's political report endorsed the 'consolidation' line. He stressed that the house had not caved in with the party's having relinquished article 6, and progress was being made in the forging of the new 'vanguard' role for the party. He used the same terms as Polozkov, 'socialist choice, Communist perspectives'. The party had 'dropped its claim to substitute for government bodies', but would now 'work to earn the mandate of the ruling party'.[33] The party had made the decision for a market economy, but this must be consistent with 'mastery by the state of the whole range of instruments of economic management'. Since the party had long since accepted the dismantling of the planning apparatus, this presumed step toward *dirigisme* promoted the illusion of an increase in its powers.

Yeltsin did not address the Congress until its fifth day, as if he sensed that a compromise solution had been worked out over his head. He had won a close election at the end of May, over intense RCP opposition, to the post of chairman of the Supreme Soviet of the Russian Republic's Supreme Soviet. The parliament also declared the Russian Federation to be 'sovereign', so he was now the President of Russia. He had sat quietly through the Congress sessions as it proceeded to ignore him. Then, in a speech that bristled with chilling and prophetic threats, he called the Congress illegitimate and conjured up the possibility, if the party did not mend its ways, of a confiscation of its property, with its leaders being hauled into court for abuse of the public confidence. More specifically, he laid out the agenda of the parliamentary anti-Communist forces for the next period, one that consisted in trying to take from the party its control over the armed forces and the KGB, in fact an imposition that was perfectly consistent with the new role of the party as outlined by Gorbachev. The delegates, who had interrupted the speeches of Medvedev and others with catcalls and shouts, listened in grim and pained silence to Yeltsin. As he had in previous party fora, he drove them into a protective rally against him. Yeltsin's intervention thus fortified the alliance of Gorbachev and the 'conservatives'. It also permitted Gorbachev stridently to assert in closing the rectitude of every step he had taken in the last five years. The party was now united behind him, no longer a party claiming a leading role, nor yet a parliamentary party, but one with a special 'vanguard' role that remained to be seen. On the eleventh day, Yeltsin announced that he was leaving the party, and shortly after, Shostakovsky, speaking for the Democratic platform, followed suit.[34] Gorbachev remarked pompously that the resignations were 'logical'. Yet he must have regretted the loss of Yeltsin, who up to the end of his career as a Communist filled the role of Gorbachev's bludgeon. He no doubt still hoped that Yeltsin could fill the function as well from outside the party.

In fact, Yeltsin was now a President of Russia who was bitterly opposed to the Communist Party of the Soviet Union and, of necessity, to the Soviet Union itself. Communists loved to speak about 'dual power' erupting in a revolutionary situation, as in Petrograd in 1917, when the Petrograd soviet vied with the Provisional Government for the loyalty of the troops. And many described the loss of party authority in the last two years as a kind of 'dual power'. Now Yeltsin had in his hands a revolutionary instrument as potent as anything seen in 1917, and he was in the most extraordinary position of leading Russia in rebellion against the Soviet Union.

Most outside observers formed the impression that Gorbachev had once again won the day against his opponents in another one of his astounding public performances. His great rival Ligachev, after being applauded ecstatically for his speech on the first day, a performance that oddly seemed more like a memoir than a call to action, failed to gain election to the position of 'deputy general secretary' of the Politburo. The remainder of those who had criticized Gorbachev were all routed as well. The Politburo itself seemed to have been reduced to the status of a ceremonial organization, being enlarged by the inclusion of the 15 first secretaries of the republics. It no longer had anyone in it with even a pretence to the making of state policy, and it now took on the visage of a sort of party think-tank. Gorbachev men such as Yakovlev, Shevardnadze, and Medvedev, who were no longer in the Politburo, continued on in the Presidential council. With his psuedo-Leninist rhetoric about a vanguard role, 'not in giving orders but in influencing minds', Gorbachev seemed to have caused the party to fold up its tent.

However, what the hard-liners lost in the Politburo, they gained in the Secretariat, now re-emerging as the main organ of daily control, if no longer of the entire state, then at least of party affairs. Its political colours were similar to those of the Russian Bureau set up in December 1989. A. S. Dzasokov, who replaced Medvedev as Secretary for Ideology, was rather more straight-laced than his predecessor. Falin and Gennadi Yanayev had the foreign affairs portfolios. Four other members of the Russian Bureau, Gidaspov, Kuptsov, Manayenkov, and Shenin, were included in the Secretariat. That body, meeting weekly, while the Politburo met only monthly, appeared to have made a comeback. It no longer maintained even the pretence of running the day-to-day affairs of the country, confining itself, it said, to offering policy options to the appropriate state bodies. One could say that the party emerged from the Congress much weakened, but that what remained of it was firmly in the hands of the hard-liners.

STAVROPOL

Gorbachev had again mastered the party. Not only had he gained assent to hobbling it still further, but he had forced it to recognize him as the leader in the move toward consolidation, which to most meant essentially drawing a line of defence between the bloc and the Union itself. One can only guess to what degree Gorbachev was sincere in this pledge. At any rate, he had crippled the party's potential objections to anything he might do in foreign policy. Only a few days after the Congress, Gorbachev met in Stavropol with Kohl for two days, after which Gorbachev finally dropped his objections to the new united Germany's membership in NATO. This had been the essential sticking point in the Two Plus Four talks of the foreign ministers held in May and June. Shevardnadze had proposed that special attention be given the 'external aspects' of unification, in particular a period of extended four-power rights. Both Kohl and Bush had quickly rejected any limitation on the full sovereignty of the new state. This was during the time when Gorbachev was still talking about German membership in both NATO and the Warsaw Pact; nevertheless he had conceded to Mitterrand on 25 May that Germany in NATO was 'one of the options'. At the end of June, Shevardnadze again proposed an extension of the four-power rights for five years, but Baker once more pushed the idea aside. However, Shevardnadze did get agreement to a five billion dollar loan that German commentators described frankly as a price for unification.

In his meeting with Kohl on 15–16 July, Gorbachev cleared up the remaining issues that were blocking the Two Plus Four talks. Germany was to be free to stay in NATO; Soviet troops were to have three to four years to return to the USSR, and were to be supported by a German fund to provide for their upkeep and to build housing for them in the USSR. Germany was to renounce the construction of nuclear weapons. There was agreement on two other points. The first was a treaty forgoing any further German claims to the lands east of the Oder–Neisse line now belonging to Poland. Kohl had refrained for an embarrassingly long time from renouncing any such claim, to the consternation of the Poles.

Not surprisingly, Poland was the only country in Europe whose public was decisively against German unification.[35] The second point must have been the promise of a bilateral treaty, for in September the two foreign ministers signed an undertaking that 'if one of the two states should become the target of aggression, then the other side will give the aggressor no military aid or other support'. Gorbachev could at least claim that a bilateral agreement of this kind was undoubtedly incompatible with Germany's NATO commitments.

In his first press conferences on his return from the July meeting in Stavropol, Kohl was asked repeatedly: was this a new Rapallo? His answer was of course an emphatic negative. But we might ask what anxieties were conjured up by this question? The reference was to the Russo-German treaty of 1922, uniting the two then pariah powers who considered themselves threatened by the essentially French security system established at Versailles. It was designed to be a key element in German recovery, in that it provided security against Poland and various aids to direct German rearmament. Prussian conservatives of the Bismarck tradition had considered it indispensable, recalling the Reinsurance Treaty of 1887–90, by which Bismarck had maintained a link to Russia despite the seemingly contradictory commitments of his alliance with Austria. In the early twenties, Bolsheviks such as Karl Radek liked to remind the German diplomats and generals of this fundamental relation, which Radek said had been revived by the Russian Revolution itself, at bottom, he liked to argue, a revolt against the imperialism of the British and French. The Rapallo pact, it was feared at the time, might be used against the states of east central Europe, then in the process of being organized by the French into a *cordon sanitaire* with the dual function of a hedge against the expansion of Bolshevism and a check on German claims in the east.

Did the question put to Kohl imply the possibility of future Soviet-German cooperation at the expense of the states between them? The very idea seems bizarre in view of the vast differences between the twenties and the present period, especially in the questionable suggestion of continuity in the 'Prussian' strategic aims of German foreign policy. Certainly Kohl had in effect promised not to intervene in the dispute over sovereignty of the Baltic republics, and left the Soviets with the temporary feeling that a line could be driven between the loss of the Bloc and the partition of the Soviet Union. In this limited sense, Stavropol was indeed a new Rapallo, and perhaps a kind of eastern Locarno.

But the implication of an analogy between the damage of Rapallo to Versailles and that of Stavropol to NATO was not so clear. Stavropol underlined the fundamental shift in the balance of power, hence also in the meaning of the NATO alliance, that had been brought about by German unity. Now that Russian power had collapsed in east central Europe, it was not easy to foresee how this region would figure in a new European balance.

It was perhaps with this anxious question in mind that suggestions, admittedly premature, were made in the West for the enlargement of NATO by the admission of Poland, Czechoslovakia, and Hungary. This would have made of NATO something far different from what its founders had envisaged. Where NATO had always been a nuclear alliance designed to deter a Soviet threat to Germany and the West, now it would have had a 'front line' on the river Bug. Yet it would also have to 'anchor' Germany. Was NATO now in the position of having to set up a general system like that of the period between the world wars? In the thirties that system eventually collapsed, and issues involving the eastern states were settled between Germany and Russia. Many Russians now thought that this was the way things had again turned out after the revolutions of 1989.

12 The Second Russian Revolution Gathers

One could congratulate Adolph Hitler posthumously on his glorious triumph over this country.

Aleksandr Nevzorov

The point is that [Gorbachev] made our country poor and contributed to our dissolution. At the same time he made your country stronger because he eliminated the danger.

Melor Sturua

Even when he speaks for the right wing, the country progresses to the left.

Andrei Sakharov

If Finland, if Poland, if the Ukraine break away from Russia, there is nothing bad about that. Anyone who says there is, is a chauvinist. It would be madness to continue the policy of Tsar Nicholas. . . . No nation can be free if it oppresses other nations.

Lenin

Stavropol closed the episode of *perestroika* in the former Soviet bloc. In the eyes of those who had been accusing Gorbachev of losing east Europe nothing had been gained from the entire chapter other than a non-aggression pact with a new German colossus fortified by support from NATO. For those who could remember what the USSR had enjoyed in 1985 this was bitter medicine. Yet they had worse to contemplate in the future, with the threatened disintegration of the multinational Soviet Union and a catastrophic economic collapse which promised civil war and worse. They resolved to 'reverse the course of events', as Suslov had once put it in 1980 in regard to actions to fight Poland's *Solidarity* movement. Just the same, despite the firmness with which Gorbachev had addressed them at the end of the 28th Congress, 'turning back with Gorbachev' still seemed to them the most hopeful possibility. Their eyes confronted the Gorbachev who vowed to make theirs a parliamentary party, yet these same eyes preferred to see the Gorbachev who described himself as a 'convinced Communist' committed to consolidation on the basis of the socialist choice.

He was no doubt perfectly sincere in professing a desire to preserve socialism and the integrity of the Soviet state, to draw a line between the fate of the Bloc parties and the CPSU. He regarded the Stavropol agreement as having secured German acceptance of the integrity of the USSR, as a kind of trade of German unity for the Baltic republics. But using force against Lithuania, even if it were to be successful, would amount to an admission that the same thing could have been done in the bloc in 1989, and that was the same as saying that everything since 1985 had been wrong. That was what his party opponents were now saying, but Gorbachev could not say it. In the face of the catastrophic results of his policies, he declared himself satisfied with the outcome as the dictate of history. In that case, would he press on with the transition to a market economy on the Western model, as the rapidly radicalizing democrats urged him to do? No, he could not do that either. Liquidating the old idea of the party was one thing, but liquidating socialism itself was quite another. He would defend the Union and the socialist choice by Political Means. It would all be voluntary. Even at the time it was not hard to see that the ultimate meaning of Gorbachev's Political Means was that the USSR would before long repeat the bloc experience of 1989.

In view of Gorbachev's now conscious and, one might say, even principled centrism, Yeltsin, who had no firm views of his own, as his autobiography amply demonstrated, went further 'left' with the marketizing democrats. Despite his militancy, he had up to 1989 given the impression of a man who believed in nothing beyond his own charisma. But, with Ligachev now removed as a foil, he had to oppose Gorbachev the man of the apparat. Usually this simply meant putting a minus where Gorbachev put a plus. Where Gorbachev upheld the idea that a leader's position must be in the centre, Yeltsin denounced wavering and half-measures. The unwavering and resolute measures implied by his criticisms were those that would create a market economy and a little Russia freed of her imperial burdens. So, in order to traverse the remaining distance to victory over Gorbachev, Yeltsin had to force the USSR to repeat the bloc experience of 1989.

TURNING BACK WITH GORBACHEV

The party hards were not at all comfortable with the post-Cold War position that Gorbachev staked out when Kuwait was invaded by Iraq on 2 August. Together with Bush, he pledged that Iraq's conquest would be reversed, even if that could not be done peacefully. To those who were upset about deserting Iraq, a former client, Gorbachev pointed out the differences between himself and Bush. He did not intend to send any units to the Gulf; and he agreed essentially with the Iraqi position on the linkage of the crisis to the Palestinian–Israeli dispute. Yet he insisted that the era in which the US and the Soviets could be sent to the brink of war over a Middle Eastern conflict, as they almost had in 1967, was definitely over. In some ways, his domestic position was improved by the fact that his opponents, in order to criticize this policy, now had to come up more prominently as supporters of Saddam Hussein.

At the same time, the logic of the abandonment of article 6 and the establishment of state institutions separate from the party led directly to the removal of the party from control over the armed forces and the KGB, as Yeltsin had advocated so forcefully. In June, while the Russian Communists were having their first Congress, the Congress of Peoples' Deputies declared, in a draft Decree on Power, the 'inadmissibility of political guidance by any party over law enforcement agencies, the KGB, the armed forces, and para-military formations'. When asked for his re-action to this, Polozkov said: 'My reaction is negative. Have you ever seen a de-politicized army, to say nothing of a de-politicized state security service?'[1] Yet the issue had been raised and the intentions of the democrats in parliament to drive out the party had been stated with painful clarity. Had they got their way, the real power of the party would have been broken. This was clearly understood by those who agitated for parliamentary control and, while they did not succeed at this point, they would continue the campaign.

Despite his fitful turn toward the policy of consolidation, Gorbachev was predictably attracted by the idea of expanding state (which to him meant personal) controls over the 'organs'. On 4 September, he issued a decree calling on the Defence and Interior ministries and the KGB to produce proposals to better align their organizations with 'state policy' in compliance with the earlier decision to end the Communist party's monopoly of power. A seemingly potent challenge to which there was a potent response! On the night of 9 September, several divisions in the Moscow area were put on alert and elements of two other divisions were flown into Ryazan by 30 transport aircraft, after which they were sent down the road to Moscow in full battle gear. The alert lasted for nine days. The explanations offered by ranking military people (that the troops were sent in to harvest potatoes, that they were practising for the parade scheduled for 7 November) were quickly discounted. An anti-Communist demonstration planned for 16 September was used as an excuse by Kryuchkov, Yazov, and others to impose the troop movements on Gorbachev.

Some claimed that he had agreed and joined in giving the order. This appears unlikely in view of the fact that he was in Helsinki on the ninth for summit talks on the Kuwait crisis with President Bush. Something similar had been done in February at a large Moscow demonstration when many thousands of troops had been brought in to keep order. This time the troop movements were timed to coincide with Gorbachev's absence, in time-honoured fashion. Whether Gorbachev was involved in the decision or not, he was in any case abruptly reminded where the real power lay. This demonstration of military might, as much as any other, confirmed Gorbachev's turn toward making common cause with the party hards.

In the event, nothing came of the decree ordering de-politicizing of the military and security organs. These remained under the supervision of the State and Law Department of the CPSU Central Committee. The parliament also had a Committee for Defence and State Security. Of its 38 members, seven were military men, 19 from military industry, two from the KGB, with five regional party secretaries. The committee was hand-picked by the same State and Law Department of the Central Committee, presumably in order to exercise self-supervision.[2] The feeling among the hard-liners was that the threat posed by the rise of state institutions could be parried effectively if the party could dominate them.

In addition to the national security apparatus and the Kuwait issue, Gorbachev also faced opposition on the matter of a transition to a market economy. The idea had gathered strength inside the party itself during the tumultuous events of 1989, and a powerful impression was made by the German elections in March 1990, with the parties virtually vying for the leadership of a movement toward the market. The Soviet Communists were committed to this project, at least on paper. No one could say for sure whether a 'market' actually meant capitalism or a 'social market economy', whatever that might mean. For the new parliamentarians, as for Gorbachev, it became another matter over which to waver and manoeuvre.

It was undeniably necessary to advocate something radical in view of the deepening food crisis. Ryzhkov complained that Soviet food and consumer goods were not seen in the stores because they were ending up in Poland and Czechoslovakia, and being sold there at much higher prices. Grain too was failing to reach the state granaries because it was being taken abroad by smugglers. On the other hand, supporters of a rapid Soviet reform leading to some sort of price parity were buoyed up by the experiment with rapid market transition, 'shock therapy', under way in Poland since January. Everyone had expected a social explosion in view of the harshness of the measures employed there, but the fact that the Poles did not rise up against the reforms bolstered enthusiasm for something similar in the Soviet Union.

This was certainly the Soviet reaction until June, when Lech Walesa broke with the Mazowiecki government, noting the spreading discontent and refusing to take further responsibility for 'shock therapy'. He declared 'permanent political war' on the government. There were two ways

for the Soviets to interpret this: on the one side, the rapid decline o'
Mazowiecki's popularity showed the political perils of shock therapy while
on the other, Walesa seemed, despite this, to be resolved to continue ir
the same vein. The Polish case seemed to be an encouragement for botl
sides in the Soviet debate over transition to the market. This would agair
be the case in November, when Walesa won the presidency. Mazowiecki
whose government had taken responsibility for economic policy, was hu
miliated by a miserable third place finish. How should the Soviets inter
pret this? Walesa's victory could be taken as an object lesson in the
predictable reaction of a new democracy to economic shock therapy – o
perhaps as proof that a strong leader, in this case, one 'with an axe is
his hand', as Walesa liked to put it, could alone guide the country to a
market economy.

In July the 'war of the banks' ensued, with Yeltsin trying to seize auth
ority over Union banks in Russia, and Gorbachev responding by decree
ing control for Gosbank until a new Union Treaty was drafted and passed
The issue of economic reform was more and more intimately connecter
to the issue of relations between the republics and the centre. It hac
been said many times that the economic crisis was exacerbated by the
disease of local autarchy. Now Yeltsin had become the unrelenting cham
pion of the most militant localism, acting from the 'locality' of Russia
with headquarters in Moscow. On his side, Gorbachev, who looked more
and more of a figurehead, in effect admitted repeatedly that the Sovie
Union as constituted was illegitimate, and in need of being newly foundec
by means of a Union Treaty. In a period of 'paradigm ecstasy' and ram
pant utopianism, the voluntary Union Treaty was perhaps the most utopias
idea. Gorbachev acted as if it would be the solution to all the nationalit
problems. He was now not only trying to assert control over financia
affairs but also over the millions of weapons that had seeped out of mili
tary armouries and into the hands of various nationalist extremists.

Despite, or perhaps because of, his weakness, radical economists presser
upon him plans for the rapid introduction of the market. Presidentia
Council member Shatalin, who had publicly professed his passion for mor
rapid reform while in the United States during Gorbachev's June visit
submitted an ambitious '500 Days' plan that envisaged the eventual transfe
out of state hands of most of the industrial enterprises and almost all of th
construction industry and retail trade. Ryzhkov tried to moderate Shatalin'
plan, arguing that it would cause a massive outbreak of strikes and othe
unrest. Yeltsin denounced Ryzhkov's moderation and called it 'trying t
mate a hedgehog with a snake'. Gorbachev split the difference and sheep
ishly declared that he had preferred the Shatalin plan. The presumer
efforts of Gorbachev to develop a compromise plan dragged on for month:
until it was finally accepted that the '500 days' idea was dead. During th

whole period, Gorbachev adhered to the centrist position, seeming to agree with the radicals and with their critics in almost equal proportion.

The supporters of market reforms foresaw severe difficulties. They cited three inhibiting factors: the popular collectivism of the Russian tradition, the experience of 70 years of influence by Soviet Communism, and the unattractive and, in terms of popular support, unsuccessful 'shock therapy' in Poland. Most judged a significant portion of the Soviet population to be opposed to a market economy. Yuri Afanasyev estimated that those speaking out against a market economy were 'not simply extremist conservatives, but make up almost a majority of our citizens'.[3] He found the country caught in a dangerous contradiction: on the one hand, the economy could not change fundamentally without breaking with the methods of administrative command, but on the other, the military-industrial complex could not be controlled without resort to those very methods. Many prominent democrats agreed with Klyamkin, Shatalin, Migranyan, and others who repeatedly insisted that a new authoritarianism was crucial to the transition to a market economy.

The opponents of the market economy also strove to win Gorbachev's soul. When the 500 Days plan had appeared in August, Ryzhkov commissioned a study of its possible consequences in respect to organized crime. For this he turned to Sergei Kurginyan and Vladimir Ovchinsky, two intellectuals who had enjoyed the patronage of Vorotnikov and Prokofiev since 1989. They and their associates in a think-tank known as the Experimental Creative Centre undertook the study, which concluded that the 500 Days plan would be an unprecedented catastrophe.[4] Privatization of capital goods would immediately raise the question of who would own the new enterprises. Ordinary Russians could not buy them. Soviet savings were very small in relation to the value of Soviet firms, except in the case of the 'mafias', whose cash holdings they estimated to be in excess of 900 billion rubles. Only these and foreign buyers would be able to participate in any real market economy, and they would immediately snatch up vast amounts of land and raw materials on the cheap. Inferior Western goods, many that failed to meet Western health standards, would flood the Soviet market, and the country would become a dumping ground for toxic wastes of every kind, tranforming the Russian countryside into an ecological disaster area. The new entrepreneurs would no longer be loyal to their own country, but instead would comprise a 'comprador bourgeoisie' acting essentially as intermediaries for foreigners. Those who advocated radical reforms, the Shatalins and Yavlinskys, thus served these masters and were, in effect, 'agents of imperialism'.

Instead of the transition advocated by those whom they called 'pseudo-democrats', Kurginyan and Ovchinsky urged an 'authoritaritarian modernization' like that of China, with an opening to market forces under

the strict control of state power. Imitating the Western economies would
only mean the ruthless exploitation of the country's natural resource
and perhaps even the destruction of the world ecological balance. In any
case, they argued, the truly dynamic countries of the future were no
those of the soulless West, but Japan, China, Korea, Singapore, and Cuba
where modern technology was not permitted to swallow up deeper spiri
tual traditions and values. Cuba's interest in a synthesis between Marxism
and liberation theology was, they thought, especially pertinent for the
USSR. Russia too must find the scheme of modernization most in tune
with its traditional spiritual culture. Gorbachev's reforms had been a step
in the right direction, and the guidance of the great leader was still indis
pensable, but now he must reshape the state to pursue a form of corpo
ratism, commercializing (rather than privatizing) a number of the industrial
ministries and transforming them into something like modern multi
national corporations in order to compete in the world market. The state
must nurture the development of high technology in the communication
and computer-related fields and keep them free of foreign control.

Thus the hard-liners, in opposing the abandonment of the country to
market forces, did not simply rely on invoking Stalinist precedent.[5] The
produced an ideological alternative to the House of Europe idea, and
they could now ground their complaints about the democrats' inroads on
collectivist traditions and practices in a conceptual model drawn in large
part from the most successful corporatist-statist economies in the world
Both they and their rivals, the prominent supporters of the market economy
had devised schemes dependent on a dramatic increase of Gorbachev'
power, indeed to dictatorial proportions. From his standpoint, the strug
gle for his soul could not have been more flattering. And in both the
divergent scenarios offered to him, he was cast as a heroic modernizer.

THE SUPREME CENTRIST

So it was not really a question of how to choose between them, but rather
how to use the alternative visions to enhance his own powers. This wa
indeed the method of Gorbachev's handling of economic reform plans
as would again be in evidence at the time of preparations of proposal
for Western aid from the G-7 London Economic Summit of July 1991
Gorbachev encourages the alternate plans, then casts about for his own
synthesis, while the plans are relegated to the archives. However, Gorbache
himself emerges from the process with enhanced powers. The Supreme
Soviet granted him emergency authority to stabilize the economy at the
end of September. There were rumours at this time of a confrontation
between Gorbachev and the military leaders who demanded more ener

getic action against the 'second economy' and the refractory republics, and there was at least the appearance of compliance on his part. Gorbachev's first decrees concerned a 'special work regime' for the railways, to accompany the appeals he had made to the leaders of the republics in September, while Moscow was under military alert, to 'redouble their efforts against the collapse of law and order'.

At the end of the month, he insisted to the Supreme Soviet that 'the transition to the market is directly linked to the political stability of our society'.[6] The parliament responded the following month by legislation attacking the 'shadow economy', to include measures against the sale of state-produced goods and price-fixing by traders of scarce goods. All the while Gorbachev insisted that he was acting in the name of Lenin's concept of the NEP as practised in the twenties. Against those comrades who refused to understand that Lenin's resort to NEP was 'not a tactical ploy', he described his own actions, not as a deviation from socialism, but a deviation from the Stalin–Brezhnev 'bureaucratic mimicry' of socialism, which itself was a deviation from the path chosen by the Revolution in 1917.[7]

Yet the collapse of the authority of Moscow was evident to the delegates of the Supreme Soviet who rose up against him in November. They complained that their legislation was ignored by the apparatus men in the republics and even in some cases by the centre as well. Gorbachev tried to satisfy them by drastic reorganization of the government. He disbanded the Presidential Council (replacing it by a Security Council)[8] and proposed that a Federation Council of all the republic Presidents coordinate action between Moscow and the republics. A Cabinet of Ministers was to replace the Council of Ministers, with the power of its Chairman, Ryzhkov, drastically reduced. Gorbachev proclaimed the need for stern action, in the face of the worsening food situation, to preclude famine and perhaps even civil war. 'Any emergency measures would be justified', he said, to ensure deliveries of cargo from the rural areas to the towns. The authorities must moreover respond to the 'justified demands for the strengthening of law and order'.[9] He pronounced himself ready to do what was necessary to defend against the division of the Union, and to this end called it imperative to bolster the strength of the armed forces and 'protect the prestige of those who guard the security of the state'.[10] 'In a word', he said, 'enough of sowing discord and panic.'

The authoritarian turn of Gorbachev's government became unmistakable even to his most passionate admirers in December. Minister of Defence Yazov addressed the nation on television on 28 November, accusing the authorities in the Baltic republics of 'grossly violating the constitution', and 'openly confronting the troops of the Baltic military district.'[11] He was referring to local measures permitting young men to avoid conscription, and moves in the direction of establishing republican armed forces. Accordingly, Gorbachev decreed on 1 December that any action in defence matters by the republics in conflict with Moscow was null and void.

The next day, Gorbachev removed Bakatin from his post as Interior Minister, replacing him with Boris Pugo, who had headed the Latvian KGB in 1980–84. Colonel-General Boris Gromov, a critic of Gorbachev's weakness on matters involving internal order, became his deputy. The impression was strong that measures would soon be taken against the refractory republics. But it was unlikely that they would be attacked all at once; the army could not stand that kind of strain. More likely, one or another of them would be subdued and brought under central mandate as a lesson to the others. Would it be one of the Baltics, or would it be Moldova, where the central power was supporting the claims of a 'Trans-Dniester' republic? Or perhaps Georgia, where Moscow was supporting the South Ossetians' efforts to resist the Georgian nationalists?

The last possibility may have prompted the resignation of Shevardnadze from the foreign ministry on 20 December, with a farewell speech warning against the 'onset of dictatorship'. He pronounced the democratic forces 'scattered'. He defended his policy on Kuwait against the misinformation spread by the 'boys with colonels' stripes'. The latter reference was to the hostile attacks in parliament directed against him by leaders of the *Soyuz* caucus, in particular, Viktor Alksnis, Nikolai Petrushenko, Yuri Blokhin, and Yevgeni Kogan. These were officers in the military comissariat who had formed the caucus in the first months of the year in response to the decision to abolish article 6. They had started with 103 members, but their ranks had grown by this time to some 500, around one-quarter of the Congress of Deputies. They made special pleas on behalf of the 70 million Soviet citizens who lived outside their 'own' territories.[12] They had launched the attack against Shevardnadze for his Kuwait policy in October, and claimed credit as well for the removal of Bakatin. Between the democrats and the proponents of the 'old way', they claimed to represent a 'third force'.

In his resignation speech, Shevardnadze asked the question: Who was behind these colonels? Perhaps it was those who had introduced a petition, a 'Letter of the Fifty-Three', in the Congress of Peoples' Deputies the day before his speech. Signed by a group of hard-liners that included Generals Mikhail Moiseyev and Valentin Varennikov, and the Patriarch Aleksei II, it called for Gorbachev to use emergency powers to rein in the secessionist republics. The Communist Party's January plenum indicated further that the colonels were backed not merely by high officials in the Army, KGB, and MPA, but by a full-scale campaign for a party revival against all the reforms of the Gorbachev era.

As with the campaign for the formation of a Russian Communist Party the previous spring, the initiative apparently came from Leningrad. In an article for *Pravda* before the plenum, Gidaspov argued that the party had gone awry since it strayed from the path of *uskorenie* in 1985–87; the country had needed scientific and technical progress rather than a dismantling of its entire planning apparatus. The Politburo had 'disappeared', and the media was in the hands of those who wanted to restore capitalism. A counter-offensive was imperative.[13] Speeches by Ivashko, Dzasokhov, and Shenin repeated the theme. The party should not renounce control over the economy. Collective property must have priority over private. The party centre must control the party periphery. The prior importance of class perspectives over those of Universal Human Values was crucial. These themes were stressed by Polozkov, and later underlined in an article for *Sovetskaia Rossiia* by Ligachev himself.

Yet even in this welter of resolute talk, there were waverings. Shenin's call for centralized party control made him in effect a rival of Gorbachev for the post of general secretary. He would be a key figure in the coup attempt in August. But some of the regional secretaries, despite the fact that it was they who had suffered most from the chaos in the centre, could not help but enjoy their gain of status in the new, albeit weakened, Politburo. They wondered about re-centralization. Moscow chief Prokofiev also showed himself a centrist, as had been increasingly evident through the last year. He spoke of a division in the Politburo. He spoke of the necessity of a market economy, as outlined in party documents for at least a year, but one arrived at through authoritarian means. As models he mentioned South Korea, 'and I do not hesitate even to say, Chile'.

Prokofiev's attitude expressed, among other things, the core of consensus in the two streams of economic reform thought. Migranyan, Klyamkin, and many others had called for westernizing the Soviet economy under a dictatorship on behalf of civil society; Kurginyan, Ovchinsky, and their school spoke of a corporatist 'Eurasian' alternative preserving the organic social unity of Russia and the Union. Prokofiev could see, as could Gorbachev, that the two scenarios, both requiring what the parliamentarian Sergei Stankevich called 'the idea of authoritarian modernization', were not divided by a great chasm. It was not at all as the famous quip had it, that 'you can't cross a chasm in two leaps'; rather this was a chasm that one could walk across, or as Gorbachev was doing, stand astride.

DRESS REHEARSAL AT VILNIUS

The division on policy toward Kuwait was rather more sharp, but in general geopolitical outlines, it mirrored the competing domestic visions. Shevardnadze and his supporters were wagering everything on a closer relationship with the West, which implied a deterioration in relations with Cuba, North Korea, Vietnam, and even perhaps eventually China, as well as Libya, Ethiopia, Yemen, and the PLO. Against them was the persistent critique of the 'Arabists', with Primakov at their head, arguing that the Soviet Union must stake out an independent position in the Kuwait crisis or forfeit its status as a superpower. Their view was that, despite the humiliations of the last two years, the USSR had no alternative to playing this role, since it was the only power in possession of all the military requisites, including nuclear weapons. It had to because it had in the past and because it could still do so now. They railed at being excluded from the decisions that had ranged the USSR on the side of the UN and the United States in opposing the Iraqi annexation of Kuwait.

However, as the crisis came to a head and the war began, it became clear that Gorbachev had characteristically decided to pursue both policy lines, backing the United States and its allies in the use of force, but still manoeuvring to make a peace settlement in the hope of preserving something of a position of honest broker in the Middle East.

Among the advantages of this dualistic approach, as against that of trying to protect Iraq from its fate, was the fact that the West would, it was thought, now be preoccupied with the Middle East for an extended period, so that Moscow could impose itself on Lithuania and the other Baltic republics while world opinion was distracted. The analogy with October–November 1956, when the Western conflict over Suez competed for headlines with Soviet intervention in Hungary, was repeatedly invoked in the Western press as the war approached. Even so, full-scale combat in the Gulf necessarily reduced the importance in Western minds of Soviet action in the Baltics.

Soviet moves against Lithuania actually began well before war broke out in the Gulf on 17 January. OMON 'Black Berets',[14] special Interior Ministry troops, seized previously nationalized Communist party property in Riga and Vilnius. On 7 January paratroopers entered the three Baltic republics, as well as Armenia, Georgia, and Moldova, and parts of the Ukraine to round up deserters and draft dodgers. Three days later, Gorbachev charged, in language similar to that used at the end of November by Yazov, that there had been flagrant violations of the constitution in Lithuania. 'Slogans of democracy', he said indignantly, were being used to cover for a policy 'aimed at restoring a bourgeois system.'

Up to this point, no one would have doubted that the intervention was on Gorbachev's order. On 13 January, MVD troops had to plough through thousands of people trying to block them from taking the Vilnius television centre. The centre was then seized by Group Alpha of the Seventh Directorate of the KGB. This was the same organization that had spearheaded the Soviet intervention in Afghanistan in 1979 by arresting and killing President Hafizullah Amin. It would be ordered, unsuccessfully, to carry out a similar mission in respect of the White House and Boris Yeltsin in August. And it would carry out the same Yeltsin orders with a massacre of his parliamentary opponents in the White House in October 1993. Sending Group Alpha against the Vilnius television station was a little like Tiananmen Square. It was a bloody action. There were 13 killed and 230 injured.

The next day Gorbachev tried to disavow responsibility, claiming that the attack had been carried out without his prior knowledge. This was not as ridiculous as some thought at the time. He may not have known every detail of the military plans. Nevertheless, even when he spoke later of how much the deaths in Vilnius had upset him, he continued vehemently to blame the consequences of the intervention on the leaderships of the Baltic republics.

There was reason for Gorbachev to think that a firm effort in the Baltics to save the Union would be appreciated. When Lithuania had first declared independence in March 1990, the Congress of Peoples Deputies had backed Gorbachev in his efforts, ultimately successful, to pressure the Lithuanians to suspend the declaration. The use of troops, however, had a far different effect. Yeltsin issued a public appeal to the troops to disobey orders. The intellectuals who had taken part in the *glasnost* campaign rose up in near-unanimity to condemn the action as criminal, a new 'Bloody Sunday'. In party history, the suppressions of worker protests by armed force on 9 January 1905, 'Bloody Sunday', marked the last time the workers would march under icons and pictures of the Tsar, and the beginning of their radicalization. Now the same moment had arrived, they said, for 'humane socialism, the New Thinking, and the House of Europe'.[15] Mournful articles announced the turning in of party cards. An opinion poll conducted in Moscow and Leningrad on 16 January showed similar sentiments: 74 per cent condemned the military action, including even 46 per cent of those who described themselves as party 'hard-liners'.[16] A massive Moscow rally of 100 000 on 20 January echoed the protest.

Gorbachev, who was ready on 15 January to follow up the action by suspending the liberal press law, relented on the 22nd. He declared himself to be shocked and upset at the killings. On the 30th he announced that the special units would be withdrawn. The coup in the Baltics was wound down and turned into a steady, almost unpublicized, series of minor skirmishes over the manning of border posts and the guarding of airports. At the time of the September troop movements, there had been talk of a plan for a coup that would begin in the Soviet far east and spread toward the capitals, testing the waters to see what the reaction was, so that it might be cancelled if necessary. The Baltic actions bore the earmarks of something similar.

Gorbachev again scurried for an intermediate position. Months later, Shevardnadze would be claiming that neither Gorbachev nor he had known anything of the plans for the action. Perhaps this means the plans were drawn up after Shevardnadze's resignation at Christmas.[17] They were at any rate probably in the nature of a reconnaissance in force, to gauge the strength of the resistance, which in the event proved to be considerable. Gorbachev had to retreat, but he did not change his mind about independence for the Baltics, as had most of the intellectuals and the bulk of urban public opinion.

Lenin once said that the revolution of 1905 was a 'dress rehearsal' for 1917. Vilnius was a dress rehearsal for the coup of August 1991. It showed clearly the willingness of the leaders of the security organs to use force, the ability of Yeltsin to call with apparent impunity for troops to disobey orders, and the general outrage from the intelligentsia and the public with which anyone attempting a coup of this sort would be faced. The troops had carried out their orders, but they were angered by the government's and even their own higher officers' disavowals of their actions. Bitter and humiliated, some denounced the wavering of those who gave them orders, and many must have concluded that failure to carry out such orders in the future would not be the worst eventuality.

KUWAIT AND 'DEFENSIVE DEFENCE'

While this was going on, the American air war in the Gulf was under way. Despite the formidable damage it inflicted in the first weeks, it was not forcing an Iraqi surrender and was not proving to be a substitute for a ground war. In this fact, Gorbachev thought he saw a small opportunity to reassert Soviet capacities as a great power. He dispatched Primakov to Baghdad on 14 February to secure Saddam Hussein's agreement to withdraw from Kuwait in return for lifting all UN sanctions. Had this been permitted to stand, Gorbachev could conceivably have posed as a peacemaker and defender of Arab interests. German Foreign Minister Hans-Dietrich Genscher complicated the picture by urging the West to trust Gorbachev.

While Primakov wondered aloud about Saddam Hussein's 'Masada complex', he nevertheless asserted that 'The Soviet flag has been shown, and it is being perceived very positively. We are a superpower; we have our own line, our own politics.' When he was asked, on the 'Nightline' television programme, why the Soviets had not cleared their 18 February peace plan with the US, he replied: 'If people believe that we have to report everything to the United States and do not have an individual line, then this is a mistake. The Soviet Union has not been reduced to the status of a secondary power.'[18]

The whole episode of the Iraqi seizure of Kuwait would have been turned to Soviet advantage. If a decision were not achieved soon, Gorbachev would have made the Gulf War his own diplomatic triumph. However, the point was that the Soviet diplomatic initiatives had made the American ground war inevitable. Bush seized control of the situation by his ultimatum to Iraq on the 22nd, giving it 24 hours to initiate withdrawal. Saddam Hussein could not bring himself to comply, and the ground war proceeded. Bush had refused to permit Gorbachev to steal this moment and to establish himself as the key figure in Middle Eastern affairs, in effect a mediator between the United States and the world. Thus the world was reminded of the decline of Soviet influence in the Middle East and its subordinate position in the post-Cold War order.

Coming on the heels of the loss of the bloc, and the unification of Germany in NATO, the Kuwait war resulted in a 'third crisis of the New Thinking', as one Soviet observer described it. The loss of international status was dramatic. Soviet influence in the Arab world could no longer be considered even a small counter to the military projection of the United States. The manner in which Bush had brushed aside Gorbachev's peace manoeuvres gave another example of how absurdly ineffectual were Gorbachev's 'political means' of influencing events against the wishes of a real power. Soviet military people had to admit that there was now only one superpower. Marshal Oleg Losik, an armour officer, estimated that the United States was now twice as powerful as the USSR. Soviet prestige in the Arab world plunged. Arab nationalists were incensed by the continued release of Soviet Jews for resettlement on the West Bank. There was a powerful echo of pro-Iraq sentiment in Soviet Central Asia, and disappointment with Moscow's seeming impotence.

The apparent military lessons were harsher than at the time of the defeat of Soviet arms in Lebanon in 1982. Soviet military officials spoke openly of the demonstrated inadequacy of Soviet air defence equipment against the electronic counter-measures employed by the United States. General Yazov was optimistic in saying that Soviet gear could only continue to suffice 'for two or three years'. The Academy of the General Staff commissioned a special study of the war, which concluded that surprise, air superiority, the use of space-based systems, and advanced conventional munitions all pointed to the advent of a new era of 'air-space war'.[19] The arguments of General Ogarkov, so roughly overridden in the development of the New Thinking, now appeared to be vindicated with a vengeance. The undeniable superiority of the offensive seemed a stinging rebuke to the defensive doctrine that had occupied centre stage since 1987.

GORBACHEV TURNS TO YELTSIN

Not everyone in Soviet public life still considered these matters worthy of concern, in view of the expectation of an approaching coup or civil war, accepted, almost fatally, by even the most militant of the democrats. Gorbachev's turn toward the line of consolidation had presaged for them the ascendancy of the military-industrial complex, eager to close the book on the era of the Gorbachev reforms. Leading democrats wanted a last stand against this presumed impending coup, but the population at large, while it was willing to support the democrats, retained little faith in the promises of *perestroika* as advertised for the last five years. An all-Union poll at the end of March asked the question: What is *perestroika*? Seventeen per cent defined it as the start of democracy or the transformation of society, while an equal number called it 'a word used to disguise the struggle for power among leaders'. Another 18 per cent called it 'an attempt to maintain power by the rulers at the expense of a certain democracy', while 14 per cent said it was an 'obsolete slogan which has outlived itself'. Another question was: 'Would you have supported the changes now under way in the country back in 1985 when they started, had you possessed your present knowledge of these changes?' Those answering 'no' were 38 per cent, while those answering yes were 23 per cent.[20]

Gorbachev himself was not prepared to say that it was time for 'post-*perestroika*', despite his turn toward the line of consolidation. He still strove to be the man of the golden mean, albeit one who defined that mean in a way more congenial to Kryuchkov, Yazov, and Pugo, the disciples of order. He had spoken of the revolutionary tasks of his reforms since 1988, but now he seemed to mean a revolutionary situation in which the Soviet state was actually reconsidering the alternatives of 1917. He warned that many of the democrats who now advocated 'neo-Bolshevik' methods, that is strikes and popular demonstrations of all sorts, were courting disaster. A coup from their quarter would certainly bring on a civil war. This was the message of his speech in March in Belorussia, ostensibly aimed at the democrats and mentioning a number of them by name.[21] But the message was meant to be heard on the other side as well, as a warning to hard-liners that they could not consolidate without him.

At the same time he asserted his opinion that the central differences between him and Yeltsin were now quite clear: he was for the preservation of the Soviet Union and the socialist choice, while Yeltsin was opposed. If one were to judge only by the statements made by these two men in March and April, this would seem to be perfectly accurate. In the perspective of the last four years, however, since Yeltsin first ran into trouble with Yegor Ligachev in the summer of 1987, one could see that this was only a brief moment of conceptual and political clarity for both Yeltsin and Gorbachev, neither of whom held any position for which he was willing to suffer the slightest loss of personal leverage.

Gorbachev found himself momentarily at the head of a movement for consolidation in international policy. The Foreign Ministry was trying to make up for the collapse of the Warsaw Pact by a series of bilateral treaties with the states of the former bloc, rather like Stalin's and Litvinov's non-aggression pacts with the Baltic states, Poland and Romania in 1929–31.[22] The Soviets of the earlier period had been eager to encourage the weakening of the French *cordon sanitaire*, which they achieved finally by a pact with France itself in 1932. The comparison shows how far Soviet power had declined. Far from threatening these smaller countries, the Soviets were now fearfully trying to prevent them from becoming a base for action against the USSR.

This also applied to Russia's southern neighbours. Gorbachev tried to take advantage of Iran's worries about its own Azeri minority, and Turkey's hostility to Armenian ambitions, by offering them pacts. Eventually Romania pledged not to encourage irredentist pro-Romanian sentiment in Moldova, on the understanding that the Soviets would not offer sympathy to Hungary about Romanian Transylvania, where there is a sizable Hungarian minority.

All the neighbours of the Soviet Union had to recognize the onset of a harder line. This did not prevent Iceland, Holland, Denmark, and the entire Nordic Council from enthusiastically encouraging the independence movements in the Baltic republics. A new Soviet intransigeance was also manifest. It could be seen in Soviet stiffening on the timetable for troop withdrawals from the former bloc, contravention of the letter of the November Conventional Forces in Europe agreement, resumption of militiary support to the MPLA in Angola. Yet there also seemed to be a sense that these signs of life were only the result of a decaying and fragmented Soviet political and military command structure, and that in the end, as long as Gorbachev was on hand, the general debility would not be reversed.

Gorbachev spoke furiously against the collapse of the Union, yet it was not even certain that he was directing the ongoing military actions to prevent it. Protracted low-intensity military operations continued against the customs posts in the Baltic republics. Soviet troops and arms accumulated in South Ossetia. Soviet helicopters aided Azeri forces in their attacks on Armenian villages in the strip of land between Nagorno-Karabakh and the Armenian border. The army and the internal security forces were apparently firmly under party control, while it was not clear whether they were under Gorbachev's. About the state of their morale, there were dark speculations.

It was alarming for the hard-liners to contemplate that troops might have to be used to quell a wave of strikes throughout the country. In March and April the miners came out again demanding, among other things, the resignation of Gorbachev. This time they were joined by a general strike in Minsk, and a threatened all-Union political general strike, all directed against Gorbachev. The President-General Secretary gloomily contemplated the menace of a civil war in Belorussia with its concentration of strategic aircraft bases and missile launch sites.

Yet, ever the optimist, he promised a renewal of the Union on the basis of a Union treaty, to be given a general mandate, he hoped, in the 17 March referendum on what he called 'the preservation of our thousand-year-old state'. The results of this vote were generally in line with expectations. Among the 80 per cent who voted, 76 per cent favoured the Union. The six refractory republics replied negatively in their various ways: the Baltics and Georgia held their own 'counter-referenda'; Moldova declared a boycott, largely successful; Armenia scheduled a legal referendum for the fall. The others endorsed the Union, most strongly in the Central Asian republics. But there was not very impressive support in the big Russian cities, a suggestion that the recent strikes had caused a certain disaffection among urban workers. The western Ukraine voted overwhelmingly against. Three *oblast*s, Lvov, Tarnopol, and Ivanovo-Frankovsk, had less than 20 per cent favourable. These were the historic regions of Galicia and Volhynia, Catholic rather than Orthodox in confessional tradition, whose people had lived under Habsburg rule until 1919, and then under Polish, but had only been Soviet since 1945.

Was the referendum a victory for Gorbachev? He had not been confronted with any major unpleasant surprises, and he could now claim a mandate for his efforts to draft a new Union Treaty. The hard-liners in the Supreme Soviet took the vote as a mandate to strengthen discipline by cracking down on nationalists and separatists. Strictly speaking, the procurator general could have brought up on charges any republic president who refused to hold the referendum. There were many violations of the constitution with which the nationalist leaders could be charged, but Gorbachev was overseeing the writing of a new constitution anyhow. Speaking for the *Soyuz* caucus, Petrushenko claimed that the time had come for 'bold, determined and consistent action', and said of Gorbachev that 'History will not forgive him if he fails to use this opportunity.'

But there was no bold and determined action. Gorbachev was tacking toward the harder line, but he appeared only to want to do for the Union what could be done by 'political means'. By contrast Alksnis, in the pose of one who thought nothing of taking on a civil war in the country, referred to Gorbachev contemptuously as a 'follower of Ghandhi', who did not accept the role of force in politics. Gorbachev's hard-line critics were fortunate in that it was not they who were called upon to do what they proposed. Nevertheless, their disaffection, added to that of the democrats whom Gorbachev had been denouncing for some six months by this time, created a broad front of those who, for their different reasons, had come to feel that they had had enough of him.

It was a broad front but hardly a coherent one. Alksnis was the most audible voice among the hard-liners, but no one could be sure what he intended for the post-*perestroika*. In an interview with *Moscow News* in February, he had added to the mystery by an amazing declaration that he was independent of 'Marshal Yazov and the whole Politburo' and even his own apparatus of political officers in the armed forces.[23] He advocated a National Salvation committee like that which had appeared in Lithuania in January, 'but not one designed to save socialism'. Instead, it would ban all parties, 'the Communist Party included', and introduce a market economy, 'forcefully if need be'. The army he said, wanted a single state, 'neither capitalist nor socialist'.

The interview was so bizarre and disturbing that Aleksandr Nevzorov, on his television show, tried to deny that it had ever taken place, only to have Alksnis himself verify its contents a few days later. To judge by what Alksnis had said, he was not a Communist fundamentalist, but a colonel with pretensions to the role of a man on horseback. Western speculation was astray in the assumption that Gorbachev's opposition was made up exclusively of diehard Stalinists. Many of his opponents, having endorsed the course of reform, but not its disastrous results, now centralized their thoughts and reduced things to the simple complaint that they could do it better. If the issue of the post-*perestroika* was not clear, neither was the issue of post-Gorbachev. Speculation on his replacement centred on Anatoli Lukyanov, Gorbachev's close friend and former protégé. This gives a hint of the thinking of Gorbachev's opponents, who had up to now never been able to contemplate ruling without him, and whose opposition had always been directed to the purpose of getting him under control. The difference between Gorbachev and Lukyanov apparently consisted in the greater prospect for their control over the latter. So now they sought finally to have done with Gorbachev at the April Central Committee plenum.

Gorbachev's hard line, toward which he had been sidling fitfully at least since his ultimata to Lithuania after the independence declaration of 11 March 1990, was now exhausted, in that it was no longer bolstering his personal power. So he proceeded to discard it like an old coat. Indeed, now he could only save himself by another abrupt turn. On the first day of the plenum, 25 April, which was widely expected to see his ouster, he had to sit and listen to a series of harsh criticisms of his weaknesses, centring on his inability to apply force to the solution of the problems of Nagorno-Karabakh and South Ossetia. While the discussion was going on, MVD troops were attacking airfields and other public buildings throughout Lithuania.

But Gorbachev was again too nimble for his enemies. He confronted them with an agreement that he had signed with Yeltsin and the other leaders of the nine republics who had voted for the Union on 17 March, arranged at his dacha at Novo-Ogaryevo the day before the plenum convened. Dubbed the Nine Plus One agreement, it pledged a new Union Treaty that would abolish all the state bodies based on the old constitution. There were guarantees of the transfer of responsibility for much state property from the centre to the republics, including transfer of the striking Vorkuta mines to the Russian republic. Gorbachev had signed on for what he called a new federal Union.

The newly emergent republic-level personalities who had been elevated to an entitled seat on the Politburo at the Twenty-Eighth Congress now found that the country was in effect to be partitioned and that they were to be heads of sovereign states. This won them over to Gorbachev's side. The Politburo came before the Central Committee to urge confidence in him. People such as Kazakhstan President Nursultan Nazarbaev, who had been attacking his weakness a few weeks before, underwent a sudden conversion and spoke up on his behalf at the second day's session of the plenum. By June, Nazarbaev would be promoting a '15 plus 0' arrangement for negotiations among the republics without any central participation. The fact that the Central Asian republics were the only remaining bastions of Communist control improved Nazarbaev's personal leverage immensely. In economic affairs, moreover, he favoured a conception remarkably similar to the post-*perestroika* plans of Kurginyan for an eastern model of corporatism.

At the same time, the anti-Gorbachev delegates also had to reckon with a petition calling for a new party, with 150 signatures gathered by a newly created faction, the Movement for Democratic Reforms. Its central figure was Colonel Aleksandr Rutskoi, not a liberal but until recently a fierce Russian nationalist who had experienced an epiphany since failing to win a seat to the parliament in 1989. Likening the Vilnius events of January to those in Tbilisi in 1989, he had concluded that Gorbachev's 'leadership' consisted in blaming his errors by turns on the democrats and on the army, and that only Yeltsin was capable of stopping the bloodshed between the centre and the nationalities. His supporters in the Supreme Soviet were called 'Communists for Democracy'. Those who held a good opinion of him and Yeltsin, or who were developing one, rallied to the defence of Gorbachev. The vote for retaining him as 'gensek' was 332–13.

THE LAST TURN

Gorbachev's manoeuvres had again prevailed: this time a bloc with the 'democrats' and what he now called the 'conservatives' (Rutskoi) against the 'reactionaries'. Moreover, Yeltsin, who only days ago had been divided from Gorbachev by a line of blood – the issues, they both agreed, were no less than the defence of socialism and the Union – was now an ally. Thus ended Gorbachev's period of communion with the hard-liners, and with it the central fiction of the turn: that a line could be drawn between liquidation of the Soviet bloc and liquidation of the Soviet Union. The logic of Gorbachev's struggle for liquidation was that his power, since it only 'increased' with liquidation, could not be built by consolidation. Moreover the liquidation must continue as long as Gorbachev could fight.

Yeltsin had made no comparable sacrifice for this bloc. He professed himself overjoyed with the Nine Plus One deal because it meant that the nine were now 'sovereign states'. Yeltsin saw the partition of the country more clearly than his supporters. There was a pervasive suspicion that these deals were not worth more than the paper they were written on. Yeltsin's friends in the 'Democratic Russia' caucus in the Russian federal parliament actually thought he had yielded too much to the centre. But Yeltsin saw the opportunities for his own parliamentary action that were thus opened up. He may also have had a vision of Gorbachev ending up like Poland's General Jaruzelski, sitting up in the high throne in the Polish Sejm watching the democrats and the ex-Communists liquidating all his former power.

Had Yeltsin got it right? Did the Nine Plus One agreement mean the end of the old Soviet Union, the CPSU, the gensek, and the rest, and the beginning of a new Russian state led by Yeltsin? The six refractory republics appeared to be defined out of the country, and the remainder of the Soviet Union would be partitioned, not for the last time. Could Gorbachev have believed that he was solving the problem? The republics given their internal independence within the Union were, in the drift of things, bound to impose more severe linguistic rules for commerce and administration. Russians would be increasingly discriminated against, as they had been in the Baltics. At some point one could imagine several million Russians deprived of full citizenship and driven out, that is, back into the Russian republic. After that, the republics could become more genuinely sovereign, casting off the Soviet skin. The processes deliberately set loose by Gorbachev in 1987, ostensibly to bolster *perestroika*, might not even yet have reached their limit.

Gorbachev in any case acted like a man pulled from a stormy sea. He chided the 'reactionaries' for their lack of vision and their dogmatism. He declared himself open to reconsideration of vast plans for economic reform, on the order of the discarded '500 Days'. Grigori Yavlinsky was dispatched to the United States to work out another version of the 500 Days with a group of Harvard economists and political scientists, a plan designed to cohere with the IMF–World Bank report on the Soviet economy and therefore to appeal to the leading industrial nations' G-7 economic summit in July. Nothing was official on either side, but the Soviet–American team was proposing annual aid of $30 billion, Soviet membership in the IMF and the World Bank, and the end of restrictions on trade, in return for Soviet market reforms on the Polish model. In particular, the Soviet economists were hoping for a sizable fund to back the transition of the ruble to full convertibility.

Soviet Prime Minister Valentin Pavlov dismissed this play for Western aid as a pipe dream. He countered by asking the Supreme Soviet for extensive new powers to enact his own economic legislation. Pavlov warned that nothing would be forthcoming from the West, and that his own measures would have to be in the spirit of the acts made since he took office earlier in the year. Price increases and other actions had been promoted by Pavlov to remove the excess cash that had accumulated as a result of the big wage increases voted by workers since 1988 when they got the right to elect their managers. Pavlov had accompanied his measures with charges of a Western 'bank plot' to corner billions of rubles and dominate the Soviet economy. This had been in line with the prognostications and warnings of Kurginyan and his staff. Pavlov was played up in the Western press as an enemy of reform, but in fact he did not oppose a transition to a kind of market economy. He was only trying to administer what democrat critics called 'shock without the therapy', that is, preparing the ground for markets and convertibility while retaining full fiscal powers to control the process from above. Now he wanted to take these powers out of Gorbachev's hands.

Gorbachev and his allies quickly defeated Pavlov's move in the parliament. Gorbachev angrily declared himself fed up with the machinations of the 'reactionaries', and declared defiantly that they could do their worst, but they did not frighten him. Did that mean that Gorbachev had made another, this time a decisive turn, and that he now endorsed massive market reform on the order of the 500 Days or the Yavlinsky–Harvard 'Grand Bargain'? One might have thought so, especially one who had not seen Gorbachev in action over the last five years. But in fact Gorbachev was even now characteristically trying to synthesize the two approaches, Yavlinsky and Pavlov.

Western economic writers who were thrilled by the vistas of the Yavlinsky programme, and who saw in him a worthy Russian counterpart to market-minded finance ministers in the former bloc, men such as Balcerowicz in Poland or Klaus in Czechoslovakia, were likely to overlook the similarity, despite the accompanying rhetoric, of the orientations of Yavlinsky and Pavlov. That is, they were so impressed by vistas of a Russian El Dorado that they tended to discount the fact that the USSR of the Pavlovs was also committed to moving toward a market with, as Pavlov had said many times, 'various forms of ownership'. The argument between the two sides in this struggle for economic policy was really an argument over pacing, with competing judgements, based on the events in Poland and other former bloc economies, of society's reaction to these extreme and in many ways draconian reforms.

In June Yeltsin, with Rutskoi as running mate, was elected to the Presidency of Russia with 58 per cent of the vote. Acting like a President of a sovereign country, Yeltsin let it be known that he would soon put in place a Russian Defence and National Security Committee and a Russian KGB. How could this be squared with the idea that under the Union Treaty such matters would be reserved for the centre? This struck at the heart of Gorbachev's power in the proposed new Union. But he acted, characteristically, as if there were no problem at all. Gorbachev did his best to share the limelight with Yeltsin, and television showed the two accepting congratulations at elaborate ceremonies.

Gorbachev was of course no Yeltsin. He had never been popularly elected to anything. By the drift of things it now appeared that Gorbachev, having accepted the democratic definition of legitimacy, had to do his best to become Yeltsin's sort of politician, thus the better to put himself in a position where he could eventually be popularly elected as President of the Soviet Union. What was Yeltsin's sort of politician? Well, no one would argue that Yeltsin had got his way because of adherence to a programme. He had only embraced a Little Russia and a market economy in the course of the last year. His popularity was gained by virtue of being the enemy of Ligachev, the opponent of the apparatus. Yeltsin had this much in common with Lech Walesa or Vaclav Havel. Since the country was following in the path cut out by the revolutions in the former bloc, which despite their being led by Communists, were anti-Communist revolutions, the only real way to gain votes was through anti-Communist gestures and acts.

So Gorbachev now tried to regain his own reputation as an opponent of the apparatus. He hailed the appearance of the Movement for Democratic Reform, especially when Shevardnadze, Yakovlev, Popov, and Sobchak declared their expectation that the movement would soon evolve into a new political party. If Gorbachev could only depart the Communist party and take the lead of this party with its roster of leaders identified with reforms, some of whom had won elections, then he himself might also win an election – if of course he were lucky enough to run against one or several of those pathetic apparatus candidates, perhaps a

Ryzhkov, or a Makashov. That had certainly been Yeltsin's trick, running against the CPSU, one which could be done by others. This kind of reasoning may have grasped the essential negative dynamic of the revolutions of 1989, the struggle against the Communist apparatus men. Since this struggle was not in the name of any particular economic or social programme, the programmes enacted by the elected regimes had a certain period of grace with which to conduct economic experiments. If these should come up short, or be especially painful, more space and time could be gained by deepening the revolution against the remnants of the *ancien régime*. Like the Jacobins of the French Revolution, they could for a time at least, offer to make a pair of shoes out of the skin of an aristocrat, or in this case, an ex-Communist.

GRINDING DOWN 'THOSE WHO CANNOT KEEP UP'

Paradoxically, while Gorbachev delivered blow after blow to the Communist party, he still clung to old and now quite outmoded reflexes of Soviet interests in Europe. When the Yugoslav civil war broke out in July, with the Yugoslav Federal Army attacking to secure customs posts in Slovenia, Gorbachev reacted to the German economic and political intervention on Slovenia's behalf by warning against the 'internationalization' of the conflict. Soviet generals no doubt watched Yugoslavia's nascent civil war with their own situation in mind.

The Brioni accords, by means of which EC diplomats effected a momentary compromise in a presumed Yugoslav internal matter, were precisely such an 'internationalization', and another measure of the shift in the balance of power as a result of the Soviet decline. Events anywhere in Europe were now the business of federated Europe. Gorbachev still tried to project an image of his country's importance, despite its discomfiture. He came to the London G-7 economic summit, not with the Grand Bargain to offer, but with warnings about the collapse of the USSR into civil war in the absence of major Western financial assistance.

He dramatized this in characteristic fashion. He had always liked to present every meeting with a *fait accompli*. In this case it was a proposed START agreement, cutting warheads on both sides to a level under 8000, the United States to about 9000 and the Soviets to about 7000. The SS-18 heavy missiles, often cited in the United States as the most pressing threat of a counterforce first-strike against the American land-based missiles, the cause of a 'window of vulnerability', were cut in half, from 308 to 154. The demand that Moscow had made since before Gorbachev's advent to power, for linkage to the SDI programme, had now been completely forgotten, as had been evident for at least two years. The Sakharov position, that the United States would relent once the Soviet threat was gone, was the attitude that showed most clearly, while Soviet work on its own ballistic missile defence, admitted by Gorbachev since 1988, would no doubt continue, although one might now wonder how it would be financed.

On one level the agreement was an anti-climax, since arms control was by this time no longer recognized as the central aspect of Soviet–American relations. Perhaps it was realized, as Aleksei Arbatov once put it, that even with START reductions in both arsenals by 50 per cent, an *increase* in instability might still result because of the decrease in targets and modernizations of those arsenals. Perhaps there was general weariness with the process of negotiation. Or, perhaps there was a realization that intentional strategic action with nuclear weapons, even at the height of periods of tension, was never the main threat of war, and that accident was always the greater menace. In view of this, the revolution going on inside the USSR probably presented a greater likelihood of unforeseen and fatal consequences than anything that had ever been produced by the arms race.

At any rate, Gorbachev was now mainly interested in the START treaty to dramatize his own achievements as the liquidator of the Cold War, so as not to appear before the Seven merely as a deserving pauper. They in any case did not see him as deserving, nor did they offer the alms he wanted. Instead, he got an advance in Soviet prestige by admission to associate status in the IMF and the World Bank. In return, he told London press conferences only that the USSR looked forward to an interface with the world economy, that he wanted increased contacts and was willing to relax various Soviet restrictions on private property, but not that he was willing to convert the Soviet Union according to the Polish model. Even in rags, Gorbachev still wanted to be a centrist and to find room to manoeuvre. Nevertheless, on his trip home, he told reporters that his efforts would still not be well received in the USSR, since in the final analysis, he was coming home 'with empty hands'.

Everyone thought that his critics were going to make another attempt to unhorse him at the 25–26 July Central Committee plenum. Again he tried to present them with a *fait accompli*, this time more precise stipulations of the modalities of the Union Treaty, including taxation parameters. He also presented to the plenum a new programme, aiming toward an 'open economy', a convertible ruble, and various forms of property ownership. In this programme, it was conceded that 'History is not simply a continuous process of building socialism.' A new campaign against Stalinism was urged, as in January 1987, to cast fire on 'those who adhere to the past'. Said to be the joint work of Shakhnazarov and Gorbachev, the programme was a basis for sorting out the ranks of the CPSU, either by a very risky purge of the hard-liners, or by a series of splits among the four platform factions. It was generally sensed among supporters of reform that a parting of the ways was imminent. Ambartsumov thought Gorbachev would resign. Melor Sturua hoped he would be content to do so, thus to preserve his place in history as a 'Great Reformer', rather than continue to participate in further tragic events.

But Gorbachev could not quit the party while it still had a chance to act against him. His new programme would have gone a long way toward the transformation of the CPSU along the lines of other ex-Communist parties in the former bloc. Yeltsin had pushed matters along impatiently by issuing a decree on 20 July against party cells in government offices. This appeared to be the primary thing on the minds of the delegates to the plenum, who sought Gorbachev's protection from Yeltsin. And he promised to annul the decree. But most could tell that he did not mean what he said. Actually Gorbachev said he would nullify *only* if the Committee on Constitutional Oversight found it illegal. This was no real defence. Nevertheless Aleksandr Yakovlev found it to be too much; he gave his support to Yeltsin's 'de-partification' and actually claimed to find the departures of the new programme to be too timid. About the Yakovlev position Gorbachev said enigmatically 'not everyone can keep up'.

At the same time he spoke of a 'convergence' between himself and Yeltsin. At a plenum in which he stressed unity in order to foil the efforts of the Movement for Democratic Reform, he also concentrated fire on the newly formed Bolshevik Platform and Communist Initiative factions. At the Soviet–American summit at the end of July, the START agreement took a backseat to the issue of the condition of Gorbachev's reforms, with the American President Bush actually pressing on Soviet audiences the need to release the Baltics, desert Castro's Cuba, and move more quickly to a 'market-oriented' economy, this as a precondition for US aid. The issue of a renewal of US attacks on Iraq's nuclear facilities, not so completely destroyed as was thought in March, was also broached between the two leaders. Gorbachev tried to make the summit an advertisement for his good relations with Yeltsin, so that some of the magic could rub off on him, but Yeltsin refused to oblige, insisting on a separate meeting with Bush.

A situation of 'dual power' could not be finessed by any of Gorbachev's stratagems. Yeltsin's Russia, acting like a sovereign Great Power, loomed as the successor to Gorbachev's Soviet Union. To the half-somnolent soon-to-be ex-Soviet intelligentsia, eagerly testing its newly articulated Russian nationalist reflexes, the scale of the conflict was barely perceived. The

Soviet Union, they were coming to think, had always been Russia, the ridiculous cosmopolitan universalism of the Communist idea merely a pretentious and ludicrous façade. The real warmth of the relation among the disparate peoples united with Russia would only be enhanced by a show of magnanimity toward their national aspirations. In the end these peoples would cling to Russia as the most advanced and most civilized culture among their number, whose language opened the door to the world. The old Communist structures could decay and fall away like a dress eaten by moths, but the Russia of Pushkin and Tolstoy would never be rejected by the former 'Soviet peoples'.

Their bedraggled hard-line opponents, watching everything collapse around them, were at their wits' end. The warnings of impending disaster, with which they had showered the party for two years, continued to fall on deaf ears. They hoped that a show of force might somehow ward off the inevitable, but who could provide it? Their leader was in league with their arch-enemy, and now had the majority of the party willing to endorse its own self-liquidation. Since Andrei Nuikin's article on the putsch of September 1990, they had been told many times that any attempt to alter the natural course of things by force would only cause their immediate destruction. But how could they pass from the historical stage without any semblance of a fight? If only they had acted three or four years ago! But perhaps even now something could be done.

13 From the Coup to the End

Nor called the Gods with vulgar spight
To vindicate his helpless Right
But bowed his comely head
Down as upon a bed.

Andrew Marvell

Was Gorbachev intimidated? Not in the least. Warnings abounded that some movement by the hard-liners was afoot, but he had been threatened many times by now and the result was always the same: they would send some delegation to him to beg him to change course and he would finesse the whole thing. He was convinced that they lacked the capacity to force anything on him, and told the press specifically after the July plenum that he was not at all frightened by them.

On the eve of that gathering, the hard-liners had published an Open Letter in *Sovetskaya Rossiia* with the heading 'A Word to the People', an attempt to counter the Gorbachev–Shakhnazarov draft programme for the social-democratization of the party. It was signed by Yuri Blokhin of the *Soiuz* caucus, the writers Yuri Bondarev and Valentin Rasputin, Generals Varennikov and Gromov, Vasili Starodubtsev of the Peasants' Union and Aleksandr Tizyakov for the industrial managers. Starodubtsev and Tizyakov would be on the Emergency Committee of the August putsch. The Letter called for a broad movement to save the country from those 'who do not love it, and who travel across the seas seeking from foreign mentors advice and blessings.'[1] Only intervention by the military, the letter said, could prevent the present slide 'into darkness and nothingness', and lead the country 'into a sovereign future without humiliation'.

Gorbachev was not much impressed by threats from this quarter. True, on the eve of the plenum, 32 of 72 regional secretaries in the Russian Federation had declared that Gorbachev should be called to account for his policies.[2] But this had happened before, and Gorbachev's account had usually been good enough to quell any trepidations. The hard-liners' miserable 13 votes in the April plenum were not such as to inspire fear, and that small cell of opposition had not grown any larger at the July plenum, in which the radical social-democratization of the party was posed even more sharply.

In fact he acted as if he thought that he might not be moving quickly enough on the transformation of the party, and that too much delay would permit someone else to split the party and take out a majority. One of the most prominent pretenders was Yeltsin ally Aleksandr Rutskoi, who had made the most serious challenge to Gorbachev as leader of the reform forces within the party by forming a faction called Communists for Democracy. Rutskoi sponsored a Moscow conference of that group on 2 August, at which it was decided to constitute a caucus in the Russian parliament under the name 'Democratic Party of Russian Communists', this to provide opposition to the 'Communists of Russia' group that acted under the discipline of the Russian Communist Party. The conference quickly won the enthusiastic endorsement of Aleksandr Yakovlev and Gavriil Popov. But *Pravda* declared the conference 'null and void', and moreover announced that Rutskoi himself had been expelled from the CPSU.[3] This could be interpreted as a threatening action by the party hards circling the wagons, but Gorbachev himself was still inside the lager as long as he did not break with the party.

With the CPSU seemingly in his pocket, Gorbachev won more ground in the Russian Communist Party, up to this time a centre of opposition, with the resignation of Polozkov, who had been championing the hard-line position. His replacement, Valentin Kuptsov, very quickly won Gorbachev's support. It seemed that Gorbachev had made a coup in the RCP. The meaning of the change in leadership became clear immediately when Kuptsov took up plaintive discussions with Yeltsin, with the aim of begging him to relent on his 'de-partification' decree of 20 July. The idea of the leadership change apparently was that Kuptsov would succeed in getting with balm and unction what could not be got by Polozkov's intransigence. But it was no good. In a meeting on 13 August, Yeltsin told Kuptsov that the 'de-partification' would be complete by the end of the year. In fact it was to be complete by the end of the month.

The expulsion of Rutskoi was quickly followed by a move to expel Yakovlev, one of the leading lights in the Movement for Democratic Reform, but the action was pre-empted on 16 August when Yakovlev resigned, warning darkly that 'a party and state coup' was in the offing. 'An influential Stalinist group' had formed in the leadership, he said, one that wished to end the political course that had been followed since 1985. Yakovlev had issued a similar alarm in January. On 21 August, in an interview with Austrian television, he claimed that he had been warning Gorbachev continuously since the Twenty-eighth Party Congress in July 1990, even supplying the names of the potential putschists. But Gorbachev did not seem to Yakovlev to have appreciated the danger, and even told the latter that his opponents 'lack the courage to stage a coup'.

This was the state of Gorbachev's mind as he departed for Cape Foros in the Crimea for a few days' repose to work on the text of a speech he was to give at the ceremonial signing of the Union Treaty on 20 August. While at Foros he also consulted with his advisor Anatoli Chernyaev, on an article outlining some broad perspectives of the reform process. In this article Gorbachev and Chernyaev sought to answer two questions: first, had *perestroika* been a mistake? And second, 'What policy is to be pursued in a situation of economic crisis, dangerous signs of disintegration and chaos, and of fear for the next day?'[4] To the first question, their answer was unequivocal; the course had been correct, one that the Soviet Union might have followed in its early days 'had it not been for the Stalinist *Thermidor* in the mid-twenties, which betrayed and trampled on the ideas of the Great Revolution'. Gorbachev distinguished his thinking from the views of 'those who seek salvation in repentance', and want to reject the whole experience of the country after 1917. He was still a defender of the socialist idea and the heritage of Lenin, now making its difficult way in a democratic environment. To those on the other side who rejected the path of reform and sought to turn back, he insisted that this would only lead to a division of society 'into reds and whites', and ultimately court a civil disaster.

He similarly rejected the critique of his opponents, to the effect that 'imperialism' had been handed on a platter what it could not win by force, that the Gorbachev revolution had squandered what had been won in the Second World War by the sacrifice of millions, and that the foreign policy positions of the state had been weakened to the point where the Soviet Union was in effect dancing to someone else's tune. To this he answered that

> A firm basis has been created for the further strengthening of the foreign policy positions of our state. Let us recall that great Russian minister of foreign affairs of the last century, Prince Gorchakov, who said 'Russia is concentrating' when referring to the revival of her authority following the reforms of the eighteen-sixties and eighteen-seventies.[5]

The essential task, said Gorbachev, was to continue to steer between 'the neo-Stalinists, the retrograde types' on the one side and 'the adventurists and ultra-radicals' on the other, in order to keep *perestroika* on a peaceful course. This could now only be done through the Novo-Ogaryevo process of voluntarily creating a new Union. What he called 'The Union of Soviet Sovereign Republics' would permit each Republic to take part in appointing one house of a national legislature and would allow them to veto new laws. They would own their own resources, establish and levy their own taxes, and conduct their own diplomatic relations. Moscow would be reduced to the role of 'coordinating' foreign policy for the Union. Even the coup itself would not dampen Gorbachev's enthusiasm for this project, probably the greatest of his many utopias. But it is note-worthy that neither of his critics from the 'two camps' of opposition were tempted by it in the least. Yeltsin indicated many times that he consid-ered the prospective Union Treaty that would soon issue from the 'Novo-Ogarevo process' to be the final act of Russia's emancipation from the Soviet Union. And the hard-liners saw the matter in exactly the same way. Only Gorbachev continued, as in the past, to have faith in a centrist course, and in his ability to hold off every challenge by steering it deftly.

THE 'COMMITTEE'

Gorbachev was at work in his study in the state dacha at Cape Foros on 18 February when, at 4:50 pm, the head of his bodyguard informed him that a group of men had arrived at the dacha demanding to see him. They had been admitted because they were led by General Yuri Plekhanov, of the Guards Service, formerly the Ninth Chief Directorate of the KGB, responsible for the security of the leadership. Before Gorbachev con-fronted the delegation accompanying Plekhanov, he tried to make a call out from his phones, only to find that all the lines had been cut. So before hearing their demands, he knew of their intentions. He took a moment to inform his wife, their daughter and grandaughter of his fears before the delegation arrived.

The group included the men closest to Gorbachev's Presidential office. Valeri Boldin was a member of the Politburo and Secretariat, who had functioned as chief of staff to Gorbachev since the days of his ephemeral Presidential Council of March 1990, and after that his Defence Council, on which Boldin also sat. Oleg Baklanov was First Deputy Chairman of the Defence Council, and a member of the Secretariat. Oleg Shenin was a member of the Secretariat responsible for organizational matters. Gen-eral Vitali Varennikov was First Deputy Defence Minister who had helped direct the attacks in Lithuania in January. We have already noted the prominence of Baklanov and Boldin, and their likely responsibility for party supervision of the security organs at the time of the troop move-ments of September 1990. So this was not a group of nonentities.

Gorbachev immediately demanded to know in whose authority they claimed to act. The reply was

'The Committee.'

'What Committee?'

'Well, the Committee set up to deal with the emergency situation in the country.'

'Who set it up? I didn't create it and the Supreme Soviet didn't create it. Who created it?'[6]

Gorbachev was not to get a satisfactory answer to this question straight-away. What was this Committee? In fact it was to call itself the State Committee on the Extraordinary Situation (*Gosudarstvennyi Komitet po Chrezvychainomu Polozheniiu*, or GKChP). Headed by Vice President Gennadi Yanaev, who was also the member of the Secretariat respon-sible for international affairs, it included Prime Minister Pavlov, KGB chief Kryuchkov, Interior Minister Pugo, Defence Minister Marshal Yazov, Baklanov, Vasili Starodubtsev and Aleksandr Tizyakov, the last speaking respectively for the kolkhoz peasants and industrial managers.

What Yakovlev had foreseen as a 'party and state coup' was in fact being directed by a group of high figures in the state rather than by a party group as such, even while its leaders were all prominent party fig-ures. Yanaev stressed that they had not acted through any party body. Indeed, they could not have secured the support of the Politburo, nor even the Secretariat, much less the Central Committee. They had ar-rived at the present stratagem only after exhausting the alternatives, having tried criticizing Gorbachev, threatening him, as in September and December 1990, attempting to remove him by a Central Committee vote in April, and trying to outmanoeuvre him in parliament in June. This was a des-perate last stand driven by the knowledge that, even if the chances for success were not good, this was the eleventh hour for Communism and for the Soviet state.

Vyachislav Dashichev told German television on 21 August that Yeltsin's entourage regarded Baklanov, Gorbachev's deputy on the Defence Council, as the real leader of the coup, and that he was seconded by Shenin and Moscow party chief Prokofiev. Ivan Silayev, the Russian Prime Minister, insisted that the real 'ideologist' of the coup was Lukyanov, the Chair-man of the Supreme Soviet, who was not on the GKChP at all, but ap-peared in the role of an intermediary between the Committee and the various parliamentary organs. Lukyanov, Gorbachev's old law school col-league, was evidently responsible for the odd psuedo-legalism of the whole operation. The Committee announced its existence to the world at 6:30am on 19 August. Claiming that Gorbachev had been removed from his pos-ition by reason of 'ill health', Yanaev pronounced a six-month state of emergency, with the aim of overcoming 'the profound and comprehen-sive crisis, the political, ethnic and civil strife, the chaos and anarchy' that threatened to engulf the nation. The emergency entailed the ban-ning of all political parties and mass movements that might hinder 'nor-malization of the situation', and suspension of the publication of newspapers, except for *Pravda*, *Sovetskaya Rossiia*, and a few others that were consid-ered reliable. Television channels broadcast only classical music and sports news, interrupted occasionally by reading of the Committee's first de-crees. These acts were defended by invoking point 14 of article 4 of the

Soviet law 'On Legal Conditions Applying in a State of Emergency.' Articles 124 and 127 of the Constitution were cited to support the establishment of the Committee as 'the supreme body of state power'.

The USSR Committee for Constitutional Review issued a statement containing the amazing opinion that the transfer of the functions of the state bodies to the authority of the GKChP could be legally justified 'if there is strict observance of the requirements of the USSR Constitution and other laws'.[7] The articles that had been cited indicated that a state of emergency could only be sanctioned by the Supreme Soviet. And Lukyanov was the chairman of the Supreme Soviet. This, apparently, was the 'ideology' of the coup, a 'lawful' putsch establishing the Supreme Soviet as the parliamentary cover for the Committee. Baklanov would be the driving force in the decisions of the Committee, and Lukyanov would act as the parliamentary-legal liaison. It was an idea that could only have been hatched in Gorbachev's Russia, with its obsession with the transition to legality coexisting alongside a quite traditional bureaucratic arbitrariness.

This was of course not known to Gorbachev as he confronted the group sent to arrest him. He knew only that they wanted him to sign a decree authorizing the state of emergency. Several days later, after the coup had failed, a welter of stories circulated to the effect that Gorbachev had from the beginning been in league with the plotters. These wounded Gorbachev grievously, and may have played some role in the discussions that accompanied his eventual ouster in December. It was reported at that time that Gorbachev, once he had decided to resign, had earnestly negotiated for six hours with Yeltsin, seeking a guarantee of immunity to any charges of complicity in the coup. Yeltsin had not yielded, saying only that 'if you did anything wrong, now is the time to confess it, while you are still President'. Was Gorbachev connected with the coup? Shevardnadze later seemed to indicate that he had anticipated it, telling CNN that 'he [Gorbachev] was worried about a coup on 17 August. Probably he was told about it.'[8] Some of the wildest claims about his role can be dismissed, for example that of the then Georgian President Zviad Gamsakhurdia, who argued that Gorbachev had helped plan the whole operation in order to strengthen his own personal powers after the coup had failed. In that case, Gorbachev's motives would be truly inexplicable. The idea was rejected as 'ridiculous', first by President Bush and then by the rest of the world.

But Gorbachev seemed most concerned to refute a story circulated by Lukyanov in an interview the latter gave on 19 August, to which Gorbachev referred enigmatically in his book on the coup, saying that he had been accused of aiming to 'sit out' the coup and then to arrive 'ready to serve'. Gorbachev reports Baklanov suggesting to him 'You take a rest and while you are away we'll do the dirty work and you will return to Moscow.'[9]

But this idea should be considered in different terms. According to the report of Deputy Oleg Borodin, Lukyanov met with ten deputies of the hard-line *Soiuz* caucus on 19 August, claiming that Gorbachev had agreed to the Committee's programme of action, on the condition that its creation be sanctioned by the Supreme Soviet. Even if this is doubted, its underlying suggestion – that Gorbachev may have had some inkling of the 'legal' basis of the plot – cannot be so easily dismissed. On his own account, he agreed with the statement of the country's woes presented to him by Baklanov and the others, and suggested 'let us raise these questions in the Supreme Soviet'. He reflects that 'I had succeeded more than once in recent years in checking or averting a dangerous turn of events. And this time I again thought that these people would understand and change their minds.'[10]

Lukyanov told the deputies further that Gorbachev had changed his own mind about the timing of signing the Union Treaty after he had received an 'extremely offensive reply from Yeltsin to his invitation to attend a meeting of the Federation Council on 21 August, the day after the Union Treaty was to be signed. Yeltsin had apparently told him that since all the organs of Union power would be abolished by the Treaty, there would be no point in attending the meeting.[11] This is well known to have been Yeltsin's opinion. Several days after the defeat of the coup, he told a television audience why he thought the plotters had sought to forestall the Union Treaty: 'Instead of the great power the centre has now, all that was left to it (under the treaty) was a small role, mostly coordinating.'[12]

Yet this shows only that Gorbachev may have immediately understood the legalistic concept of the coup and tried to finesse the matter as he had done in the past. It might also indicate that the plotters envisioned a role for Gorbachev even in the event that he did not cooperate and agreed only after a few weeks or months of 'illness' to return to Moscow to play his centrist role again. Kryuchkov later told his interrogators that 'It wasn't a matter of completely depriving the President of power. This is very important. In no conversation did we talk about that.'[13] It may have been that the same technique was employed as was used to pressure Gorbachev in the coup of September 1990. Of course, however plausible Gorbachev's explanation, he spoke in the knowledge that, in the overheated atmosphere of a future political trial, these things might count heavily against him.

It is likely that mere pressure on Gorbachev was the first option, and his failure to cooperate necessitated that the Committee proceed with the military operations on its own. These were organized by Kryuchkov for the KGB, probably at least several weeks in advance.[14] The lack of a major campaign at the July plenum probably indicates that the hard-liners had given up on the idea of a party rising against Gorbachev's policies and were resolved to use the means eventually employed. Kryuchkov ordered the KGB troops based in Sevastopol to encircle the dacha at Foros and keep the Presidential aircraft from taking off. Marshal Yazov later told his interrogators that he had only been contacted by the plotters on Saturday, 17 August, after which he hurriedly called together Colonel-General Nikolai Kalinin, commander of the Moscow military district and Lt-General Pavel Grachev, who commanded the airborne troops. Grachev was to move an airborne division into town from its base at Tula. Kalinin was to bring in the 'Taman' mechanized infantry division and the 'Kantemirov' armoured division. These troops were assigned to purely defensive positions at various key points, the Kremlin, the State Security offices, the bank, the telephone exchange, and the approaches to the Russian parliament building, the 'White House'. The KGB troops succeeded in silencing the media, despite some slip-ups: CNN and some Russian news agencies were able to get information to the outside and Gorbachev himself at Foros to receive information from an unjammed BBC; the radio station Moscow Echo was shut down, but it sent bulletins by a clandestine transmitter. The Chinese coup in May 1989 had been much more thorough in this regard. But the actions of the KGB and military were effective enough, and insured victory for the first phase of the enterprise, the installation of the Committee.

A COUP WITH A LEGALIST 'IDEOLOGY'

This was, however, only the first phase. Even after the plotters had gone on to silence the potential opposition, something that they did not in the event manage to accomplish, the last phase, the restoration of order in the nation at large and in the republics, would have loomed before them. They were certainly thinking in these terms. On the first day of the putsch, they had troops take over the radio and television station in Kaunas. The telephone exchange in Vilnius was raided, and the port of Klaipeda was put under blockade. Alfreds Rubiks, of the Latvian Communist Party, announced plans for an emergency committee for that republic. Armoured troops took up positions around Riga and a tank battalion entered Tallinn. General Fyodor Kuzmin, acting in the name of the Committee, declared himself the *de facto* ruler of the Baltic republics.

Had the Committee managed to establish itself in Moscow, the first order of the day would have been the opening up of protracted military and police operations in the Baltics. Nothing less than that would have served to bring them into line. The men of the Committee could probably count on getting their way in Belorussia and the Central Asian republics, but would most likely have to face a bitter defiance in Moldova, the Caucasian republics, and most importantly, Ukraine. One had to assume that the democrats in Russia would also be in opposition, at least to the degree that they had been in January. So the troops and security forces would have been engaged in skirmishes and confrontations with crowds all over the Soviet Union. And should any major units rebel against this onerous duty, as is highly probable when conscripts are used in such situations, there would have ensued a full-scale civil war, in a country with tens of thousands of nuclear warheads.[15] Quite a formidable programme of 'political struggle'. In view of this long-run agenda, seemingly dictated by the necessities of the situation, the plotters were in a sense lucky that they failed in the next step, the silencing of the opposition. Boris Yeltsin was to be their 'saviour'.

Somehow they had overlooked Yeltsin. He got the news of the coup from the official announcements at 6:30am on 19 August at his dacha at Usovo, just outside Moscow. Every newspaper account of the coup professed astonishment that Yeltsin had not been arrested. KGB units were watching his dacha but did nothing. Possibly this was part of the 'ideology' of the coup, that is, a proclamation of emergency in the Soviet government to which the republican bodies, including the Russian government, were expected to knuckle under. Yanaev said explicitly at the Committee's first press conference that he sought the cooperation of the Russian leadership. Yeltsin was not considered an admirer of Gorbachev's—perhaps he would be tempted to see some profit in the great reformer's political incapacity. Kryuchkov told his interrogators on 22 August: 'Our GKChP did not take any actions that were directed in any way against the Russian republic leadership or Russia.'[16]

On the other hand, Gorbachev, speaking of the confrontation at Foros on 18 August, reported that 'Baklanov said that Yeltsin had been arrested. Then he corrected himself: he would be arrested on the way.'[17] Several sources reported that Yeltsin had been warned of his impending arrest, either by an associate of the GKChP or by a KGB official. Leningrad mayor Anatoli Sobchak said that he got a call at 6:30am Monday at his Moscow apartment from friends in Kazakhstan saying that they had received documents from an 'Emergency Committee' hours earlier. He drove to Yeltsin's dacha at Usovo to meet with Yeltsin and some other supporters, including Ruslan Khasbulatov, chairman of the Russian Supreme Soviet. There was a brief conference on strategy, with Sobchak assigned to go back to Leningrad, where General Samsonov, speaking for the GKChP, had already declared himself in charge. Yeltsin's daughter told him: 'Daddy, pull yourself together. Now everything depends on you.'[18] According to Sobchak, troops arrived at the dacha to arrest Yeltsin just ten minutes after he and the others departed for Moscow. It is not clear how Sobchak knew this. If an arresting party did come, they may have lost time by going first to Yeltsin's Moscow residence. He had been in Kazakhstan the previous day and had only returned the night of 18 August. He may have been informed from Kazakhstan in the same way as Sobchak. Most curious, and doubtful, is the idea that this vaunted intelligence force had simply lost track of the President of Russia. The thought persists that the arrest of Yeltsin was not included originally in the plans for the coup, but was improvised at the last minute.

Yeltsin at any rate arrived at the Russian parliament building, the White House, to establish resistance to the GKChP. A statement by Lukyanov was being circulated denouncing the Union Treaty and calling for a debate on it in the USSR Supreme Soviet. Lukyanov had the backing of the Presidium of the Cabinet of Ministers, which on Saturday, had given a similar opinion, advising Gorbachev and the republican leaders to amend the draft, in order to strengthen the prerogative of the centre. Yeltsin began his campaign with a flurry of phone calls. An oddity of this 'legal' putsch was the good communications it had with the centres and leaders of resistance. This made it possible for Yeltsin and his forces to anticipate the actions of the Committee, and even to negotiate various defections from their ranks. Yeltsin first talked with Yanaev, and got little satisfaction. But in a conversation with General Grachev, who was under the orders of Yazov and the Committee, he requested that a token force of his men be sent to the White House to take up positions there to defend the Russian government.[19] Grachev agreed to this, and sent a group from the Taman division commanded by Major-General Aleksandr Lebed, with ten armoured personnel carriers. The fact that it was a parade unit, lacking ammunition, did not detract from its impressive appearance. Once at his post, the commander of the platoon, a Lt Nikolai Kotlerov, was invited in to speak with Yeltsin, and was quickly won to the cause. Thus by these timely acts did the Yeltsin revolution acquire its first armed forces.

At 9:00 Yeltsin, Silayev, and Khasbulatov made an appeal against the removal of 'the lawfully elected President of the country' by a 'right-wing and anti-constitutional coup'.[20] 'Constitutional' would be the most contested idea for the next three days. The appeal complained of the impositions of 'unconstitutional bodies, including party ones'. It would be hard to maintain that the constitution of 1977 did not provide for a role for these 'party bodies'. And those who were now calling them unconstitutional were themselves on the verge of abolishing the state that had provided the constitution, with 'all its structures of power', as Yeltsin's formula had put it.

But the idea of a Russian sovereignty apart from the USSR was put into mothballs for the moment. The grand deception of saving the Soviet

Union now became the central theme of Yeltsin's struggle for power. The leadership of Russia described itself as the champion of 'the unity of the Soviet Union and the unity of Russia'.[21] It urged a return to 'normal constitutional development', calling on the armed forces to disobey the orders of the GKChP and the workers to stage a general strike.

At 12:30 Yeltsin issued a Presidential decree calling the establishment of the GKChP 'a coup d'état which is nothing else than a state crime' against the 'legally elected authority' in the Russian republic, and punishable under the Russian republic's criminal code.[22] Again an interesting formulation that seemed to make the actions of the Soviet Union's officials subject to the higher authority of the Russian republic's laws. But the whole matter was not going to be decided in an impartial court, rather in the court of world history. And Yeltsin had done about everything a great revolutionary could do to win the day for his cause. He had stated the case for his side's legitimacy both as a 'constitutional' and, what is not necessarily the same thing, an 'elected' authority. He had called his enemies criminals and threatened them, as he had when he had left their party one year previously, with arrest and punishment. He made these ringing statements before several thousand people who had come to the White House and, standing on one of the vehicles so obligingly sent him by General Grachev, appealed to all the citizens to respond. Shaking the hand of one of the soldiers, who must have been wondering what he was doing there, Yeltsin said dryly to the delight of the crowd, 'Apparently they are not going to shoot the Russian President just yet.'

Several thousand Yeltsin supporters gathered outside the White House during the rest of the afternoon and evening, throwing up makeshift barricades with overturned cars and tramcars. At 5:00pm the Committee held a news conference to explain the 'illness' of Gorbachev and attempt to get the media used to their faces, but the cascade of insulting questions to which they were subjected by sceptical reporters only made the eight members of the new government cringe. The televised sight of Yanaev's hands shaking as he spoke was particularly demoralizing. He tried to assure the public that both political and market reforms would continue, and that at some point 'my friend President Gorbachev' would be rejoining the government. 'He is not a state criminal', said Yanaev 'this is a man who did everything to put us on the path of democracy.'[2] Yanaev stressed that the setting up of the Committee was not the work of either the Politburo or the Secretariat. As to the question of who would replace Gorbachev as general secretary of the party, Yanaev said Deputy General Secretary Vladimir Ivashko was ready to work. He also announced that the Committee would seek the approval of the state of emergency by a two-thirds majority of the Supreme Soviet when it convened on 27 August.

The committee at this time may have been unsure of itself, and in the case of some members even fatalistic, but it was not confronted with an avalanche of counterforce. Coal-miners at Vorkuta and the Kuzbass were to go on strike on the morning of 20 August, and would be joined by some other scattered strikes. But there was to be no general strike. Resistance in the Baltic republics and in Moldova gathered force as was expected. The deputy chairman of the Sverdlovsk soviet identified the emergency as a 'military coup d'état' and urged opposition, as did authorities in Tyumen, Tomsk, Khabarovsk, and Kamchatka.

But in Novosibirsk, the party propaganda chief, Anatoli Sabanov, supported the Emergency, complaining that 'What could be called a coup is the signing of the Union treaty and the forthcoming changes in the Russian government bodies.'[24] In Kazan, Donetsk, and Alma Ata, the situation was described as normal.[25] Soviet troops succeeded in sealing off Moldova's border with Romania. Georgian television was appealing for calm, discipline, and order. Most important, Ukrainian President Leonid Kravchuk was not opposing the Committee. Yeltsin had claimed, not entirely truthfully, to have his support, but in fact Kravchuk had been paid a visit by General Varennikov and two other generals in support of the Committee and had been assured that the Emergency would require no special measures of him. So he acted with caution, referring the matter, according to the Lukyanov 'ideology', to the Ukrainian Supreme Soviet. In a law-based state, he said, 'everything must proceed on the basis of the law, including the announcement of a state of emergency'.[26]

NOT MILITARY ART BUT POLITICAL MEANS

The resistance did succeed in gathering some support from the military. Yeltsin enjoyed the advantage of having assiduously cultivated military support in the course of his recent electoral campaigns. He had prior assurances of support in the event of a coup from General Grachev, commander of the Tula Division paratroops.[27] Grachev was likely one source of warning to Yeltsin about the coup, since he had already been brought into advance discussions by Kryuchkov and the other conspirators. Yeltsin and his Vice President, Rutskoi, had spoken only recently to the troops in the units now on duty in Moscow. This was not to be important for promoting a mass defection from the ranks, but for establishing for Yeltsin a certain legitimacy with the commanders. One of the oddities of the burgeoning Soviet parliamentary life was that, unlike other countries, it permitted uniformed officers to run for office and sit in parliament. In effect, because the officers were not kept out of politics, a certain *civisme* was promoted among them. Although many accounts in the Western press

spoke about 'an old-style Soviet coup', or of a 'Stalinist coup', in fact
the USSR had no tradition whatever of military coups, and may be said
to have been in many respects a model of civilian control over the military

General Konstantin Kobets, chief of the Russian Republic Defence
Committee, spoke to the crowd in front of the White House on the first
day, telling them that 'Just because these officers and generals are wear
ing uniforms, does not mean that the soldiers will support them.'[28] Along
with Rutskoi, Kobets kept up contact with Yazov's 15-member collegium
at the Defence Ministry. They did not recruit so many 'defectors' as was
thought originally. What was often described in later press accounts as
defection sometimes meant simply the withdrawal of a unit back to its
base, as when General Lebed took a group of vehicles from the White
House watch and brought them back to the barracks. In fact the first
orders given the units supporting the coup had not involved taking any
one or anything under fire, but simply taking up 'protective positions' to
guard against assault. And 'defecting' to the side of Yeltsin was not a
great contrast from this mood for the responsible commanders; it did
not mean attacking anyone, but merely staying in contact with the Russian
President and trying to argue his case for caution, reflection, and ulti
mately, inaction with the excessively communicative general staff of Yazov
As long as no one did anything Yeltsin and the democrats stood to gain

Since the plan, 'the ideology', of the GKChP was so delicate and sub
tle, so devoid of the violence and military audacity that are usually asso
ciated with such actions, the inaction of the plotters on the first night
while the Yeltsin forces dug in and made their appeals, may be explained
not just in terms of their moral paralysis and malaise, but also by their
complacency. They were apparently waiting to see how much the resist
ance would grow.

But they lost ground on the night of the 19–20th in Leningrad. Before
Sobchak left Moscow for Leningrad, he gave an airport interview strongly
condemning the 'coup d'état by a group of adventurers', and urging an
immediate general strike and disobedience to orders by the troops. The
Leningrad City Council had already expressed support for Yeltsin's early
decrees, despite the action of General Samsonov in declaring martial law
Sobchak said later that his own arrest had been left to the airport KGB
who inexplicably bungled the job.

Once he arrived back in Leningrad, he confronted General Samsonov
and his staff with the same sort of appeals that were being made to the
officers in Moscow, but with one special element. Sobchak told them that
they were all plotters 'from the legal point of view', and that if they were
to follow the orders of the Committee they would all end up on trial
'like the Nazis were tried at Nuremburg'. Sobchak reminded Samsonov
of the suppression of nationalist protests in Tbilisi in April 1989, and his

1aving wisely avoided obeying orders and having 'kept in the shade' at
hat time. He tried to convince the general that, as in recent actions of
his sort, the campaign would not be carried through, and after its col-
1apse, others would be blamed, as in Vilnius in January 1991. 'Don't you
,ee, General', he told Samsonov, 'what nonentities these people are? They
vill not retain power even if they seize it!'[29] Sobchak finally managed to
:ase the General's reversal of position by having Yeltsin put him, by
lecree, under the authority of deputy mayor Admiral Vyacheslav
3hcherbakov, who was against the coup. Sobchak was able to assure the
3eneral that in any future inquiry he would be covered, by virtue of
1aving obeyed lawful orders. Sobchak could then announce that troops
n Leningrad would not obey the GKChP. The Committee suddenly dis-
overed that they had lost their Leningrad bastion.

On the 20th Yeltsin's campaign continued to gather momentum. He
ried to proclaim the criminal liability of the putschists as often as poss-
ble and to plead with them to turn back before it was too late. They
vere more and more suffused with a sense of hopelessness as a result of
heir contacts with Yeltsin forces. These latter, in their public utterances,
vere not walking on eggshells as they themselves were but howling for
•unishment of the Committee. At noon Yeltsin addressed a massive crowd
•f over 100 000 in front of the White House, predicting the collapse of
he coup, while Rutskoi declared menacingly that they would have 24
1ours to close the affair down. Yeltsin claimed, probably untruthfully,
hat Lukyanov had turned against the coup, and Rutskoi said that Lukyanov
1ad admitted the illegality of the the GKChP. Yakovlev, Popov, Shevardnadze
.ddressed the rally, hurling their denunciations at the putschists. The
entral committee of the all-Union Komsomols (the League of Young
3ommunists) denounced the coup. In Leningrad, Sobchak addressed
nother huge rally estimated at 200 000.

That answered an important question for the members of the Commit-
ee, who had been working under the delusion that their action against
3orbachev would not meet with much resistance. Now they knew that
hey had something to overcome and the coup would not be bloodless.
'hey seemed to have been shocked and defeated by the idea, as was
hown by the wave of medical infirmities that swept through their ranks.
Kryuchkov, Yazov, and Pavlov were all reported ill and 'resigning' their
osts on the GKChP. News of this convinced many that the coup had
>st, although it was hard to believe that those who commanded the armed
orces could not get any of them to overpower the very modestly de-
ended White House. Ukrainian President Kravchuk's commitment to
:gality, which had sounded favourable to the plotters on Monday, now
egan to sound more like the anti-coup position of the Ukrainian nationalist
ukh organization. Kazakh President Nazarbaev also now decided, after

36 hours of deliberation, to support Yeltsin and attack the GKChP.

No one had thought that the coup could be defeated so easily. The indefatigable Elena Bonner told the people of Moscow that the greatest test was yet to come, and that now was the time for them to show themselves 'worthy of the name of residents of the capital and the state – and not simply a mob interested in sausages'. For months it had been rumoured that dire economic conditions had been imposed on the populace by the hard-liners, who were withholding food stocked in warehouses. After they had cracked down, food would suddenly appear on the shelves to placate the masses. Yazov himself seems to have believed this. Afterward, he told his interrogators: 'We were counting on the theory that somewhere there were some reserves and stockpiles . . .'. He was crushed to find that they did not exist. 'There are no such reserves', he was told by a staffer, 'and in five days we'll be howling like wolves.'

The test came on Tuesday night, 20–21 August, during which the GKChP made its last attempt to overpower the forces defending the White House. Strictly speaking, these defences were not much. Inside the building Yeltsin had only around 300 men equipped with small arms, many of them from a private security agency. Despite their intrepid character, these men would not have been much trouble for troops. But apart from the hardware of material forces, Yeltsin had some considerable advantages. Throughout the days of the coup he enjoyed superb intelligence reports from KGB leaks and defectors, often receiving copies of written orders and plans. This could not do much more than give him an idea of his potential fate. But he and his supporters were in contact with the leaders of the coup and warning them not to shed blood; he knew how demoralizing it was for them that he knew everything they were planning, and that he knew how badly they were wavering. Yeltsin himself did not waver. He told all his supporters that they would either win or die in the fighting. CBS's Diane Sawyer got one of his aides to take her through the building at a time when the assault was expected, and finally to a conference room where Yeltsin sat with his associates. As she left, he told her cheerfully 'Don't look so worried, we might win this after all!'

During the day, large crowds milled about in front of the building but they dissipated once night fell, more pronouncedly in view of the fact that it was raining all evening. Their thousands had dwindled into hundreds. They put the finishing touches on their barricades, actually just tank obstacles, in the form of overturned cars, tramcars, trolleybuses and rail ties. People could lurk in and among these obstacles and throw Molotov cocktails at the tanks and other armoured vehicles. They could also spread terrifying rumours among the GKChP troops in the vicinity that the obstacles were loaded up with various bombs and devices to start fires on vehicles that might attempt to bridge them.

That was about the extent of the defences. To attack the building with troops, some tank dozers would have been needed (which were surprisingly never in evidence) and some people would have had to be shot at. Assaulting the building itself would have produced casualties, but for troops with heavy weapons it would not have been difficult. Commanders of the units who were eventually asked to do the job supposed that the whole thing would take 20 minutes. The GKChP's orders to the troops had been obeyed well enough for them to have expected good chances in this battle.

After the failure of the coup, it seemed that every commander interviewed said that his was the mutiny that foiled the plans of the plotters. Yet all the 'military mutinies' that were so trumpeted in the press had by no means impeded the ability of the plotters to get their way by the use of force. Commanders who were later said to have 'gone over' to Yeltsin in reality had never had any orders to fire their weapons or attack anyone; if some indeed tried to evade their orders, they were far from showing any inclination to march their units into battle against the GKChP, which continually moved out units that had been close to the crowds and replaced them with fresh ones, usually from KGB divisions. In the end the presumed defections of commanders served more to weaken the *morale* of Yazov and his collegium than their actual ability to command force. In the end the Committee suffered a moral failure.

The Committee failed to realize that they had gone beyond Political Means and that now it was a matter of military art, which involves violence and killing, as with Yeltsin's actions in taking the same White House under fire in October 1993. Instead, the greatest fear among the military people under the command of the plotters was that blood might be shed, and perhaps also that afterward someone would be blamed for it. While they may have believed passionately in their cause, they were already wondering what to say in the event that it should fail. Yeltsin, by contrast, was ready to risk death in order to win this test.

'RUSSIA HAS SAVED THE UNION'

The last gasp came that night, 20–21 August. Fresh units were brought up into position. A number of forays into the obstacles were made by tanks, with the intention of showing force by clearing them and scaring off the defenders at the outside of the building. Tank dozers were not used and tank crews ended up ramming their tanks ineffectually against the overturned vehicles. In some cases they managed to flatten some cars and toss some other vehicles out of the way, but youths with Molotov cocktails also managed to set some of them on fire. One armoured personnel carrier was set aflame and forced to reverse downhill into an underpass pursued by the youths until it used its machine guns to disperse them, killing one. At least two other deaths occurred in similar confrontations. Afterwards the obstacles were built up again. A column of about two tank battalions lined up for an assault on the building awaiting orders to move.

The assault was to be a replay of the attack on the Vilnius television tower in January 1991, with much of the same dramatis personae. Kryuchkov was in charge overall. According to Moscow military district chief of staff General Zolotov, 'oral orders' were given by Deputy Defence Minister General Vladislav Achalov on the basis of coded telegrams from General Moiseyev. Achalov had been appointed to his post in December 1990, with special responsibility for emergency situations. He had directed the operations at the television tower in Vilnius, using MVD troops and paratroops to clear the way, and dispersing much of the crowd by having tanks fire some 270 blank rounds. With a path thus cleared, a special unit, Alpha Group, had entered the building, shooting out doors and windows, establishing control, and leaving just as quickly, to turn things over to the MVD troops. There had been 14 killed and over 300 wounded. Lithuanian KGB Major General Stanislav Tsaplin said later that the orders had come from General Achalov and that they only awaited confirmation by a decree from Gorbachev introducing Presidential rule.[30] However, they ended up carrying out the operation without the decree, and afterward, Gorbachev actually disavowed the attack. The officers and men soon found that their action was not treated as something heroic, but rather as a crime for which the unit was guilty. As a trial run, the Vilnius operation did not offer very promising prospects for the troops who would have to repeat it. In addition, this unit was composed of Russians who in Vilnius were acting against non-Russians. Now they would have to act against their own.

Available accounts of Group Alpha's mission against the White House differ on the actions of the principals, on times, and even dates. But some outlines of the story are clear enough. The commander of the 200-man group was Major General Viktor Karpukhin, who had been in charge since its founding by Yuri Andropov in 1974. Karpukhin later claimed, of course, that it was he who had refused the order to attack the White House. His second in command, Mikhail Golovatov, said, however, that Karpukhin had placed the unit on alert on Sunday 18 August, and had carried out the order to prepare the assault at 3am the night of 20–21 August.

The order to begin the attack came at midnight. Golovatov said that Karpukhin commanded him to execute it and repeatedly threatened him with court-martial when he showed reluctance. Golovatov said that at this point a poll (!) was taken among the troops, with every man in the unit agreeing that the order was illegal. It is likely that Karpukhin permitted the order to be debated and discussed, knowing that it would not be accepted, and that Golovatov and the other commanders of the unit thought the same way. As a rule, commanders resolved to carry out orders do not send the order down to the ranks for debate and discussion. 'We believe our refusal to obey the unlawful order saved the country from civil war,' Golovatov later said, 'If Alpha began to storm the White House, combat operations would have broken out between military units taking different sides.'[31] Afterwards, Yeltsin gave credit to Group Alpha, citing its courage as the main reason for the failure of the attack that night.

It was symbolically fitting that it had fallen to the KGB to save Soviet power. The coup plotters did not see themselves as Stalinist reactionaries determined to go back to the old ways, as their opponents naturally wanted to paint them. They were followers and protégés, as was Gorbachev himself, of Yuri Andropov, who personified the hope of the gradual and measured reform of the system. Even now, the plotters stood basically for a return to the reform course laid out by Andropov in 1983–84 and followed by Gorbachev in 1985–87. To them the KGB was of course a mighty secret police force, but also a repository of intelligence for national security in the nuclear age. They thought it logical that the succession to Brezhnev had brought forward in Andropov the Politburo figure thought to be the most authoritative in intelligence and national security matters. In the first years of the Gorbachev era they had wanted to stay on Andropov's course. The men who sat on the GKChP now no longer had Andropov, but they still had the KGB. And they had staked everything on its effectiveness. There was no backup unit on hand, according to GRU head General Vladen Mihailov, whose spetsnaz troops, commanded by the leaders of military intelligence, had been active in South Ossetia and Karabakh. These units were never activated during the days of the coup. The KGB alone had proved to be the last refuge of the defenders of Soviet Communism.

At midnight the White House defenders heard that helicopter crews had refused to assist in the assault by landing troops on the roof. At about one o'clock, sniper fire was heard, and it was returned from the darkened windows of the White House. Yeltsin was in touch with Yanaev at this time and was promised by the latter that no forcible measures would be used. Kryuchkov also told Yeltsin's associate, Gennadi Burbulis, that he knew nothing about the movement of KGB troops. This at about the time that Group Alpha was debating its orders, and reports were coming in that described large-scale movements of KGB divisions into the city. Shevardnadze arrived at about 2:00am, prepared to share the fate of the defenders. They waited a seeming eternity for the attack. But the attack never came. At 4:00 a political officer with a paratroop unit leaked it to the defenders that 'H-hour has passed', and Burbulis, after having been told over the phone by Kryuchkov that he 'could go to sleep', decided himself that 'a turning point has been reached'.

The Committee of plotters had indeed reached a turning point. Faced with the necessity to begin again with preparations of new units for another attempt at a night assault, Yazov's general staff began to yield to the opinion of Air Force General Shaposhnikov, expressed repeatedly for days, that the only 'way out' was to go to Gorbachev and have him sort matters out. Shaposhnikov was less a military factor than a diplomatist of revolution, as Lukyanov had wanted to be for the Committee. He pressed repeatedly on this weak spot among the hard-liners, their reliance on Gorbachev. Yeltsin's position in telephone conversations with Kryuchkov was similar: go to Foros, get Gorbachev, and there will be a negotiated agreement among us. Yeltsin later described this as a brilliant trick to lure Kryuchkov away from Moscow into a position where he could be arrested. This may be the case. But it is more likely that the Committee knew it was finished, and was looking merely for the best party to whom to offer its surrender.

The Communist Party now weighed in against the coup. The Secretariat and the Politburo had hardly moved a muscle for two days, taking a stance rather like the Presidents of the Central Asian republics and Ukrainian President Kravchuk. At midday on the 21st, however, a meeting of the Secretariat suddenly came out publicly against the coup, and demanded that Yanaev meet with Gorbachev. Speaking for the Secretariat, Deputy General Secretary Vladimir Ivashko said that without such a meeting, the party would not know how to instruct party units to respond to the state of emergency. But had the party not been supporting the coup before this? This was not easy to determine. When he resigned from the party on the 23rd, Nazarbaev said that in Kazakhstan, a constant flow of communications from the Secretariat had been sent to party offices urging cooperation with the GKChP. In Ukraine, the parliament's Human Rights Commission released documents showing that similar instructions were received on the 19th in the name of the Secretariat. Others were apparently sent to the secretaries of the republics under the signature, dated 20 August, of Oleg Shenin.[32] This was the most direct connection between the party and the GKChP. Yet Shenin and Prokofiev resigned from the Politburo in protest of that body's lack of support for the coup. Could Shenin have issued communications for the Secretariat on his own? Gorbachev himself later seemed convinced that the party was not to blame for the coup and that the Secretariat was under the control of 'healthy forces'.

Ivashko's call was taken up by the general staff at 3:00pm. Shaposhnikov's repeated urgings to Yazov to contact Gorbachev had become a majority position. Arkadi Volsky, head of the Union of Science and Industry and Central Committee member, called on the party to condemn the coup. Central Committee Secretary for Ideology Dzasokhov insisted that the Secretariat had from the beginning been urging a meeting with Gorbachev, and had succeeded only on the 21st. At any rate, Ivashko, now himself playing a key diplomatic role, headed a group including Kryuchkov, Yazov, and Lukyanov on a plane that left for Foros at 2:00pm. A second plane with Rutskoi, Volsky, and others left three hours later.

Gorbachev had been following events as best he could by monitoring radio reports, but he could not be sure what to expect, although he said later that he could see that the coup was unravelling on the second day. When the delegation arrived at Foros at 5:00pm, Gorbachev refused to see them and immediately ordered their arrest. Then he softened and permitted Ivashko and Lukyanov to come to his office, along with Bakatin and Primakov. Seeing now that the plotters had come as supplicants, he tried to gather in the reins of power as best he could. He called General Moiseyev and put him in charge of the armed forces in place of Yazov. Moiseyev would later relate that he had been in charge of the nuclear codes throughout the coup.

Gorbachev got in touch with various foreign leaders: Bush, Major, Kohl, and Mitterrand (who had taken almost as long to condemn the coup as Nazarbaev). Gorbachev and Yeltsin spoke by phone to confirm their consensus that the plotters must all be arrested and eventually tried for their attempt to seize power. Gorbachev knew only that he would quickly be returned to Moscow to resume his office. But he did not understand, as he tells in his account of the coup, what a different country he was returning to. In particular, he did not yet understand that his 'saviours' were no more committed to him than his recent captors had been. There was no more cruel illusion and no more artful deception than that contained in the phrase of Yeltsin, to the effect that 'Russia has saved the Union'.

GORBACHEV AND KERENSKY

Gorbachev wrote later that, on reflection, perhaps he had foreseen some attempt to remove him. No impartial observer could call the August putsch a bolt from the blue. Countless warnings had been mulled in the Soviet and foreign press for at least a year. The model they usually invoked was that of 1964, when a Central Committee plenum had removed Khrushchev. But an incident from the last Russian Revolution was a much closer fit. In 1917, General Kornilov had sought to put an end to the domestic chaos that he saw as the root of Russia's weakness in the war effort against Germany and Austria. He marched his troops on Petrograd to depose the government of Kerensky, who had no choice but to seek the military support of the Bolsheviks, thus to stake his fate on the very Bolsheviks whom he knew intended no less for him than Kornilov. On their side, the Bolsheviks knew that once they had 'saved' Kerensky, he would be theirs. Boris Yeltsin, so often accused in the past two years of neo-Bolshevism, understood no less clearly that once he had saved Gorbachev from the putschists, the fate of the hapless Soviet President and CPSU gensek would be his to dispense.

A moment had been reached in which those who had yesterday been giving the orders were now under arrest. Yazov was surprised when the order was given by Gorbachev to arrest him. Kryuchkov was said to have been confident that Gorbachev had saved him until just before his own arrest. The truth was that from the moment that they decided they needed Gorbachev after all, the plotters were lost. In his *August Coup* he claimed that he did not then know that they were coming to Foros on the agreement of Yeltsin and the Russian leaders. On the contrary, Yeltsin said later that he had told Gorbachev explicitly by phone not to negotiate with them. Rutskoi said that Gorbachev cooperated in this and that 'he was a good boy'.

On his arrival back in Moscow early Thursday morning, the 22nd, he announced to an incredulous greeting party that 'This is a serious victory of the *perestroika* process.' He promised that those who had organized the coup would be arrested and punished. 'I will do anything', he said, 'to purge the Communist party of reactionary forces.' This formula shows how little he realized, or, giving him more credit, perhaps how little he was reconciled to, the sweeping victory of Yeltsin. He added pathetically: 'I do not think we should have a witchhunt and act as we did in the past.'

In fact he had been overthrown with the plotters. A spiritual change had come over the entire country. For several days the crowds had been chanting 'Yel-tsin!' in two syllables and 'Ros-si-ya!' in three. A new flag had been unfurled, the Tsarist tricolor (without the double eagle) in white, blue, and red – for Belorussia, Ukraine, and Great Russia. Russia had been reborn in a traditional sense, the rebirth coming with its own illusion that the Slavic republics would remain united with Moscow. The Russian republic's military leaders had overthrown the Soviet Union's general staff. The new 'Russian' idea had emerged in place of the ideas of *glasnost, perestroika*, and democratization. And the new idea was powerfully personified in the revolutionary figure of Boris Yeltsin. While the Gorbachev revolution had depended, whether Gorbachev understood it or not, on the reform of the Communist party, the Yeltsin revolution depended no less on the rout of that party, now in Yeltsin's description, 'the organizing and inspiring force of the coup'.

These two impulses had a final brief contest in the following 48 hours. Gorbachev told a press conference late Thursday: 'I shall fight to the end for the renewal of the party.' The Cabinet of Ministers and the Supreme Soviet duly declared their loyalty to him. While the leaders of the coup were being arrested, he announced his new appointments to key posts. V. P. Trushin would head the MVD in place of Pugo, who had killed himself a few hours earlier. L. V. Shebarshin was put at the head of the KGB. He had been a close associate of Kryuchkov's, and had written his speeches warning about the influence of the Western intelligence agencies.[33] General Moiseyev was made Defence Minister.

On the same day, the 22nd, Yeltsin supervised the arrest of the coup's organizers, among whom he was careful to count Lukyanov. The Supreme Soviet cooperated by stripping the latter of his parliamentary immunity. Chairmen of regional soviets in Rostov, Krasnodar, Lipetsk, and Samara, who had supported the coup, were removed. Yeltsin decreed that a system of 'presidential envoys', reporting to a special Presidential control authority, was to replace the leadership of the local soviets. It was rather like the system of military governors that had ruled Russia under Peter the Great. Yeltsin's 'counter-coup', as Roy Medvedev would later call it, had for its rationale and vehicle the arrest of the plotters and prevention of another coup. Yeltsin described their motives in a Gorbachevist way that nevertheless boded ill for Gorbachev:

> Objectively, the new Union Treaty would strip virtually each of the architects of the putsch of their offices, and herein lies the secret of the conspiracy and the main motivation behind the action of the parties to it ... Their demagoguery about the fate of the homeland is little more than trickery concealing their personal selfish interests.[34]

A crowd of several thousand went to the headquarters of the KGB Thursday night to tear down the statue of Feliks Dzherzhinsky, the first head of the Cheka. Dzherzhinsky had died in 1926, so the crimes of the Stalinist period cannot be laid at his door, but in fact the symbolism was a revolt against the very forces that had fought for Communism in the earliest days of the revolution and civil war that had followed the events of 1917. A crowd that gathered the next day looked like breaking into the KGB building to ransack it as had been done to the East German Stasi headquarters. Yeltsin drew the line there. He was not to permit any wholesale attacks on the KGB, which in his revolution was to undergo a very graceful reorganization. To placate the crowd, Yeltsin offered instead a coup against Gorbachev, rescinding the latter's appointments of the day before and appointing Viktor Barannikov to head of the MVD in place of Trushin, Bakatin to the head of the KGB in place of Shebarshin, and General Shaposhnikov to the Defence Ministry to replace Moiseyev. The first deputy was to be Grachev. Gorbachev's first acts on his return had been quickly overturned.

The next step in the humiliation of the father of *perestroika* was taken before the Russian Supreme Soviet on Friday, the 23rd, to which gathering he had been invited ostensibly to speak about his struggle against the plotters. It was a replay of the humiliation of Yeltsin before the Central Committee plenum of October 1987, with the positions of the principals reversed. Now Gorbachev stood in the dock fending off the accusations and catcalls of the assembled delegates eager to leap onto the bandwagon of the Yeltsin revolution. The gathering was in a mood rather like that of central committee plenums of the past, when a recital was being made against someone who had erred.

Gorbachev rambled on in his own familiar way about the terrible ordeal that he and his family had been subjected to. But one delegate offered the opinion that Gorbachev had been talking too much about these things and too little about punishing the guilty. From the floor came question after question to Gorbachev about how he would prevent another coup. One speaker even questioned Gorbachev's zeal in the matter in view of Lukyanov's story that he, Gorbachev, had actually been involved with the plotters. In the coming days and weeks, Yeltsin continually found ways to allude to the story, obviously enjoying the way it made Gorbachev squirm. Now Yeltsin forced a fretting Gorbachev to reverse himself and endorse the Yeltsin selections for the key posts over his own of the previous day. At one point Gorbachev rebelled against the demeaning process, starting to say that he was not going to go through the whole list of appointments. Yeltsin's sardonic reply was: 'I have just mentioned these things because I didn't want you to forget.'

Yeltsin forced Gorbachev to read the minutes of the Council of Ministers meeting of the 19th, in which it endorsed the coup. This in order to have him admit that the government must be replaced. For this task Yeltsin insisted on and got the agreement of Gorbachev to appoint his own candidate, Ivan Silayev, to the post of Prime Minister of the USSR. Silayev eventually took over control of the important Interim Committee for the Management of the National Economy. Yeltsin's man was to run Gorbachev's own government.

Repeatedly the question of outlawing the Communist party was posed by the delegates and by Yeltsin. Gorbachev tried to convince them that, while the leaders should be punished, 'to declare the millions of people, workers and peasants, to be criminals, that I will never go along with'. He was thinking of all those who had supported him through thick and thin against the hard-liners, thinking that is, of the overwhelming majority of the party.

The task of driving socialism from the territory of the Soviet Union, nobody has a right to ask such a question or make such a demand. It's another type of utopia.[35]

A smiling Yeltsin reminded Gorbachev that he, Yeltsin, had decreed during the coup the nationalization of all property on Russian soil, and that would of course include Communist party property. He reminded Gorbachev of his promise made that very day to sign a decree confirming all of Yeltsin's decrees. While a flustered Gorbachev was attempting to talk his way around that, Yeltsin said dryly 'On a lighter note, shall we now sign this decree suspending the activities of the Russian Communist party?' Whereupon Yeltsin scribbled his signature with a flourish before a protesting Gorbachev. At that moment Yeltsin must have thought back to the end of the 1987 plenum at which Gorbachev and the rest of the delegates had humiliated and demoted him. Gorbachev at one point had said contemptuously to Yeltsin, 'Are you so illiterate politically that we must organize elementary school studies for you here?'[36] It was not the sort of drubbing that a public person forgets. Now a smirking Yeltsin asked the jeering assembly to show Gorbachev some more respect. 'I know some of his answers to questions might not be entirely complete but . . . maybe he's gotten tired.'

On the face of it, Gorbachev's position was not ridiculous. The coup had been carried out by a group of party people who had long since lost their support in the party. They did not have the participation of a single party body, as Yanaev had plainly stated on Monday.[37] The Committee of Eight had not mentioned Communism, Lenin, or anything but the interest of the state. This had led some to suggest that they had in fact staged a coup against the party, perhaps in the same sense that the Polish coup led by Jaruzelski, the army, and security forces in December 1981 had been a coup against the Polish party. For his part, Gorbachev was defending, not the conspirators, nor even the minority hard-line wing of the party to whom he had turned from September 1990 to April 1991, but those who had supported him in his fight with Gromyko, Ligachev, Chebrikov, and others, those who had passed the measures adopted at the Nineteenth Conference, those who had voted an end to the party's leading role, who had supported the Novo-Ogarevo process since April, and who had outvoted the hard-liners at the last two plenums. They were the reform-minded majority who had unflinchingly accepted the social-democratization of the party as described in the new platform of July. Why should they have to be put under a ban directed primarily against those they had fought for five years?

A touching complaint that would, one might think, normally be considered with respect in the parliament of a country seeking to break with

Stalinism and to rejoin the rest of civilized humanity. However, the fact was that the coup had made a break with the genteel norms invoked in the reform period and had itself been overcome by a democratic revolution. At this point the surge of revolutionary passion to oust the remnants of the old regime could perhaps best be compared to the revolutions of the past in England, France, and Russia itself. The Russian parliament that hurled its insults at Gorbachev was in a certain sense a revolutionary Long Parliament like that which formed the core of the English revolution of the seventeenth century. And the Russian democrats amply showed that they were capable of treating Communism as an *ancien régime*.

As if to prove this, Yeltsin was already on the warpath. On Friday and Saturday he suspended *Pravda*, reorganized TASS and Novosti, sealed the party headquarters and archives, along with those of the KGB and MVD, and drove the commissar apparatus out of the armed forces. Outright bans of the Communist party, with confiscations of property, were carried out in Latvia and Moldova. These were soon to become general throughout the Soviet Union. A reeling Gorbachev gave his approval, confirming the acts by decree. Then he himself resigned from the party. Two days earlier he had said: 'I will fight to the end for the renewal of the party.' Perhaps he had meant to the end of the week.

Saturday was also the day that the Soviet Union fell apart. The Ukrainian parliament declared independence, setting 1 December as the date for a referendum to ratify the decision. Without the Ukraine, the democrats' idea of *Rossiia* was just as unthinkable as the Communist idea of a Soviet Union. This appears to have been as clear to Yeltsin as to Gorbachev. On Tuesday the 27th, Yeltsin nevertheless tried to hold off the inevitable by threats. He made the perfectly reasonable suggestion that, if the Ukraine should go its own way, there should be an adjustment of the borders of the new state to reflect the fact that eastern Ukraine was populated by some ten million Russians. This provoked angry responses from Ukraine and a near panic among many of the new ex-Communists who had gone against the coup. President Nazarbaev of Kazakhstan intervened hurriedly to head off what he feared would be a potential civil war. Yeltsin had to back down. Russia and Kazakhstan found that they had in common a desire to prevent a too-rapid collapse of the Union. But that had been exactly the premise of Yeltsin's campaign against Gorbachev of the last year. Now, however, that he was coming to power and getting free of Gorbachev and the CPSU, Yeltsin began to realize his other responsibilities as the head of a historic state that had included the Ukraine since the seventeenth century. An historian of the school of E. H. Carr might observe that power was bringing forward in his personality the ascendancy of the statesman over the revolutionary.

Gorbachev too did what he could. In the event, all he could do was to threaten to resign if no new Union could be salvaged. He warned Yeltsin that increasing Russian control over domestic affairs did not mean that he had relinquished the right to conduct foreign policy. He announced the appointment of Boris Pankin as his Foreign Minister. Yeltsin responded by saying that 'The disintegration of the central authorities is not the disintegration of the country, especially not of Russia.' An odd formula that proved wrong on all counts. The collapse of Communism meant the collapse of the Soviet Union – and was moreover part of the same process that was threatening the collapse of Russia itself.

As he had in the past, Gorbachev faced internal crisis by pursuing foreign policy victories. American Secretary of State Baker visited in September, permitting a more positive impression of his attitude toward Yeltsin. President Bush was widely thought to be unhappy about the rise of Yeltsin and the eclipse of Gorbachev, with whom he had enjoyed a close rapport. This was a position that drew a great deal of criticism in the United States from many who argued that unfavourable impressions of Yeltsin and reports about his personal instability had been unduly stressed. True enough, Bush and his advisors had a certain affection for Gorbachev. But their attitude toward the Yeltsin revolution was probably more affected by the tableau of national disintegration that stretched out before them as they contemplated the post-coup Soviet Union.

The breakup of the Soviet Union meant on the face of it that Ukraine and Kazakhstan would become major nuclear powers, perhaps not superpowers, but nuclear powers on a level with the next rank, England, France, or China. Other former Soviet nuclear weapons in Estonia, and especially in Belarus, would be factors making for instability and grave dangers. If the economic and ethnic chaos were to deepen, these weapons might fall into the wrong hands, as had already happened with so many Soviet weapons in the Caucasian republics. The straitened conditions facing the military were as well prompting the idea that the sale of weapons could be an important means of support. Small arms, including automatic weapons, were as likely to be found in urban flea markets as old uniforms and insignia. Yeltsin encouraged this by explicitly authorizing air force units to sell planes to raise cash. He himself offered the MIG-29 for sale in European markets, along with the latest T-82 tanks.

Most of the nuclear weapons were in Russia. But there were as many as 5000 warheads in the republics that were becoming independent. In Kazakhstan and Ukraine there were silos for the SS-18s, the 'heavy missiles' that had for some time been judged as posing the most potent threat of a disarming counterforce strike against American ICBM silos. Belarus had many mobile ICBMs. There were bases for strategic bombers in Estonia,

Belarus, Ukraine, and Kazakhstan. Then came the tactical weapons attached to the armour and mechanized infantry divisions, of which there were 86 in the non-Russian republics.[38]

Prior to the coup, Ukraine, Belorussia, and Kazakhstan had all declared an intention to make their republics nuclear-free zones. And afterward, most of the same pledges were repeated, but with various qualifications, sometimes tied up with their general claims to parts of the Soviet armed forces on their soil. A flood of statements from Yeltsin and Gorbachev assured the West and plans were immediately announced to put all these weapons under Russian control, wherever possible, moving them into Russia. But nuclear weapons are not easily moved around like a pallet of oil drums. The ICBMs were considered the least worrisome, owing to the elaborate procedures necessary for their arming and use, including authorizing codes from the centre. At the other end, tactical weapons might actually be stolen or taken in mutinies in the same way that helicopters, tanks, artillery pieces and other gear had already disappeared from Soviet bases in the Caucasus.

As to the effect on the previous Soviet position on non-proliferation of nuclear weapons, which in recent decades had been on the whole quite satisfactory to the West, the picture was now alarming. Republics pressed for hard currency could not help considering sales of fissionable material, if they had it. Fifty per cent of the plants involved in the nuclear fuel cycle were in Central Asian republics. Tadzhikistan was reported to have offered to sell uranium and plutonium. Most distressing was the situation of scientists and engineers who had worked in the nuclear industry and, with the chaos of the past several years, were now either reduced to penury or trying to make a living on the free market. An estimated two thousand or so of these actually understood the engineering problems involved in making nuclear weapons. It was possible for Western organizations to absorb them in various projects in order to supply them with a living, but any nation seeking to build nuclear weapons could easily lure them away by offering better salaries.

By far the most portentous result of the breakup of the Soviet Union was the emergence of the newly independent Ukraine. Possessing an industrial base of proportions hardly inferior to that of Russia (producing, for example, around half of the Soviet tanks), and an agriculture normally producing four or five times its own consumption of grain, Ukraine had, except for oil, all the economic elements to aspire to be at least a middle-rank European power. Both Gorbachev and Yeltsin tried their best to bring Ukraine back into some kind of relation with Moscow, either as part of the Soviet Union or as a sovereign member of a new federation or confederation. In this effort, the closest Gorbachev was able to come was his plan for an economic union of eight Soviet republics.

At the same time, however, Ukrainian politicians were striving to put their 24 August declaration of independence on a firmer footing in preparation for the all-important vote on independence scheduled for 1 December. Ukrainian Prime Minister Vitold Fokin railed against the excessive 'centralism' in the plan. A bitter Gorbachev listened to the demands of Fokin about the division of hard currency and gold reserves, and remarked darkly: 'Well then, how about the matter of borders?' This threat, however, had lost its bite. On 16 October, the Ukrainian Supreme Soviet officially rejected the plan.

While Gorbachev strove against impossible barriers to cling to an economic union with Ukraine, the men who had established Yeltsin's power in the Soviet military tried in their own way to hold on to the same link. Soviet Defence Minister Shaposhnikov issued a demand to his Ukrainian counterpart General Konstantin Morozov to report to him in Moscow for orders. This was at a time when Ukraine had already staked a claim to its own army of 325 000 men from the million-plus Soviet troops serving on its territory. Despite the Soviet military's pressure, however, Morozov defied the order. Their backbones stiffened by the Soviet threats, the Ukrainian leaders now resolved to raise a national army of 400 000. More than this, they demanded control over the nuclear weapons on their soil, until of course these were destroyed pursuant to a disarmament agreement with the West. This evoked no panic in the Western press. Generally it was thought that the Ukrainians were simply using the 180 strategic missiles on their soil as a bargaining chip, and that they were sincere about not wanting to be a nuclear power. In truth, no one could then say what the nuclear future of Ukraine would be. One Ukrainian delegate to the economic talks reflected that in view of the military tensions, 'We should sign anything. By 1 December, it will all be academic.'

Ukraine was emerging as a nation state for the first time, not in the wake of a peasant revolt led by a Cossack hetman, as in the seventeenth century, nor from the chaos of Russian defeat by Imperial Germany, as in 1918, but with all the trimmings of democratic legitimacy as recognized throughout Europe. The geographic and historical existence of the new state made it necessarily an enemy of Russia. The democratic intellectuals in Russia who had hailed Yeltsin's victory and the Novo-Ogaryevo process were slowly realizing that this meant going on without Ukraine, and were beginning to repent their visions of a peaceful 'Little Russia'. And every bumptious remark made in Moscow in the spirit of more confident times only quickened the fears expressed in Kiev of 'Russian Imperialism'.

The issue was sharpened up by the little pockets of opposition to the Yeltsin revolution and its probable consequences. During the French Revolution, La Vendée, a district in the west of France rose up in defiance of the Revolution's claim to conscript the peasantry for its army, and, led by refractory priests, fought a bitter counter-revolutionary struggle against the new order for two years. Now there were Vendées of the second Russian Revolution that adjoined the territory claimed by Ukraine. Russians in the industrial area on the left bank of the Dniester river refused to accept the Moldovan claim on their territory in view of their anticipation of a probable future annexation of Moldova by Romania. General Gennadi Yakovlev committed the Russian Fourteenth Army to the defence of the 'Trans-Dniestrian Republic'. When his superior, General Yuri Kuznetzov was asked how General Yakovlev could have acted thus, he replied: 'It's his affair.'

The Ukrainians assumed, not unreasonably, that this military rising could not have happened except as it fit the plans of Moscow. They also feared that the rebel republic could stir a war with Russia, as could other issues in other regions. The Crimea or the eastern, mostly Russian *oblast*s in Ukraine might also present severe problems in the future. These were not small disputes easily solved by vague minority treaties. At the moment, to be sure, neither Russia nor Ukraine wanted to fight about them. But in future years, perhaps only a few years, that could change. The new Little Russia, with millions of its nationals in the neighbouring newly independent countries, like the French *pied noirs* after the loss of Algeria, could only live at peace with these neighbours by the unlikely act of accepting the expulsion and return of the millions of Russians, who had always thought they were 'Soviets', back to Russia. Neither General Rutskoi, nor any other of the Yeltsinian politicians were talking in this voice. They were on the contrary giving assurances that 'Russia will defend Russians' – a formula that bore an eerie and threatening resemblance to the attitude of Serbia toward the Serbians in the ex-Yugoslavia.

In this atmosphere of rising menace and spiritual crisis, Gorbachev had reached the point of being written off as a leader. Nevertheless he persuaded himself that if he could cobble together some kind of new Union, on almost any domestic foundation, he would preserve and even enhance his international status. He had long since lost his taste for wrestling with the vast economic problems of the Soviet states. Foreign policy was his real *métier*, and the area of his greatest historical achievements. As head of a new post-Soviet confederation of states, he could be rather like the head of the EC, or of the United Nations, a position suitable to a world-level diplomatist of his stature and universalist ideas. For this last contrivance, Gorbachev was not depending on any deft manoeuvre; he was simply trying to get the leaders to agree on the advantages of staying together. In fact, this had been his nationality policy all along. It was really quite simple and lacking in guile.

Even so, how could Yeltsin be recruited to cooperate with Gorbachev's plans? Had the hero of August come all this way, braving the opprobrium of the party hard-liners, defeating them at the polls, and finally standing up to their tanks and guns, in order to carry Gorbachev's bags as the domestic manager of the limping Russian economy? This while Gorbachev remained the centre of focus in the international life of the new state? Did Yeltsin owe this to Gorbachev in return for previous services?

Yeltsin's ineffectual actions in regard to the new Union thus need no interpreting. He was, of course, on record as being determined to fight to save the Union. But not a Union led by Gorbachev. His prestige was certainly great enough by now to push Gorbachev roughly aside in the spirit of the 23 August parliamentary session. This would hardly, however, enhance his prestige in international affairs where Gorbachev had semi-divine status. He would have to fit into the role of successor and supporter of Gorbachev who only picked up the great leader's baton as the latter fell of his own weight. At any rate the destruction of Gorbachev's power did not have to be accomplished by Yeltsin. As the polls showed clearly enough, Ukraine was going to vote for secession on 1 December, and everyone knew there could be no Union without Ukraine. Gorbachev understood this stance of Yeltsin's well enough. 'I realized the Russian President was dissembling', he wrote later in regard to last-minute discussions on the Union Treaty, 'and that the secessionist position taken by the Ukrainian leadership was actually a godsend for him, because it played into his own hand as he stalled over the Union Treaty. He clearly had his own plan in mind.'[39]

Gorbachev had assurances of the same kind from Ukrainian President Kravchuk that every effort was being made to resist nationalist pressures to interpret the vote for independence as a complete secession. Kravchuk would have to play the same double game with Gorbachev (and even with Yeltsin) that Yeltsin was playing with Gorbachev: he would have to maintain that Ukraine's drive for real sovereignty was not what it appeared and that Ukrainians still supported the efforts being made to defend some sort of Union. As expected, the referendum results on 1 December overwhelmingly endorsed the 24 August decision for independence. The choice was made by 90 per cent of voters and included majorities in the Russian *oblast*s in the eastern Ukraine. Kravchuk himself got 61 per cent of the vote and the clearest of mandates. But he took pains to assure Gorbachev that cooperation was still possible.

Yeltsin moved quickly to follow up on the vote. Russia extended its recognition on 3 December. The United States, which had not been eager to see Ukrainian secession for the obvious reasons, had already reconciled itself to the fact and urged only that Ukraine deliver its nuclear weapons to Russia and observe human rights. On 4 December the USSR's

Soviet of the Union approved a Gorbachev draft Union Treaty. The Presidents of the republics in question were still stringing Gorbachev along even at this late date. Gorbachev was still being led to believe that he was engaged in a process of negotiation for the Treaty. On 6 December, the Ukrainian parliament voted not to sign any Union treaty with the Soviet Union, and declared that the treaty of 1922 establishing the Soviet Union was henceforth null and void.

Gorbachev spoke to Yeltsin briefly to ask what he planned to do in response, and was told by the latter that he planned to go to Minsk to meet with Presidents Shushkevich (of Belarus) and Kravchuk, and that a 'Slavic Union' might be discussed there. Gorbachev responded that this was 'unacceptable'.[40] It turned out to be acceptable enough to the parties, however, who in declaring their Commonwealth of Independent States (*Sodruzhestvo Nyezavisimikh Gosudarstv*) on 8 December also announced that 'The Soviet Union, as a subject of international law and geopolitical reality, ceases to exist.'

Their agreement, said Gorbachev, who was in an extreme state of denial, was 'just a sketch'. Secretary Baker told him, for a different reason, that it was 'just a shell'. Baker had met with Yeltsin and continued to doubt, correctly, that the CIS was anything more than a device to break up the Union. There could be no unity among the three states on foreign and defence policy. By 12 December the Commonwealth was joined by Kazakhstan and the other Central Asian states, and by the 21st, by Moldova, Azerbaijan, and Armenia. Gorbachev could not believe that it was happening. 'If the USSR were to split up', he later recalled in the anguished tone of those days, 'we'd have to divide our armed forces. Ours was a unique defence space, with strategic forces and early-warning systems that monitored the global military situation. There was nothing like it in the world except in America. Were we really going to rip it apart?'[41]

Yeltsin and his team had had more opportunity to think about it, and had decided that such an end of empire would not be the worst case. Gennadi Burbulis was said to believe that Russia's exit from imperial status could be rather like that of Britain after 1947, even to the point of maintaining a certain middle-level power internationally and a modest nuclear deterrent.

Gorbachev, angry at the systematic policy of deception practised by all those with whom he had been attempting to deal honestly, described the formation of the Commonwealth as 'a coup', and resolved not to go gently into the night. He spent what were to be his last hours in office engaged in frantic meetings and phone calls with military men. He correctly refused to believe that they could consciously agree to the breakup of their multinational state and armed forces. Both he and Yeltsin now vied in their *lutte finale* for the loyalty of the military men, both promising them, with varying degrees of sincerity, the maintenance of unity in this 'unique defence space'.

Burbulis remarked that the question of 'Who rules the army?' would be resolved in the next few days.[42] In the unlikely case that Gorbachev had won real military support, a real civil war might have ensued. But Yeltsin knew that the question of 'Who rules the army?' also depended, in the present circumstances, on the question of who pays it. Here he could claim, unlike Gorbachev, that in the future he would most likely be in charge of a government with a defence budget. Yeltsin also understood the desire of the military to keep the common defence space. So he promised exactly that to the generals, appointing a CIS military staff to be headquartered at Minsk. A mirage of course, but it proved to be enough to keep them out of any future 'Soviet' plans of Gorbachev's.

Thus Gorbachev did not have his last stand and had to go quietly. He got into contact at the end with foreign leaders, Bush, Major, Kohl, Mitterrand, who were at first concerned at the dire possibility of the outbreak of civil war and who, when they saw how the matter would actually be resolved, offered gentle and appreciative words to the great reformer. Yeltsin was also constantly in contact, trying to disarm by requesting Gorbachev's support for vague future projects. In this same spirit, Burbulis even made ambiguous hints that there would still be a role for Gorbachev in the new state. One was reminded of Khrushchev at the plenum that removed him from office in 1964, after an eloquent recital of his inadequacies, meekly asking if he could stay on as Minister of Agriculture. Yeltsin pronounced with his own unique sense of decorum: 'We do not want to follow the tradition that has taken shape since 1917 of burying each former head and leader of the state and subsequently reburying him or regarding him as a criminal.' He gave assurances that Gorbachev would have a dignified exit, while stressing at the same time that it would indeed be an exit.

Gorbachev collected himself, addressed the nation on television to bid his last farewells, urged in a generous way that all who heard him support the Commonwealth and Russia, and of course the leadership of Boris Yeltsin. 'In the end', he told Major, having power was not so important and losing it would not be so bad – 'life will take care of the rest; everything will fall into place.' One could almost hear in this remark an echo of the maxim that he had loved to quote in his student days, Hegel's 'The truth is always concrete.' In the early days of his rise to power, this remark had been recalled by his classmate Zdenek Mlynar, with the implicit suggestion that here was a man who understood hard facts, the better to master them, a man who, in a sense, took reality by the throat. Now, viewed from a different perspective after the seven lean years that had followed, the maxim, invoked by Gorbachev, would have seemed to be a mark of resignation, in a spirit of fatalism that suggests the image of a sinner in the hands of an angry God.

On his departure there was a small observance for the transfer of the codes and the authority to use nuclear weapons, but no other formalities. For some reason Gorbachev was struck by this and could not help but remark on the oddity of there being no ceremony to bid farewell to a head of state, as in other civilized countries.

14 Conclusion: Utopia and Repentance

I'll never tell you everything I know.
                                    Mikhail S. Gorbachev, 22 August 1991

Question: Were you able to see at the time how things would turn out?
Gorbachev: I'll keep that to myself. I've already said more than I usually say . . . who knows Gorbachev's designs?
                                    Press conference, 11 December 1992

Caesar knew the Gauls better than modern Europe knows the Russians.
                                    Alexander Herzen

Does wisdom appear on earth as a raven inspired by the smell of carrion?
                                    Nietzsche

Epochal events inevitably produce, alongside the expected elation, a certain ambivalence. The last partition of a vast multinational empire, that of Habsburg Austria-Hungary at the end of the First World War, was also greeted with mixed sentiments. The collapse liberated some of the same peoples whose fate was touched by Gorbachev's revolution. Even in the midst of the post-war elation, with the demise not only of the Habsburg Empire, but the Hohenzollern, the Ottoman, and the Romanov as well, many thoughtful people living in the victor countries could see a tragic potential in the 'Balkanizing of Central Europe'. In fact fewer new states emerged then than in the events of 1989–91. The economist John Maynard Keynes lamented the demise of a functioning economic unit, a 'free trade zone' ruled from Vienna and Budapest. Winston Churchill later called the breakup of the Habsburg state 'a cardinal tragedy', and even a necessary cause of the Second World War.

The Balkanization of southeastern Europe proceeded apace, with the constant relative aggrandizement of Prussia and the German Reich, which, though tired and war-scarred, was intact and locally overwhelming. There is not one of the peoples or provinces that constituted the empire of the Habsburgs for whom gaining their independence has not brought the tortures which ancient poets and theologians had reserved for the damned.[1]

Naturally pessimism was not the predominant sentiment. No one lamented the collapse of the institution of absolute monarchy, nor the fall of empires that denied freedom to smaller nations. Most agreed with the American President Woodrow Wilson that the independence of the many peoples living in the area between Germany and Russia was the *sine qua non* of democratic, and therefore peaceful, development. Austria-Hungary's unity was retrospectively regarded as a mere device for the exploitation of the subject Slavic peoples by Germans and Magyars. While the Balkanization of the region might not be optimal, it was thought to be preferable to the alternatives offered by history – their Ottomanization, or their domination by either Germany or Russia.

THE NECESSITY OF FOLLY

So too with the Gorbachev revolution, which was greeted with a rapturous rejoicing that was only cut by fears of a looming orgy of nationalist passions and conflicts in the wake of the collapse of the Soviet bloc. The worshipful audiences who came to see Gorbachev on his American trip four months after his resignation heard him blithely describe the recent revolutions as the anticipated result of his struggle for *perestroika*. As the process had deepened, he told his foreign well-wishers, he had realized that more drastic measures had been called for. This had even been the case with the subject nationalities. 'When we began *perestroika*', he explained, 'we wanted to reform the unitary state by a real federalism.'[2]

Even among the most enthusiastic listeners, who wondered at the words of the man who changed the world, there were some whose sense of recall told them that this was not so. Gorbachev had not begun with any idea of altering the relation of the Union and the republics. Yet, after his fall he wanted to take credit for every change since he had assumed power – even those that everyone knew were made against his will – all of them products of his programme of *perestroika* and the inexorable logic of history.

Many others shared the interpretation of his revolution as the result of inexorable logic. There was a desire to overlook the terrain of the Gorbachev reform course over seven years, and to see the demise of the Communist idea as something foreordained and foreseen. Bloated colonial empires in our century, it was said, have simply proven to be impossibly anachronistic. Moreover, Communism, like Nazism, had been out of step with the fundamental drive of the last two centuries for democracy and personal freedom. The idea of the Gorbachev revolution putting an exclamation point on the End of History expressed the quite unexceptional notion that the alternatives to the democratic republic had failed historically. Excited optimists proclaimed the end of an international politics of power balance and found instead that the universalism of liberal democracy would henceforth be the key to permanent world peace. Looking beyond the national and ethnic particulars of east central and eastern Europe, some argued that democracy and the free market would now quickly solve all the residual problems of the area.[3]

A variant of this view held that thinking about international life in terms of power and national interest was now anachronistic. Democracies did not make war against each other. European and world federalism were the order of the day. Not national power but the idea of collective security – a system of restraints on nations employing a legal definition of aggression and general enforcement by permanent international organizations – would now provide security for a former Soviet world undergoing a rapid transition to a system of interlinked free market economies. In the first few years after the fall of Soviet power, The End of History idea was the ideological accompaniment in the West to a programme of economic 'shock therapy' in the former Soviet bloc. It was 1919 again. The viewpoint of Wilsonian democratic internationalism was back with a vengeance.

In this atmosphere it was easy to suggest that in that last analysis the Gorbachev revolution must have occurred because Gorbachev had arrived with a recognition of the futility of maintaining the Soviet empire and its global reach, which he rightly saw as an encumbrance, and which he liquidated in a spirit of rational retrenchment.

The allure of this suggestion is easy to see, but viewing Gorbachev's revolution this way involves a denial of its history. In fact the liquidation of the bloc was never decided or accepted by Gorbachev, who was still optimistic right up to 18 March 1990, when the East German voters rejected not just Communism, but Social Democracy to boot. The line on the graph of Soviet economic performance – one that indicates a decline in rates of growth in the early eighties – cannot be used to argue, except in the grip of passion, for the inevitability of the catastrophic economic collapse of the late eighties. Indeed, when Gorbachev came to power, there were no such predictions of breakdown. Similarly, the evident necessity for a change in foreign policy to a line of *recueillement* – recognized by Gromyko and the rest of the leaders in the early eighties – is far from showing the fatal inevitability of the partition of the bloc and the country in the following years. To see the events of 1989–91 as a foregone conclusion would mean that there would be no need to ask how it happened.

The preceding narrative is called upon to explain an apparent paradox: how was it that the man of destiny who transformed the world did not intend any of the consequences of his acts? How can the immense triumphs of his revolution be squared with the failure of his most cherished projects? Is this modern history's first case of revolution by sheer inadvertence? The common sense of many observers of the process told them that Gorbachev was indeed a great reformer who had meant well in hoping to contain the Cold War, but who, to the general benefit, had lost control of things and ended up presiding over the demise of Communism. True enough, he was a bungler, they might reason, but he was at any rate 'our bungler'.

It is certainly tempting simply to consider Gorbachev a bungler. Even if we accept his desire to change the world as genuine, and there is no reason to doubt it, we do not see in his acts any clear policy moving on a straight line. The landscape of his reforms is dotted with abrupt, seemingly fitful changes of direction. It will be recalled that at first, he strove to end alcoholism and smoking, but he soon dropped the idea, leaving the state finances that had depended on these revenues badly shaken. Then he earnestly strove to combat unearned incomes, only to turn to the opposite policy of encouraging cooperative business. Tighter controls over production, and a radical expansion of the administrative apparatus intended to handle them, yielded in the same way to market decentralization and the closing down of state agencies. Up to July 1988, he fought for 'Acceleration of the Perfection of Communism', then for its 'Reconstruction', then after March 1990 for 'Consolidation behind the Socialist Choice and Communist Perspectives', and finally, after April 1991, for the 'Novo-Ogaryevo Process' that dismantled the Soviet Union. Within these phases there were still more turns. His zigs-zags were so abrupt and frantic (and so unsuccessful) that one must assume that they could not have been part of any reform blueprint.

The zig-zags are inexplicable unless we see them against the background of the struggle for power that engaged him to the very end of his period in office. His unpredictable policy turns usually had their source in manoeuvres to outrun opponents. They were frequently part of an 'escape forward', to call up support from outside in order to weaken his rivals within the leadership. Gorbachev was certainly a unique figure who overturned all the rules and conventions of Soviet politics, but he was in the most crucial sense a product of that system of power struggle, with its lack of normal constitutional limits and its passion for ideas. These are features that leap out from the pages at anyone who studies the previous succession struggles that provide the essential rhythm of Soviet history.

Every succession struggle in the Soviet past tossed up a maximum leader whose victory was codified in a new history of the party and the revolution. Gorbachev's rise to power was no exception. History gave him his intellectual legitimacy. The debates of the *glasnost* campaign in 1987–88 were expected to furnish an ideological underpinning for all his domestic reforms. The reforms in turn were to demonstrate to the West how genuine was his call for a new approach in foreign policy. This campaign had its effect. At the end of 1987, Reagan and his advisers began to reconsider Soviet history and to accept that Gorbachev was truly different from past Soviet leaders precisely because of the fresh ideas produced by *glasnost*.

Following in the footsteps of Andropov, Gorbachev advertised national security and arms control as his strong suit. In this area, his initial ideas were by no means tragic or misbegotten. Detente was needed if the country were to have room to reform. But in matters of theory and party history, he lacked the intellectual credentials of a Lenin, or even of a Stalin, so he depended on the *glasnost* intellectuals to provide him with the material to undermine the 'Stalinism' of his opponents.

Thus the Soviet intelligentsia was given a unique opportunity to shape the ideological course of the reforms. Writers who had made their careers by supplying the appropriate phrases to Khrushchev, Brezhnev, Andropov, and Chernenko were now invited to craft a new anti-Stalinist credo for Gorbachev. For this they were absurdly ill-equipped. They had survived up to then on their ability to determine what the line was and on whom to focus the requisite attacks. But once they set about to produce more than ideological manipulations, it soon became clear that even under the new rules their 'double think' could be pointed in almost any direction.

Foreign policy was also presumably *their* strong suit. Yet one result of their liberation under Gorbachev was that the propaganda language of the previous peace offensives was transformed by turns into a language of reform. When 'de-ideologization of foreign policy' was invoked at first, it was intended to caution the American adherents of the Reagan Doctrine against anti-Communist fanaticism and attempts to turn back the historical clock. Eventually it came to mean that the USSR itself had to repent of ideological fanaticism. When 'interdependence' was first adapted from Western parlance, it was meant to describe a community of fate of the superpowers in their possession of nuclear weapons and their common stake in preventing a nuclear conflagration. Eventually 'interdependence' came to signify the futility of the USSR in trying to develop its economy apart from the world market.

Reform ideas grew more radical in their implications as the struggle with the opposition intensified. In 1985–87 Gorbachev followed a course of which Andropov would have approved, promising that his return to detente would defuse the threat of the American SDI programme. When this failed at Rejkyavik, he made an 'escape forward', deciding that current Soviet problems went back to the time when theory had been distorted under Stalin. Thus the *glasnost* campaign of 1987 with Rybakov's novel *Children of the Arbat* as its centrepiece. When this too ran aground

as a result of the backfire of the Yeltsin affair and the ideological defence of Stalinism by academician Vaganov and his associates, Gorbachev
made another leap forward, encouraging redoubled debate on the Stalin
question, this time ultimately centred around Shatrov's play *Onward,
Onward, Onward*.

Glasnost was a splendid weapon in the struggle against the opponents
of Gorbachev, but not a useful guide to action. Out of the two press
campaigns came only a revival of the ideas of Bukharin, a debunking of
myths that created its own myth. Everything done since 1928 was discovered to be a dirty tarpaulin that could supposedly be easily cast aside
to reveal a 'multiform' social and economic structure evoking the halcyon
days of NEPist Russia – a time when Soviet Russia was an agrarian country.
This conception turned *perestroika* sharply away from reshaping the existing
industrial management toward dismantling it as a remnant of Stalinism.[4]

When it became necessary for Gorbachev to dodge responsibility for
the conflict between Armenia and Azerbaijan over Karabakh, the *glasnost*
intellectuals discovered that Lenin, in contrast to Stalin, had actually been
very kind to the nationalities and had permitted them every option, including secession. In the search to mobilize grass-roots struggle against
the conservative party apparatus, the national Communists and even the
nationalists of the Union republics were sought as allies.

The *perestroichiki* found themselves in the position of the sorcerer's
apprentice. The protracted discussion of the crimes of Stalin had by 1989
filled the public at large, as well as many intellectuals, with a feeling of
national self-loathing that could only be remedied, it was widely judged,
by 'repentance' and renunciation of the 'sins of October'. Thus *glasnost*
confirmed the old adage about the conservatism of revolutions and their
desire to return to an idealized past. The struggle for power, in calling
up popular support against Gorbachev's Politburo opponents, also inadvertently called up a sweeping, revolutionary anti-Communism among the
intelligentsia.

The apparent paradox – Gorbachev's destruction of Communism in
his attempts to strengthen it – may seem less onerous if it is considered
that this was actually the *only* way that Communism could have been
destroyed. Had Gorbachev set out in 1985 with a conscious intention to
replace Communism with democracy he would surely have been stopped.
Without his legion of Communist supporters, consistently comprising an
overwhelming majority at every plenum and congress, convinced that
Gorbachev was leading them to new positions of strength in a democratic system, he would never have had the backing to establish the institutions that, despite him and them, ended by destroying Communism.
This may have been bungling, to be sure, but bungling of a particular
Soviet type, bungling in the course of entrenched and bitter struggle,

bungling in the name of newly discovered orthodoxies, bungling driven by slogans promising a heaven on earth as the result of the next historical adventure. Thus the bizarre spectacle of Gorbachev's confidently arrogating more and greater powers to himself as President of a regime that was crumbling to powder.

So, the utopia of Soviet power ended in repentance of the sin of hubris. An odd but, some might say, quite Russian fate for the ex-Soviet intelligentsia. In the spirit of renunciation of ill-gotten gains, repentance called for the liberation of the subject peoples. But what power was to remain for the ex-Soviet Russian successor state? The breakup of the Soviet Union seemed to condemn Russia to another *Smutnoe Vremya*, a Time of Troubles, comparable to the confused and turbulent period at the end of the sixteenth and beginning of the seventeenth centuries, between the end of the reign of Ivan the Terrible and the establishment of the Romanov dynasty. In those decades of chaos and civil war, Muscovy, defending a western border similar to the border of 1991, was invaded and occupied by Polish and Swedish troops. Nevertheless a great national rally led by the Church managed to expel the foreigners and elect a new Tsar, Michael Romanov.

The Revolution of 1917 was also a kind of Time of Troubles, with the traditional Russian lands having scattered, and a civil war spreading to every corner of the country, with foreign troops in action on several fronts. This time the national rally was led by the Bolsheviks who, despite their coming into power with slogans that made them seem to be anarchy personified, were still able to rally many Russian nationalists, especially military men, on the basis of patriotic appeals. Much of the territorial patrimony of the tsarist regime was recovered. The civil war had been a godsend for Bolshevism.

The revolution of 1991, however, would not be blessed by a civil war and foreign intervention and, lacking that, would not only have to accept the loss of Ukraine and White Ruthenia but the possible future disintegration of Russia proper. The intelligentsia had to consider the idea that the *Rossiia* of the Yeltsin revolution – one that included White Ruthenia and Ukraine – might also prove to be a utopia. They were in the ludicrous position of the Austrian Germans, who in the middle of the First World War and on the eve of their state's breakup declared, as their revenge on the subject nationalities, that the Habsburg Empire must in the future become a unitary *German* state. Many of the same Soviet intellectuals who rejected the slogans of Stalinism had embraced a Russian nationalism of the most innocent and dangerous kind. At its centre was a Russia that dispenses with its traditional claim to lead the many peoples of the former Union, while assuming that these same peoples will nevertheless follow the lead of Russians voluntarily because *this* Russia is not the Russia of Lenin and Stalin but that of Tolstoy and Pushkin.

The behaviour of the intelligentsia in the events of 1989–91 at least provides us with the answer to an interesting question that has persisted in the West for some decades. It has always been wondered what was, in the broadest sociological sense, the intelligentsia's position in regard to the socialist idea. Asking the question required one to accept the Soviet definition of the word 'intelligentsia' – that is, white collar workers, civil servants, administrative and technical personnel, professionals of various kinds – the class that the German socialists earlier in this century had called a *neue Mittelstand*, a 'new middle estate'. It had been argued persistently by left and right opponents of the Soviet regime that socialism in practice was not the rule of the workers but of the intelligentsia, a New Class that had come to inherit the position in a modern industrial society once held by the now displaced bourgeoisie.

Andrei Amalrik's *Will the Soviet Union Survive until 1984?*, a famous analysis of the chances for a democratic movement, a work written in 1970 that still rewards the attention paid it today, considered it likely that the intelligentsia would be the main support for Soviet democracy.[5] Amalrik assumed that the democratic movement would not attack the foundations of nationalized industry. Yet the intelligentsia came out increasingly after 1989 in favour of a full market economy on the Western model. As well as the intelligentsia had been treated under the Communists, they wanted something more. They drew themselves up to their full height and collectively rejected the self-image of a gaggle of cowed state employees. The question of where the New Class stood on the question of socialism was answered clearly and unequivocally.

At the same time the peculiar circumstances of the intelligentsia's market choice now need to be borne in mind. The mood at the time of Communism's fall was repentance of the crimes of Stalinism, alongside a passion to re-enter the civilized world and to emulate western Europe. The values of western propertied civil society, of social and political experience dating from the Magna Carta, medieval town charters and free universities, Canon law – these are lacking to the quite traditionless ex-Soviet intelligentsia, which was only created in the course of the purges and campaigns of Stalin. Despite their seemingly middle-class instincts of the moment, their decision for the market was not so much an expression of the traditional values and interests of a real civil society as it was a moody and pragmatic reflex. For them, the market will have to deliver prosperity. It must tend to their care and feeding or the mood may change as abruptly as it did when the slide into repentance began in 1988.

This would hardly be a fair test for the idea of market economy, which will have to produce results that it has never generated in its long history. Optimists said in 1992–93 that everything would depend on the degree of help that would be forthcoming from the outside world. Yet a rescue of this sort has never been made. The miracle that led West Germany back to pre-war economic levels between 1945 and 1949 was performed in conditions far different from those of Russia in the Yeltsin revolution. In a sense the intelligentsia is in the throes of a utopia of repentance, and one that threatens a fierce backlash.

Gorbachev was originally led astray in 1988 by a different utopia of 'soviet power' in the economy, in the name of which he tore down the existing system and its decades-old division of labour, thinking its destruction would let loose constructive socialist energies. Captive of the regime's traditional propaganda of Soviet Man, he was 'unleashing the spontaneity of the workers in production'. But he really unleashed *mestnichestvo*, unbridled local particularism, first in the republics but then also in Russia, a regime of local embargoes, breaking down the exchange of products necessary for industry to function. Gorbachev found that he had to make emergency purchases of needed commodities abroad with hard currency earnings. By one estimate, the majority of the USSR's estimated 2000 metric tons of gold had been sold off by 1991, with gold production dropping 50 per cent in 1989–92. Oil exports, the other major hard currency earner, had dropped by two-thirds.[6] The market plans of his most hidebound opponents could not have been worse than what actually was done. The chaos in which the Russian economy now finds itself, and will find itself for some time to come, is not so much the result of the legacy of Stalinist rule, nor even of the stringencies of the free market choice under Yeltsin and Gaidar, as it is the result of the Gorbachev reforms themselves.

If the anticipated transition to an economy like that of the United States or Sweden should falter, the intelligentsia will find its faith in Western constitutional and democratic ideals sorely tested. As it contemplates prices of millions of rubles in everyday transactions, as it faces economic horrors of the kind that were seen in the early days of the Weimar Republic, the response will no doubt be to pin hopes on a Russian Pinochet. This idea has emerged so many times in the previous pages of this volume, voiced by both the democrats and their Communist opponents, that we may be permitted to suppose that it, along with democracy, represents something spontaneous in the Gorbachev-era political culture. The same academic, Andranik Migranyan, who in 1989 called for a dictatorship of Gorbachev to forestall Yeltsin's neo-Bolshevism, also called, in the wake of the coup in 1991, for a Yeltsin dictatorship supplanting Gorbachev, saying 'Yeltsin should rule by decree', and urging the suspension of all constitutional structures. The West would do well to understand that the ex-Soviet intelligentsia that has turned toward liberal democracy in the Gorbachev years is not itself a civil society, but a class that has decided the matter only as a pragmatic course of action, in a confused pursuit of immediate results. If this course should fail, it will no doubt turn to another saviour, probably toward a new dictatorship.

TOCQUEVILLE AND SPENGLER

Gorbachev, in overthrowing his government, his empire, and the existing European balance, has as well overthrown all our history, which will now have to be rewritten. Prior to 1989, the consensus of the most sophisticated and thoughtful studies was that the division of Europe first crafted at Yalta was not ephemeral, because it served the interests of many states, particularly the most powerful ones.[7] The division of Europe was thought to be permanent, going beyond its Cold War origins, and reflecting in fact the satisfaction of Germany's neighbours and other powerful states with the balance achieved by German defeat in the Second World War.

It was common for studies of the Cold War to argue, in homage to the European historians of the balance of power, that the security of the nations of Europe had always depended on restraining any one state that might attempt to unify Europe. The vitality of the European community was seen to have been assured by its national and even political diversity, and thus protected, oddly enough, by mutual antagonism and balance of strength. In the sixteenth century the potential universal power was the Austria of Charles V, then the Spain of Philip II; in the seventeenth, it was the France of Louis XIV; in the eighteenth and early nineteenth, it was the revolutionary France of Napoleon. Balance was preserved by wars of continental coalition against the power-seeking European unity.

Yet a traditional continental balance, it was argued, had not been enough to contain the power of Germany after its unification by Bismarck in 1870. For that, the outlier powers, America and Russia, had to be called in to restore the equilibrium in the two world wars of our century. In this reading, the outcome of these two wars had been foreseen by that wise observer of the 1830s, Alexis de Tocqueville, who noted the following in his study of American Democracy:

> There are now two great nations in the world which, starting from different points, seem to be advancing toward the same goal: the Russians and the Americans. Both have grown in their obscurity; and while the world's attention was occupied elsewhere, they have suddenly taken their place among the leading nations, making the world take note of their birth and their greatness almost at the same instant... Their point of departure is different and their paths diverse; nevertheless each seems called by some secret design of providence one day to hold in its hands, the destinies of half the world.[8]

Tocqueville was thought to have seen the rise of the superpowers and the mid-twentieth-century balance of power in a vision of genius. The two crude outlier powers, both culturally shaped by their contacts with Western civilization (and despite their both lacking a great cuisine) would come to dominate comely Europe, and therefore the world.

Before 1989, no one believed in the permanent division of Europe more firmly than Gorbachev:

> Let me say quite plainly, [he wrote in 1987] that all these statements about the revival of German unity are far from being *realpolitik*, to use the German expression. They have given West Germany nothing in the last forty years... There are and will be two German states with different social and political systems... one should proceed from the existing realities and not engage in incendiary speculations.[9]

Gorbachev's utopia, shared by many, was a reformed USSR in a renewed detente process of such depth and range as to amount to a liquidation of the Cold War, without disturbing the existing balance of the states of Europe. A reform-minded Soviet Union could be imagined to be malleable and open to the influences of Western culture and market forces. The Western model could be seen to have its best opportunities in a politically diverse Europe, once of course, the Soviets had abandoned their mischief in the Third World. That would have been the optimum outcome, in line with the visions of Tocqueville, for maximizing the influence of both the Soviet Union and the United States. Despite their intense rivalry, the two powers would still find themselves in the curious position of *frères-ennemis*.

All this has proved to be ephemeral. Instead of the permanent division of Germany and Europe, we now have a united Germany and a Europe restored from what will have to be regarded as the temporary divisions of the post-war era. Our history must be rewritten. Where it used to be said that the national revolts in Berlin in 1953, Budapest, Warsaw, and Poznan in 1956, and Prague in 1968 only demonstrated the impermeability of the blocs as spheres of influence, these risings will now be seen as indelible evidence of the inevitability of the breakup of those blocs.

It was also once argued that the east–west partition of Europe along an Iron Curtain had deep historical roots. The historian Arnold Toynbee saw a rough correspondence to the original division between Latin Christendom and Byzantium.[10] He considered the 'dwarfing' of the European powers by Russia and America the result of an inevitable process that was only speeded up by the two world wars, an ongoing historical tendency toward the unity of the world that was thought to be enhanced by the United States and Russia having been drawn into Europe. It was once argued, especially in the United States, that a politician like de Gaulle, who attacked the Yalta settlement as an artificial separation of the continent into imperial spheres of influence, was himself an anachronism and a throwback to bygone days. Now many will be tempted to see him as a statesman of vision who fought for a dream of eternal Europe and eternal Russia, beyond Communism and beyond the Cold War.

In place of Tocqueville's vision of the United States and Russia overshadowing Europe, we will now have to view the history of the twentieth century differently. The last two attempts at German hegemony may come to be seen as mere preliminaries showing the inexorability of European unity under one or another kind of German leadership, coming to the fore despite all the most formidable obstacles. The defeat of Germany in the two world wars of our century will no doubt be seen by the twenty-first century as a feeble attempt to ward off the inevitable. To be sure, no one today supposes that German reunification means a return to the scenarios of the Third Reich; the return of Europe for many observers means a boost for the cause of European federalism. If there is any hegemony, one wag has remarked, it is that of the Bundesbank and not the Wehrmacht.

Optimists go on from there to tell us that Europe will now be peacefully organized from the Atlantic to the Urals through the miracle of federative polity. A federated West Europe will now serve as a magnet for a federated East Central Europe and a federated Commonweath of Independent States. Democracy will prove a solvent of the old quarrels. It is certainly incontestable that, if everyone is happy and no wants to fight, there will be no unhappiness and no fights. Invoking the word federalism is only another way of stating this unexceptional proposition.

Despite the migration of millions of people, the poisonous ethnic conflicts that have been reborn, the chaos and dislocation due to the application of a new economic model, democracy may indeed solve the major problems. But the experience of the period before the Second World War, the last time, and in many cases the only time, when the nations of east central Europe were independent, leads us to expect a more turbulent time ahead. As we make our estimate of what to expect, we should bear in mind the essential facts of the balance of power between the two world wars. In that period peace was to have been guaranteed by the idea of Collective Security, as embodied in the League of Nations. A legal definition of aggression bound all the members to common efforts to preserve peace in the general interest. Germany was relatively integrated in her territories, but weakened by defeat and required by the Versailles treaty to pay reparations and remain disarmed. Yet everyone knew that whenever Germany decided to defy the French and the other powers, she could threaten to dominate Europe. The French were thought to be the great continental power, with their Eastern alliance system comprising the Baltic states, Poland, and the countries of the Little Entente, Czechoslovakia, Romania, and Yugoslavia, all of whom had taken lands from Hungary in the breakup of the Habsburg Empire.

The bloc that Clemenceau called a *cordon sanitaire* was assigned a dual task of hemming in Germany and walling off Soviet Communism. But these states all had outstanding territorial and minority conflicts with one another and turbulent domestic politics as a result of the hard economic conditions attending the organization of so many new national economies. Only one state, Czechoslovakia, went through the entire 20-year period as a democracy; the others all ended up as one or another species of dictatorship.

The leading Revisionist powers – that is, powers unhappy with the Versailles Treaty and hoping for territorial changes – were Italy, Germany, and Hungary. Italy, Hungary, and Austria did their best to undermine the Little Entente and to encourage those who undermined France. Under the leadership of a dynamic Germany, and with no substantial opposition from an intimidated and recoiling Russia, the Revisionist powers had by 1941 succeeded in the partition of Poland. The Little Entente proved to be no obstacle to Hungary in her pursuit of the partition of Czechoslovakia, Romania, and Yugoslavia. East Central Europe collapsed like a house of cards.

Today it seems far from such a fate. Yet similar patterns have resurfaced. The Baltic states are tending to gravitate, as they did then, toward the Scandinavian states. Poland cannot help but worry about the lands in Pomerania, Silesia, and East Prussia given her by Stalin. No one expects immediate German claims, but how long can Germans look eastward at cities like Szczecin, Wroclaw, Gdansk, and Kaliningrad without recalling that they were once called Stettin, Breslau, Danzig, and, in the case of the birthplace of Kant, Königsberg? Nor has Germany forgotten about the expulsion of millions of Germans from the Sudetenland. It will never regard the borders of the Czech republic as just. Hungary has to be interested in South Slovakia, Sub-Carpathian Ruthenia, the Banat, and Transylvania, with their substantial Hungarian minorities, lands she regained under Hitler, only to lose to Stalin.

The Balkans are seething in the way that they did before 1914, with many of the same issues, such as the fate of Croatia and Macedonia, still in question. In fact the fighting in Croatia in 1991 suggested the spectacle, not only of the impotence of the federated European Community, but also of Germany and Austria defying France, Holland, Britain, and the EC as well, in recognizing and supporting Croatia. The Serbs complained about German interference in lands that had once been part of the Habsburg Empire, and got the moral support, at least implicitly, of Britain,

France, and Russia, in a ghostly echo of the diplomatic lineup of 1914.

Perhaps we cannot make too much of these things. History does not repeat itself. It is bizarre to speak today, as the Serbs do, of a hostile ring of Catholic, German-sponsored states in a 'Teutonic Bloc' in central Europe. Germany is a beacon of economic success and social progress whose domestic policies might well be envied by others in the West. Yet no state has been strengthened in its potential by the revolutions of 1989 and 1991 as has Germany. Even assuming that Germany continues to be a model democracy, she may find herself in the coming decades the only European state capable of acting with immediate economic and political effect *vis-à-vis* conflicts in the countries to her east. Not much of this is visible now, but German ability to act will emerge when she throws off, as she must one day, the nuclear and other military abstinences from her defeat in the Second World War.

German authority now derives in no small measure from a position as the potential European federator. Tocqueville, who predicted a Europe balanced by the United States and Russia, has proven to be a less reliable guide than Spengler, who wrote the following in his *Decline of the West*:

> A United States of Europe, realized through Napoleon as a founder of a romantic and popular military monarchy, is the analog of the realm of the Diadochi [the heirs of Alexander the Great]; when realized as a twenty-first century economic organism by a matter-of-fact Caesar, it will be the counterpart to the *imperium romanum*.

Gorbachev's policy in the events of 1989 had originally contemplated a new Central European bloc of neutrals between east and west, but in the end he was faced with a return to the old European state system with Germany as its natural centre of gravity. In that case, what is the natural position for the new, more modest post-*perestroika* Russia? Gorbachev saw Russo-German cooperation as essential, and expressed pride in the Stavropol agreement as perhaps his greatest foreign policy achievement. Shortly after the coup, while he was still struggling for the Novo-Ogaryevo process, he rejected the idea of 'a united [Soviet] Union as a counterweight to a united Germany', advocating close collaboration between the two countries 'to avoid shocks and chaos'.[11]

Just prior to the coup, in his 'Crimea article', he compared his own foreign policy with that of Prince Gorchakov and the line of *recueillement*. Gorbachev wanted to have his own era compared with the era of the Russian reforms of the 1860s. Sure enough, as Gorchakov's policy had led indirectly to Bismarck's unification of Germany, Gorbachev's had led to its reunification. In the subsequent era of Bismarck, that is, from 1871–90, Russia refrained from participating in any combination openly hostile to

Germany, in a period when war might have broken out over a number of different disputes. Russia was extremely aggressive then, while she is utterly demoralized today. Yet Gorbachev in defeat continued to look to Germany for succour, even going as far as to express the desire that Ukraine's actions within the Commonwealth of Independent States should take for a model the status of Bavaria in Germany.

At the time of the signing of the Stavropol agreement, it was understandable that journalists and other observers should immediately ask if Stavropol was in fact another Rapallo. The fragmented and contentious landscape between Berlin and Moscow would seem to call for some kind of understanding between the two powers. They both remember that the First World War resulted from the prolonged disintegration at the end of the nineteenth century of another multinational empire, Ottoman Turkey, and the subsequent struggle among its successor states in the Balkans, in which Germany and Russia ultimately had to choose sides. The Second World War originated in the disintegration of the multinational Habsburg Empire in 1918, and the birth of another group of successor states in East Central Europe. This area again turned into a battleground between Germany and Russia. Now the collapse of the multinational Soviet empire has left behind a destitute group of new states with outstanding border disputes. Germany has recovered its unity and Russia will at least survive as a nuclear power, so they both have to consider the political landscape between them and its seemingly endless potential for conflict.

This is especially difficult because Russia will never accept the boundaries of 1991 as permanent. And precisely this is put to her by the West as a requirement of post-Gorbachev revival. Russia must act as a good citizen of Europe, a supranational actor sponsoring various disinterested peacekeeping enterprises, in Karabakh, for example, or Moldova, while at the same time assisting the transition of Ukraine and the other nations to full sovereignty, in a spirit of resignation and self-abnegation consistent with the mood of 1989–91.

Some Soviet writers then compared Russia's situation with that of Germany after the Second World War, no doubt looking toward the same kind of moral reformation that the Germans underwent, similarly eased by a new Marshall Plan. But it would be more apt to compare Russia's post-Soviet condition with that of Germany after the First World War, a nation ruined, partitioned, and humiliated, but not really defeated, nevertheless sullen, revisionist, and pervaded by the sense of having been stabbed in the back by traitorous leaders.

These themes will no doubt continue to be sounded by many of the leading Russian politicians. Russia will be called an Eurasian power whose most natural ties are with Iran, India, and China. It will be said to have no choice, in the face of the eastward expansion of the Western alliance, but to scuttle arms control and help to arm Asia. It will be set the task of reassembling the Union by one means or another, by those of a Stresemann or a Hitler. No less a friend of the West than Boris Yeltsin himself subscribes to the idea of Russia as a Eurasian power and has said many times, referring to Russian minorities in the Near Abroad, that, one way or another, 'Russia will defend Russians.' Russia is now, however, only one actor among several in a new configuration of Eurasian balances and her fate will depend to a great degree upon what Gorbachev's children, the Germans and the Ukrainians – not to mention the Romanians, Turks, Iranians, Indians, and Chinese – will decide to do.

However future developments unfold, history will trace their beginnings to Gorbachev, according to Gorbachev credit for progress and blame for misfortune. One is reminded of the historians' controversy around Bismarck's intentions in his last years: one school held that he meant to support Russia, and another that he leaned toward Britain. Of this, historian Bernadotte Schmitt remarked that 'Bismarck's own utterances support each view.' As will those of Gorbachev. So he cannot fail to have his place in history as the centre of all the changes of this era.

When Hegel spoke of the World Historical Individual, he meant a Caesar, an Alexander, a Napoleon. This will not do for Gorbachev. And he will not be compared to the great figures in Russian history, Peter the Great, Catherine the Great, or Lenin, all of whom gathered up power and wielded it without diffidence. Even Yeltsin will no doubt appear to be more forceful and purposeful than the prodigal Gorbachev. In fact no figure from history offers a worthy comparison, and from literature, perhaps there is only King Lear, drawn by vanity to seek affection in ritual obeisances, misled by a foolish credulity into the partition of his kingdom among those who did not love him, finally banished by ingrates, yet at the end of it all strangely lifted by his losses to a higher wisdom.

Gorbachev is not gone; he will say and do more before he quits this world. In respect to all the Russian and East European revolutions of the twentieth century, posterity will no doubt make him the central figure, and a figure of paradox. It was he who proved the work of Lenin and Stalin to have been a detour. A statesman who played the key role in liquidating one revolution and in making another, he is likely to be thought at the same time a traitor at home and a hero abroad, a man who ended up by disowning conquerors and conquests, claiming to see in them a source of weakness rather than strength. Duff Cooper said of Talleyrand that 'He could surrender without qualm possessions he had not coveted.'[12] Gorbachev too will be remembered by admirers as one who may have acted out of love for his own country, but whose animating spirit was broader and drew more deeply on a core of European and Atlantic belief, one that attracted him irresistibly to rejoin the larger community, and thus the world.

Notes

1 A GORBACHEV EPIPHANY

1. Interview with Reagan in *Washington Post*, 26 Feb. 1988, A18.
2. A. W. DePorte, *Europe Between the Superpowers: The Enduring Balance* (New Haven and London, Yale University Press, 1979), xii.
3. With some exceptions, for example, Michael Howard, 'The Gorbachev Challenge and the Defense of the West', *Survival* (Nov.–Dec. 1988).
4. *Time*, 4 Jan. 1988, 16.
5. *New York Times*, 16 May 1989, A1, A10.
6. *Nezavisimaia gazeta*, 8 July 1994, 1. Gorbachev's account of the 'political trial of the ex-President' is in his *Memoirs* (New York and London, Doubleday, 1995), 681–3.

2 THE OLD REGIME OF THE SOVIET COMMUNISTS: FOREIGN POLICY IN THE COLD WAR

1. Quoted in B. H. Sumner, *Russia and the Balkans, 1870–1880* (Oxford University Press, 1937), 98. V. P. Potemkin (ed.), *Istoriia diplomatii* (Moscow, Gosudarstvennoe sotsialno-ekonomicheskoe izdatelstvo, 1941), 470–71 translates Gorchakov's original French *se recueille* as *sosredotochivaetsia* (to be concentrated, fixed, focused).
2. Arkady N. Shevchenko, *Breaking with Moscow* (New York, Knopf, 1985), 156. Jon Jacobson, *When the Soviet Union Entered World Politics* (Berkeley, Los Angeles, University of California Press), 104.
3. Andrei Gromyko, *Memories*, trans. Harold Shukman (London and Sydney, 1989), 342–3.
4. *Voprosy istorii*, No. 1 (1945), 4.
5. *Khrushchev Remembers: The Last Testament* (New York, Little, Brown, 1976), 393.
6. A. W. DePorte, *De Gaulle's Foreign Policy, 1944–1946* (Cambridge, MA, Harvard University Press, 1968), 29–44. Georges Bidault, *Resistance* (London, Weidenfeld and Nicolson, 1965), 147–9.
7. E. S. Varga, *Izmeneniia v ekonomike kapitalizma v itoge vtoroi mirovoi voiny* (Moscow, 1946), 226–68.
8. E. S. Varga, *Anglo-amerikanskie ekonomicheskie otnosheniia* (Moscow, 1946), 17–18.
9. E. S. Varga, 'Demokratiia novogo tipa', *Mirovoe khoziaistvo i mirovaia politika* (March 1947), 14.
10. Ibid., 10.
11. Ibid., 13.
12. Milovan Djilas, *Conversations with Stalin* (New York, Harcourt, Brace, 1962), 113.
13. Zhdanov's speech in Leningrad, 6 Feb. 1946, text in *Pravda*, 8 Feb. 1946, 3.
14. Ibid.
15. Text in *Bolshevik*, No. 3 (Feb. 1946), 2.

16. 'Prikaz' of the Ministry of the Armed Forces, signed by Stalin, text in *Pravda*, 7 Nov. 1946, 1–3.
17. Zhdanov's speech of 6 Nov. 1946, text in *Pravda*, 7 Nov. 1946, 1–3.
18. Ibid., 2.
19. Stenographic record of the discussion in a supplement to *Mirovoe khoziaistvo i mirovaia politika* (Dec. 1947).
20. K. V. Ostrovityanov, 'O nedostatkakh i zadachakh nauchno-issledovatelskoi raboty v oblasti ekonomiki', *Voprosy ekonomiki*, No. 8 (1948), 71–2.
21. Varga's testimony in a discussion in *Voprosy ekonomiki*, No. 3 (1949).
22. E. S. Varga, 'Protiv reformistkogo napravleniia v rabotakh po imperializmu', *Voprosy ekonomiki*, No. 3 (1949), 83–4.
23. Speech by Grosz on Hungarian television, *FBIS-EEU* (88–172), 5 Sep. 1988, 19.
24. See Boris Nicolaevsky, 'SSSR i Kitai', *Sotsialisticheskii vestnik* (May 1955), 79–82. Also Boris Souvarine, 'L'Ordre des préséances au Politbureau soviétique', *BEIPI* (16–30 June 1953), 1–5.
25. Wang Ming, *Mao's Betrayal* (Moscow, 1979), 63–4.
26. Molotov's speech to the Moscow Soviet, 6 Nov. 1948, text in *For a Lasting Peace, For a Peoples' Democracy*, 15 Nov. 1948, 1.
27. *Khrushchev Remembers* (New York, Little, Brown, 1970), 400–5.
28. Acheson's Testimony to the Senate Foreign Relations Committee, 11 Sep. 1950 (Washington, USGPO, 1974), 374.
29. Sherman Adams, *Firsthand Report* (New York, Harper and Bros., 1961), 124.
30. J. V. Stalin, *Economic Problems of Socialism in the USSR*, supplement to *New Times*, 29 Oct. 1952, 14.
31. K. V. Ostrovityanov, 'Sotsialisticheskoe planirovanie i zakon stoimosti', *Voprosy ekonomiki*, No. 1 (1948), 30.
32. Edward Fursdon, *The European Defense Community: A History* (New York, St Martin's, 1980), 272–3, reviewing the ample evidence for this widely held view.
33. Mohamed Heikal, *Cutting the Lion's Tail: Suez through Egyptian Eyes* (New York, Arbor House, 1987), 104.
34. E. S. Varga, in a discussion in *Voprosy ekonomiki*, No. 9 (1948), 54.
35. George Modelski, *Atomic Energy in the Communist Bloc* (Melbourne, 1959), 181–4.
36. *Khrushchev Remembers*, 498.
37. See Richard Lowenthal, *World Communism: The Disintegration of a Secular Faith* (New York, Oxford University Press, 1966), 79–80.
38. Anwar el-Sadat, *In Search of Identity* (New York, Harper and Row, 1978), 147.
39. See Tito's letter of 19 April 1957, text in Steven Clissold, *Yugoslavia and the Soviet Union, 1939–1973: A Documentary Survey* (London, Oxford University Press, 1975), 273.
40. *People of the World Unite! Strive for the Complete Prohibition and Thorough Destruction of Nuclear Weapons*, text in *Peking Review*, No. 32 (9 Aug. 1963), 11.
41. Maurice Couve de Murville, *Une politique étrangère, 1958–1969* (Paris, Plon, 1971), 32–3.
42. David Schoenbrun, *The Three Lives of Charles de Gaulle* (New York, Atheneum, 1966), 291–4. Also, Alfred Grosser, *La politique extérieure de la République* (Paris, Editions du Seuil, 1965), 142–3.
43. For a colourful thesis of a Franco-American 'Cold War', in the currency markets, see Paul Einzig, *The Destiny of Gold* (London and Basingstoke, 1972), 84–90.

44. Apparently seeking US clearance to attack them. See H. R. Haldeman, *The Ends of Power* (New York, Times Books, 1978), 90–91; Henry A. Kissinger, *White House Years* (Boston, Little, Brown, 1979), 183–5.
45. Gromyko, *Memories*, 282.
46. Kissinger, *White House Years*, 1049.
47. Speech by Suslov in Tashkent, text in *Pravda*, 23 Oct. 1974, 2.
48. Ibid.
49. T. Timofeyev, 'Znamia revoliutsionnoi bor'by proletariata', *Kommunist*, No. 6 (April 1975), 99. Timofeyev was at the time director of the Academy of Sciences' Institute of the International Workers' Movement. Stalin's formula had been: 'As a result of the first crisis of the capitalist system of world economy, there issued the first world war; as a result of the second crisis of the capitalist system of world economy, there issued the second world war.' Speech of 9 Feb. 1946, text in *Bolshevik* (Feb. 1946), 2.
50. A. Mileikovskii, 'Bluzhdanie v potemkakh', *Izvestiia*, 17 Dec. 1974, 3.
51. Timofeyev, 'Znamia revoliutsionnnoi bor'by', 105.
52. L. Agapov, 'Uspekhi sil demokratii', *Izvestiia*, 30 April 1975, 2. See also the interview with Cunhal in *Pravda*, 14 May 1975, 5.
53. S. Salychev, 'Revoliutsiia i demokratiia', *Kommunist* (Nov. 1975), 220–21.
54. 'Znamia revoliutsionnoi bor'by', 102. The slogan of the Offensive had been used by Bukharin, Bela Kun, August Thalheimer, and others in Germany before the disastrous March rising of 1921.
55. See Terry McNeill, 'The Specter of Vosnesenskii Stalks Suslov,' *Radio Liberty Research*, 338/74 (11 Oct. 1974).
56. L. Brezhnev, report to the Twenty-fifth Congress, *Pravda*, 25 Feb. 1976, 2.
57. Ibid., 4.
58. See William G. Hyland, *Mortal Rivals: Understanding the Pattern of Soviet-American Conflict* (New York and London, Simon and Shuster, 1987), 131–5.
59. Korotich's memoirs, as excerpted in *Nezavisimaia gazeta* (July 1991), 15.
60. Mikhail S. Gorbachev, 'No Time for Stereotypes', *New York Times*, 24 Feb. 1992, A13.

3   HERO OF THE HARVEST

1. Martin Walker refers to Gorbachev's conversations in Britain on his 1984 visit in which he told reporters that his father had died in the war. *Waking Giant: Gorbachev's Russia* (London and New York, Pantheon), 1988, 2. Dev Murarka, *Gorbachov: The Limits of Power* (London and Melbourne, Hutchinson, 1988), 40, maintained, correctly, that he died in 1976. Gorbachev, *Memoirs*, 33, says that he was only thought to be dead in the Carpathians in 1944.
2. Interview of a classmate of Gorbachev's with the author, 1 June 1989.
3. 'Mikhail Sergeevich Gorbachev,' *Pravda*, 12 March 1985, 1; *Ezhegodnik Bolshoi Sovetskoi Entsiklopedii* (Moscow, 1987), 558.
4. Moscow Radio Home Service, 19 Sep. 1986, quoted in Murarka, 41.
5. Zdenek Mlynar, 'Il mio compagno di studi Mikhail Gorbachev', *L'Unità*, 6 April 1985, 9.
6. Quoted in *Wall Street Journal*, 12 March 1985, 28.
7. Zhores Medvedev, *Gorbachev* (New York and London, Norton, 1986), 54–5.
8. Mlynar, *loc. cit.*
9. Valery Boldin, *Ten Years That Shook the World: The Gorbachev Era as Witnessed by His Chief of Staff* (New York, Basic Books, 1994), 175. Gorbachev, *Memoirs*, 97, says simply that he died of heart failure.

10. On the decline of Kirilenko as 'Crown Prince', see Abdurakhman Avtorkhanov, *Dela i dni Kremlia: Ot Andropova k Gorbachevu* (Paris, YMCA, 1986), 61–2. Also Myron Rush, 'Succeeding Brezhnev', *Problems of Communism* (Jan.– Feb. 1983), 3.
11. For the effects of the bombing, see Olivier Roy, *Islam and Resistance in Afghanistan* (London, New York, Cambridge University Press), 1986, 197–8.
12. Viktor Golyavkin, 'Iubileinaia rech'', *Avrora*, Dec. 1981, 75. According to Elena Klepikova, who was an editor at *Avrora* before 1982, Romanov exercised extremely tight control of content at the review. See Vladimir Solovyov and Elena Klepikova, *Behind the High Kremlin Walls* (New York, Berkeley Books, 1986), XVII–XVIII.
13. Interview with Kuzichkin, *Time*, 22 Nov. 1982, 33–4.
14. Gorbachev, *Memoirs*, 142–3; Amy Knight, 'Soviet Politics and KGB/MVD Relations', *Soviet Union*, Vol. 11, part 2, 1984, 157–81.
15. For example, Joseph Kraft, reporting on discussions with Roy Medvedev, in 'Letter from Moscow', *The New Yorker* Jan. 31 1983, 106.
16. According to a statement of negotiator Iuli Kvitsinsky, on Brezhnev's death. See Strobe Talbott, *Deadly Gambits* (New York, Vintage, 1985), 159.
17. *Foreign Report* (1766), 3 March 1983, 5.
18. According to reports reaching Seweryn Bialer. See *The Soviet Paradox: External Expansion, Internal Decline* (New York, Vintage, 1986), 95.
19. With Gromyko and Ustinov exercising a 'dual regency' (*dvoekratiia*) over the country, according to Gorbachev adviser A. S. Chernyaev, *Shest' let s Gorbachevym* (Moscow, Kultura, 1993), 16.
20. Yegor Ligachev, *Inside Gorbachev's Kremlin* (New York, Pantheon, 1993), 30–31. Boldin, op. cit., 45–6.
21. James M. McConnell, 'SDI, The Soviet Investment Debate and Soviet Military Policy', *Strategic Review* (Winter 1988), 54.
22. These are described more fully in my *Soviet Succession Struggles* (London and Boston, Allen and Unwin, 1988) Chs. 3–5.
23. Ibid., 176–7.
24. Medvedev, *Gorbachev*, 146.
25. Arkadi Shevchenko cites a 1939 report of F. T. Gusev, party secretary of the Peoples' Commissariat, to this effect. *Breaking with Moscow*, 146–7. According to him, Molotov shouted to subordinates: 'Enough of Litvinov liberalism. I am going to tear out that kike's wasp's nest by the roots.'
26. Apparently as an agent of the KGB, Zhirinovsky worked for a brief time in the Jewish organization, Shalom. In 1990, with KGB help, he founded the first non-Communist party, the Liberal Democrats, and despite failing to receive the requisite 100 000 signatures to run for President as its candidate in June 1991, he was put on the ballot by a special vote of more than 50 per cent of the Russian Federation Congress of Peoples' Deputies. He surprised all by finishing third with over six million votes.

4 'ACCELERATION OF THE PERFECTION', 1985–87

1. *Current Soviet Policies X* (1988), 80.
2. Dominique Dhombres, 'M. Gromyko a lancé une mise en garde', *Le Monde* (15 March 1985), 2.
3. Ligachev, *Inside Gorbachev's Kremlin*, 73–5. This account is confirmed by Boldin, *op. cit.*, 60–61.
4. K. S. Karol, 'Le Gamin de Moscou', *Nouvelle Observateur* (15 March 1985), 33.

5. G. A. Aliev, 'Istoricheskaia pravota idei i dela Lenina', *Pravda* (23 April 1985), 2.
6. Ibid., 3.
7. V. Chebrikov, 'Sveriaias' s Leninym, rukovodstvuias' trebovaniiami partii,' *Kommunist* (June 1985), 53.
8. Ibid., 55. Chebrikov appears almost to echo the phrase of Zbigniew Brzezinski, *Ideology and Power in Soviet Politics* (New York, Praeger, 1962), 113–35.
9. Mikhail Gorbachev, *Selected Speeches and Articles* (Moscow, Progress, 1987), 52.
10. Ibid., 67.
11. Raymond Garthoff, introduction to G. D. Wardak and G. H. Turbiville (eds) *The Voroshilov Lectures: Materials from the Soviet General Staff Academy* (Washington, DC, National Defense University, 1989), 8.
12. Michael MccGwire, 'Update: Soviet Military Objectives', *World Policy Journal* (Fall, 1987), 724.
13. McConnell, *op. cit.*, 48; Mary C. Fitzgerald, 'Marshal Ogarkov and the New Revolution in Soviet Military Affairs', *Defense Analysis* (Vol. 3, No. 1, 1987).
14. N. V. Ogarkov, 'Zashchita sotsializma: opyt istorii i sovremennost'', *Krasnaia zvezda* (9 May 1984), 3.
15. Ibid., 4. So called for a character in Gogol's *Dead Souls*, suggesting mediocrity and complacency.
16. The contempt shown by the military for Brezhnev in his last months meant that a new civil–military relationship had to be worked out between Andropov and Ogarkov. Jeremy Azrael argues that the contrast in their views cannot be entirely ascribed to a strategy of presenting a hard and soft face to the West, but involved a struggle over allocation of resources. See *The Soviet Civilian Leadership and the Military High Command, 1976–1986*, RAND paper R-3521-AF (June, 1987), 30–31.
17. 'In our time', write the authors of the 1973–75 General Staff course, 'the expression "delay is the cause of deaths" or "delay is similar to death" has a deep meaning.' *Voroshilov Lectures*, 184.
18. *Strategic Survey, 1985–1986* (London, International Institute for Strategic Studies, 1988), 91.
19. Moscow Radio in Finnish, 12 April 1984, quoted in Roland Smith, *Soviet Policy towards West Germany* (Adelphi Paper No. 203, 1155, 1985), 32–3.
20. M. S. Gorbachev, *Selected Speeches*, 101.
21. Alexandre Adler, 'Gorbachev, l'irrésistible ascension du petit dernier', *Libération*, 12 March 1985, 4. Tatyana Zaslavskaya, 'Novosibirsk Report' [Text in *Survey* (Spring, 1984), 88–108].
22. Seweryn Bialer, 'Gorbachev's Road to Power', *Washington Post Weekly* (1 April 1985), 23.
23. M. S. Gorbachev, *Selected Speeches*, 222.
24. For example, in his interview of 8 January 1989: 'The new political thinking, as is known, presupposes the de-ideologization of interstate (*mezhgosudarstvennye*) relations. But this does not at all mean – as some want to interpret it – the de-ideologization of international (*mezhdunarodnye*) relations.' *Pravda* (8 January 1989), 1.
25. M. S. Gorbachev, *Selected Speeches*, 319.
26. Ibid., 267.
27. Ibid., 353. Gorbachev referred to 'Left Communists' and 'Trotskyites'. Bukharin and others had indeed taken this position. Trotsky had called instead for the famous compromise slogan, 'Neither War nor Peace', in order to avoid the campaign westward, and to clothe a policy of war avoidance in a customary ringing phrase.

28. On Stalin's use of the Brest peace as a precedent for the Hitler–Stalin pact, see Robert Slusser, 'The Role of the Foreign Ministry', in Ivo J. Lederer, *Russian Foreign Policy: Essays in Historical Perspective* (New Haven and London, Yale University Press), 226–7.

29. For example, Charles Krauthammer, 'The Poverty of Realism', *New Republic* (17 Feb. 1986).

30. Aleksandr Bovin, 'Istoriia neobratima', *Izvestiia*, 8 and 10 March 1986.

31. Interview with *Révolution Africaine* in *New Times*, 23 April 1986, 6.

32. Something like this seems to be the reluctant conclusion drawn at this time by a number of Soviet experts. See N. Simonia in *Pravda*, 18 Jan. 1989, 4. Also A. Vasiliev in *Izvestiia*, 3 Feb. 1989, 6; and V. Maksimenko, 'Sotsialisticheskaia orientatsiia: Perestroika predstavlenii', *Mirovaia ekonomika i mezhdunarodnye otnosheniia* (Feb. 1989), 93–4.

33. Andropov's speech to the Presidium of the Supreme Soviet, text in *Kommunist* (October 1983), 14.

34. For the argument that Ponomarev was an influence behind the Anti-Zionist Committee of 1984 and a backer of Romanov, see Mikhail Agursky, 'Soviet Rejectionists', *Jerusalem Post*, 10 Feb. 1985, 8.

35. G. Kh. Shakhnazarov, 'Logika politicheskogo myshleniia v iadernuiu eru', *Voposy filosofii* No. 5 (1984), 63–74.

36. A. Dobrynin, 'Za bez'iadernyi mir, navstrechu XXI veku', *Kommunist* (June 1986), 22.

37. Ibid., 28.

38. Ibid., 27.

39. Ibid., 31. One searches in vain for a reference to this pivotal speech in the memoirs of Dobrynin, *In Confidence* (New York, Random House, 1995), despite the fact that the opposed sides in the struggles of the next five years would line up according to its etiquette, describing themselves as adherents either of the 'class struggle' or the 'universal human values' positions. Dobrynin now disavows his role in the fall of Communism.

40. Shakhnazarov, 'Logika politicheskogo myshleniia', 73.

41. Ibid.

42. M. S. Gorbachev, *Selected Speeches*, 280.

43. Moscow Television Service, 12 Nov. 1985.

44. M. S. Gorbachev, *Selected Speeches*, 499, 500.

45. Ibid., 590.

46. Ibid., 559.

47. Walter Pincus, 'After SALT II, Can Moscow Afford a Big Arms Buildup?' *Washington Post*, 7 July 1986, 19.

5 GORBACHEV BOUND: THE EMERGENCE OF THE LIGACHEV OPPOSITION

1. M. Odinets and M. Poltoranin, 'Za poslednei chertoi', *Pravda*, 4 Jan. 1987, 3. Chebrikov's apology in *Pravda*, 8 Jan. 1987, 1.

2. Yegor Ligachev, *Inside Gorbachev's Kremlin* (New York, Pantheon, 1993), 318, for the view that the January plenum was the key point at which discussion of economic reform went over into discussion of remaking the political structures.

3. See the interview with the historian Serge Afanasyan, *Le Point*, 7 March 1988, 42–5.

4. Alexandre Bennigsen, 'Mullahs, Mujahidin, and Soviet Muslims', *Problems of Communism* (Nov.–Dec. 1984), 44.
5. A. Lapin, 'Gorkii urok', *Komsomolskaia pravda*, 10 Jan. 1987, 2.
6. Kolbin's speech in *FBIS-SOV*, 14 Jan. 1987, R2.
7. Yu. Bromlei, 'Natsionalnye protsessy v SSSR: dostizheniia i problemy', *Pravda*, 13 Feb. 1987, 2–3.
8. Excerpts from a meeting of social scientists in Moscow, *Pravda*, 18 April 1987, 2.
9. M. S. Gorbachev, *Speeches and Writings*, Vol. 2, 129.
10. Ibid., 106–7.
11. Ibid., 331.
12. On Gorbachev's relations with his speechwriters, see Boldin, *op. cit.*, 208–10.
13. *Lenin's Last Letters and Articles* (Moscow, Progress Publishers, 1956), 8.
14. Yegor Yakovlev, 'Farewell', *Moscow News* (hereinafter *MN*), 18 Jan. 1987, 13.
15. From the minutes of the session, text in L. Trotsky, *The Stalin School of Falsification* (New York, Pioneer, 1937), 105. My own views on this tangled question of historical interpretation are in *Soviet Succession Struggles*, chs. 3–5.
16. G. L. Smirnov, 'Revoliutsionnaia sut' obnovleniia', *Pravda*, 13 March 1987, 2.
17. Interview with V. A. Tikhonov, 'Tak kuda zhe nas zovut?' *Literaturnaia gazeta*, 8 April 1987, 12.
18. *Pravda*, 9 Aug. 1987, 1.
19. Ibid., 2.
20. Andrei Sakharov, 'Of Arms and Reforms', *Time*, 16 March 1987, 42.
21. See Robert Legvold, 'Gorbachev's New Approach to Conventional Arms Control', *Harriman Institute Forum* (Jan. 1988), 1.
22. See, for example, Horst Afheldt, 'The End of the Tank Battle', *Bulletin of Peace Proposals*, Vol. 8, No. 4, 1977, 328–31; Anders Boserup, 'Non-Offensive Defence in Europe', in Derek Paul (ed.), *Defending Europe: Options for Security*, London and Philadelphia, Taylor and Francis, 1985, 194–209; Anders Boserup, A. Karkoszka, A. von Muller and Robert Nield, 'Force Structures Specialized for Defence as a New Approach to European Security', *Pugwash Newsletter* (April 1986), 113–15.
23. Jaruzelski's speech to PRON, Patriotyczny Ruch Odrodzenia Narodowego (Patriotic Movement for National Regeneration), *Trybuna Ludu*, 9-10 May 1987, 3. This was timed to coincide with the 30th anniversary of the Rapacki Plan, a disengagement scheme put forward by Polish Foreign Minister Adam Rapacki in 1957. See Zbigniew Lesnikowski, 'W interesie bezpieczeństwa i odprężenia', *Trybuna Ludu*, 2 Oct. 1987, 5.
24. Kissinger, *White House Years*, 217.
25. Vitaly Zhurkin, Sergei Karaganov and Andrei Kortunov, 'Reasonable Sufficiency – or How to Break the Vicious Circle', *New Times* (12 Oct. 1987), 14. 'Asymmetrical Response' as a description of a continuing option in American national security strategy, subsuming the idea of linkage, is found in John Lewis Gaddis, *Strategies of Containment* (New York and Oxford, Oxford University Press, 1982).
26. Lev Semeiko, 'Vmesto gor oruzhiia: O printsipe razumnoi dostatochnosti', *Izvestiia*, 13 Aug. 1987, 5.
27. Evgenii Primakov, 'Novaia filosofiia vneshnei politiki', *Pravda*, 10 July 1987, 4.
28. See for example 'A Document of the Executive Committee of the PCI on Security Policy for Italy and Europe' (written by Alessandro Natta), *The Italian Communists* (Oct.–Sep. 1986), 29–45.
29. When Gorbachev came to Italy for Enrico Berlinguer's funeral in 1987, he

told reporters that Berlinguer's ideas had assisted his own thinking. See the interview with *L'Unità*, 18 May 1987, in *Speeches and Writings* (Oxford and New York, Pergamon, 1987), Vol. 2, 206. 'A thought-provoking lesson about a different political culture was taught me at Berlinguer's funeral.' (Gorbachev, *Memoirs*, 159)

30. A. Kokoshin and V. Larionov, 'Kurskaia bitva v svete sovremennoi oboronitelnoi doktriny', *Mirovaia ekonomika i mezhdunarodnye otnosheniia*, No. 8, 1987, 32–40.

31. As, for example, in D. Yazov, 'Warsaw Treaty Military Doctrine – For Defence of Peace and Socialism', *International Affairs*, No. 10, 1987, 3–8.

32. See Theodore Karasik and Thomas Nichols, '*Novoe myshlenie* and the Soviet Military: The Impact of Reasonable Sufficiency on the Ministry of Defence', RAND paper, P-7521, April 1989.

33. Stenographic record of the October 1987 plenum, *Izvestiia TsK KPSS*, No. 2, 1989, text in *Political Archives of the Soviet Union*, Vol. 1, No. 1, 1990, 85–7.

34. Yuri Afanasyev, 'S pozitsii pravdy i realizma', *Sovetskaia kul'tura*, 21 March 1987, 3.

35. Soboleva was an authority on Lenin's relations and conflicts with the Mensheviks and Socialist Revolutionaries in 1917, on the Legal Marxism trend in Russia at the turn of the century, and on Lenin's quarrel with the 'Economist' tendency in 1902–3. Shirikov had written about the party and the intelligentsia. Murashov was the author of a monograph on the *Iskra* period. Vaganov was the leading authority on the turn toward collectivization of agriculture, on whose work many Western historians depended. See *KPSS v bor'be za uskorenie tempov sotsialisticheskogo stroitel'stva, 1927–1929gg.* (Moscow, 1967); and *Pravyi uklon v VKP(b) i ego razgrom, 1928–1930* (Moscow, 1970).

36. An old chestnut of the rewritten history of the Revolution. Oddly enough, Stalin's speech on Lenin's 50th birthday in 1920 even maintained that the subtle tactical path taken by the party had been *against* the urgings of Lenin for quick action and that Lenin, on 25 October, had admitted that the party (this presumably including Stalin himself) and not he, Lenin, had been right. 'The Fiftieth Anniversary of V. I. Ulyanov-Lenin', 1920, text in *The Stalin School of Falsification*, 200–1. That also seems to argue unwittingly against accusations of Trotsky's criminal temporizing. Apparently everyone temporized.

37. P. Soboleva, A. Nosov, L. Shirikov and S. Murashov, 'Apropos of Yu. Afanasyev's Article', *MN*, No. 19, 1987, 11.

38. For the historiography of this dispute see my *Soviet Succession Struggles*, chs. 1, 5.

39. Yu. Afanasyev, 'Talking about the Past, We Must Keep the Future of Socialism in Mind', *MN*, No. 19, 1987, 13.

5 THE THOUGHT OF MIKHAIL GORBACHEV: A TREATISE AND A SPEECH

1. Ligachev, *Inside Gorbachev's Kremlin*, 285–7.

2. *Oktiabr' i perestroika: revoliutsiia prodolzhaetsia*, text in *Izvestiia*, 3 Nov. 1987, 2.

3. V. I. Lenin, 'On Cooperation', *Lenin's Last Letters and Articles* (Moscow, 1956), 31.

4. Radicals who did object soon found that there was no place for them in the Communist Party, as G. Miasnikov, of the Workers' Group was to find when

he called the NEP the 'New Exploitation of the Proletariat'. See V. Sorin, *Rabochaia gruppa (Miasnikovshchina)* (Moscow, 1924); and Roberto Sinigaglia, *Mjasnikov e la rivoluzione russa* (Milano, 1973).

5. 'Oktiabr' i perestroika', 2.
6. See Chapter 2.
7. *K Sotsializmu ili kapitalizmu?* (Moscow, Gosplan, 1925). This document was written before Trotsky had entered into opposition to the Stalin–Bukharin bloc and their theory of Socialism in One Country, and its views are compatible. Concluding (p. 60), he entertains the prospect that capitalism might not plunge into crisis but find new paths of development, in which case it would be time to admit that 'we had been mistaken in our fundamental historical judgments', and that the Soviet experiment would be in for difficult times.
8. 'Oktiabr' i perestroika', 2.
9. Ibid., 3.
10. Ibid., 3.
11. Text of Khrushchev's speech in *Crimes of the Stalin Era*, annotated by Boris Nicolaevsky (New York, New Leader, 1962), S13.
12. *Pravda*, 14 Oct. 1932, 3.
13. V. M. Molotov's self-criticism in *Kommunist* (Sep. 1955), 127–8.
14. 'Oktiabr' i perestroika', 3.
15. Michel Tatu, 'Une critique mesurée de la période stalinienne', *Le Monde*, 4 Nov. 1987, 4.
16. *New York Times*, 4 Nov. 1987, 4. Daniel Singer actually thought the main weakness of the speech to be its 'un-Marxist' way of seeing the central fault of the system in the 'psycho-pathology of one man'. 'On Recapturing the Past', *Nation*, 12 Dec. 1987, 716.
17. Mikhail Gorbachev, *Perestroika: New Thinking for Our Country and the World* (New York, Harper and Row, 1987), xiv.
18. M. S. Gorbachev, 'Leninizm – zhivoe tvorcheskoe uchenie, vernoe rukovodstvo k deistviu', *Pravda*, 23 April 1983, 1.
19. *Perestroika*, 26.
20. Ibid., 30.
21. Ibid., 66.
22. Fedor Burlatskii, 'Uchitsiia demokratii', *Pravda*, 18 July 1987, 3.
23. Boris Kagarlitsky, *The Thinking Reed: Intellectuals and the Russian State from 1917 to the Present* (London and New York, Verso, 1988), 17.
24. *Perestroika*, 80.
25. Ibid., 155.
26. Ibid., 157.
27. Ibid., 174.
28. Ibid., 179.
29. Ibid., 206. 'They think that if the USSR is afraid of SDI, it should be intimidated with SDI morally, economically, politically, and militarily. This explains the great stress on SDI, the aim being to exhaust us.' (220) For other references to an American 'exhaustion strategy' and a Soviet 'asymmetrical' response, see 226.
30. Ibid., 185.
31. See Anatolii Cherniaev, 'Gorbachev and the Reunification of Germany: Personal Recollections', in Gabriel Gorodetsky, *Soviet Foreign Policy, 1917–1991 A Retrospective* (London, Frank Cass, 1994), 30–44, for a purely fictional apologia suggesting that Gorbachev actually intended to unify Germany.

7 BETWEEN YELTSIN AND LIGACHEV

1. Boris Yeltsin, *Against the Grain: An Autobiography* (New York and London, Summit, 1990), 25. Perhaps Yeltsin was drawing on the traditional Jewish tale of Zlateh the goat.
2. Ibid., 138–9.
3. *Khrushchev Remembers* (Boston, Little, Brown, 1970), 262–3.
4. Yeltsin, *op. cit.*, 119.
5. As quoted by Ligachev at the October plenum. *Izvestiia TsK KPSS*, No. 2, 1989, text in *Political Archives of the Soviet Union*, Vol. 1, No. 1, 1990, 87.
6. Text of the declaration in *Across Frontiers* (Winter 1988), 8–9.
7. *Pravda*, 11 Sep. 1987, 3.
8. The text of the 12 Sep. letter to Gorbachev is in Yeltsin, *op. cit.*, 178–81.
9. Ibid., 128–30. At the meeting of 21 Oct., however, Yeltsin declared himself satisfied with the historical portions of the speech.
10. *Izvestiia TsK KPSS*, No. 2, 1989, text in *Political Archives of the Soviet Union*, Vol. 1, No. 1, 1990, 85.
11. Ibid., 86.
12. The text published in 1989 confirmed the earlier leak by a 'Soviet official' to Philip Taubman ('Aide Who Assailed Gorbachev's Pace Ousted in Moscow', *New York Times*, 12 Nov. 1987, 4) to the effect that Ligachev had openly blamed Gorbachev for promoting Yeltsin.
13. *Izvestiia TsK KPSS*, 125.
14. See Jacques Amalric, 'Un avertissement pour le numéro un soviétique', *Le Monde*, 29 Nov. 1987, 4.
15. Text in *Le Monde*, 2 Feb. 1987, 6.
16. Bill Keller, 'Soviet Marshal Sees Star Wars as Edge for U.S.', *New York Times*, 30 Oct. 1987, 1.
17. The evidence for this view is presented in Bruce Parrot, 'Soviet National Security under Gorbachev', *Problems of Communism* (Nov.–Dec. 1988), 1–36.
18. As he told a Politburo session on 17 Dec., he was taking advantage of the 'human factor' (*chelovecheskii faktor*) in international politics that permitted his personal contacts across 'class' lines with conservative representatives of the military-industrial complex, such as Ronald Reagan, to achieve real results. Chernyaev, *Shest' let*, 188–9.
19. Gennadi Gerasimov, 'Dithyrambs and Dissonances', *MN*, 6 Dec. 1987, 3.
20. *Izvestiia TsK*, 111.
21. Alexandre Adler, 'URSS: Gorbachev place un précieux allié à la tête du parti communiste moscovite', *Libération*, 12 Nov. 1987, 48.
22. *Pravda*, 28 Dec. 1987, 1.
23. Nikolai Shmelyov, 'Peril of Another Kind', *MN*, 10–17 Jan. 1988, 3.
24. 'Un entretien avec le numéro deux soviétique', *Le Monde*, 4 Dec. 1987, 6.
25. The language of this speech prefigures that of Gorbachev's historical speech of Nov. 1987. See 'Nam nuzhna polnaia pravda', *Teatr*, No. 8 (Aug. 1986), 3, 7, and *passim*.
26. *New York Times*, 13 Dec. 1987, 10.
27. *Pravda*, 15 Dec. 1987, 1.
28. *Washington Post*, 26 Feb. 1988, 18; Lou Cannon, 'Reagan: No Pact By Moscow Summit', Ibid., 1, 18.
29. Information from a member of the NSC staff, July 1988.
30. *Pravda*, 13 Jan. 1988, 1.

31. *Pravda*, 19 Feb. 1988, 1.
32. *Pravda*, 21 Feb. 1988, 1.
33. *Pravda*, 21 Feb. 1988, 2.
34. A television documentary, 'Time of Hope', was shown in December, with details of Kirov's good qualities and his fight *against* the opposition. In a review, Aleksandr Plakhov wrote: 'He was fated to be shot, and the shot that killed him did not come from a rifle with a telescopic sight raised by Josef Stalin in the famous film sequence, but rather from the opposition, blind with envy at such human beauty, and fearing it.' 'It Makes You Want To Go On Living', *MN*, 20–27 Dec. 1987, 11.
35. Anatoli Rybakov, *Children of the Arbat*, Eng. trans. by Harold Shukman (Boston and Toronto, Little, Brown, 1988), 636–7.
36. See for example, Aleksandr Latsis, 'S tochki zreniia sovremennika', *Izvestiia* 17 Aug. 1987; and Olga Kuchkina, 'Shto by ni bylo', *Komsomol'skaia pravda* 14 June 1987, 4.
37. 'Istoriko-partinaia nauka: puti perestroiki i dalneishego razvitiia', *Voprosy istorii KPSS* (July 1987), 146.
38. *Pravda*, 10 Jan. 1988, 4.
39. This despite news in *Pravda* (20 Jan. 1988) of what was called a 'second Berkhin Affair', in which KGB and police officials had assisted Odessa First Secretary Nochevkin in framing a troublesome senior police officer, A. Malyshev. The latter was released in August 1987 as a result of investigation by the party's control commission.
40. V. M. Kuritsyn, '1937 god iz istorii sovetskogo gosudarstva', *Sovetskoe gosudarstvo i pravo*, No. 2, 1988, 111. See also G. S. Popov, 'Tochki zreniia ekonomista', *Nauka i zhizn*, No. 4, 1987, 62.
41. Ibid., 110. Citing the lack of Soviet military preparedness in 1929, Kuritsyn gave figures showing deficiencies in weapons and matériel by comparison with Germany and the United States. However, at the time these were two states with which the Soviets thought they had common interests. The external threat in the present and future was expected from England, France, and Japan.
42. Fyodor Burlatsky, 'Khrushchev: Shtrikhi k politicheskomu portretu', *Literaturnaia gazeta*, 24 Feb. 1988, 14. In fact, in his struggle with Malenkov in 1954–5, Khrushchev accused his opponent of Bukharinism.
43. David Gai, 'The Doctor's Case', *MN*, 14–21, 1988, 16.
44. Lev Voskresensky, 'You Should Know, Comrades', *MN*, 6 Dec. 1987, 12. Anna Larina, *This I Cannot Forget* (New York: W. W. Norton, 1993).
45. N. I. Bukharin, 'Surovye slova', *Izvestiia*, 22 Dec. 1934, 2.
46. Yevgeni Ambartsumov, 'The Truth Triumphs', *MN*, 21–8 Feb. 1988, 3.
47. Joseph C. Harsch, 'The Relevance of the Past', *Christian Science Monitor*, 10 Feb. 1988.

8 ANOTHER ESCAPE FORWARD, 1988

1. On the basis of estimates by Goskomstat, CIA, and PlanEcon for the appropriate years. See *New York Times*, 16 Sep. 1990. Official figures put the 1988 growth figure at six per cent.
2. *Izvestiia*, 6 Dec. 1987, quoted in Anders Aslund, *Gorbachev's Struggle for Economic Reform* (Ithaca: Cornell University Press, 1991), 79–80.
3. Abel Aganbegyan, *The Economic Challenge of Perestroika* (Bloomington: University of Indiana Press, 1988), 65.

4. Ibid., 196.
5. See Chapter 7.
6. Kurashvili pleads for a legal status for civil society centred on reform of the soviets in 'K polnovlastiiu Sovetov', *Kommunist* (May 1988), 28–36; and 'Kakoi byt' strukture vlasti?', *Izvestiia*, 28 July 1988.
7. Yuri Feofanov, 'Daughters of the Arbat', *MN*, 10–17 Jan. 1988, 13. Feofanov wrongly describes them as defendants in the trial of the 'Trotskyite-Zinovievite Terrorist Organization', which would indicate the 1936 trial of Zinoviev, Kamenev and their associates. Bukharin, Krestinsky, Rakovsky and others were tried in 1938 as the 'Anti-Soviet Bloc of Rights and Trotskyites'. Lomov was shot in 1937 and Smilga, who was part of the Zinoviev–Trotsky opposition of 1926–7, was arrested in 1932 and disappeared sometime after.
8. The exception was Genrikh Yagoda, head of the political police until 1934, who probably had a hand in preparing the cases against the defendants in the first two trials, and who was earlier said to be politically allied with Bukharin.
9. Bukharin biographer Stephen Cohen, in an extensive interview with *Moscow News* (28 Feb.–6 March 1988, 13), made no mention of the foreign policy positions of Bukharin that played a prominent role in Cohen's biography. Cohen even seemed to acknowledge the difficulty of Bukharin's opposing Stalin's decision to end the NEP in 1928: 'True enough, what can you do when told that war is in the offing and that there is no time to industrialize the country along NEP lines? Many economists in the west think that in this context, Bukharin's concept was naive.'
10. Anatoli Latyshev, 'Bukharin – izvestnyi i neizvestnyi', *Nedelia*, 21–27 Dec. 1987, 6–7.
11. Vyacheslav Dashichev, 'Vostok–zapad: poisk novykh otnoshenii', *Literaturnaia gazeta*, 18 March 1988, 14.
12. Aleksandr Samsonov, 'The Whole and Bitter Truth', *MN*, 14–21 Feb. 1988, 12.
13. Interview with Mieczyslaw Rakowski in *MN*, 31 Jan.–7 Feb. 1988, 6. Chernyaev relates that letters from the Poles inquiring about the Katyn massacre helped prompt the 1987 discussion of 'white spots' in Soviet history. *Shest let*, 172.
14. Meeting in the CPSU Central Committee, *Pravda*, 13 Jan. 1988, 1–3.
15. Ibid., 2.
16. *Pravda*, 21 Feb. 1988, 2.
17. *Pravda* editorial, 21 Feb. 1988, 1.
18. *Pravda*, 19 Feb. 1988, 1.
19. V. Glagolev, 'Khudozhestvennaia istoriia i istoricheskie sud'by', *Pravda*, 10 Jan. 1988, 4.
20. M. Kim, 'Leninizm i istoricheskie sud'by sotsializma', *Pravda*, 5 Feb. 1988, 2.
21. Dmitri Kazutin, 'Everyone is under the Jurisdiction of History', *MN*, 17–24 Jan. 1988, 12.
22. See Chapter 5.
23. Vadim Kozhinov, 'Pravda i istina', *Nash sovremennik*, No. 4 (1988), 162.
24. Aganbegyan, *Economic Challenge*, 126.
25. Albert Nenarokov, 'At the End of 1922', *MN*, 3–10 Jan. 1988, 8.
26. Timur Pulatov, 'Running Ahead of the Cart', *MN*, 10–17 April 1988, 10.
27. Interview with V. Maamagi, *Pravda*, 9 Feb. 1988, 3.
28. Dmitri Kazutin, 'Three Years Later – In Tallinn', *MN*, 18 Feb. 1988, 13.
29. Round table discussion, 'Natsionalnye protsessy v SSSR: Itogi, tendentsii, problemy' in *Istoria SSSR* (Nov.–Dec. 1987), 94.
30. See Chapter 4. See Richard Hovannisian, 'Caucasian Armenia between Imperial

and Soviet Rule', in Ronald Grigor Suny (ed.), *Transcaucasia* (Ann Arbor, U. Michigan, 1983), 259–93.

31. Between 1959 and 1979 the Armenian population in Karabakh declined from 85 per cent of the total to 75 per cent, while the Azeri population increased from 14 per cent to 23 per cent.

32. Natalia Kraminova, 'The 130th Catholicos of All the Armenians', *MN*, 31 Jan.–7 Feb. 1988, 16.

33. Interview with Boris Kurashvili in *Izvestiia*, 16 Feb. 1988, 1.

34. Giulietto Chiesa, 'Storia segreta del manifesto anti-Gorbaciov', *L'Unità*, 23 May 1988.

35. Nina Andreyeva, 'Ne mogu postupatsia printsipami', *Sovetskaia Rossiia*, 13 March 1988, 3.

36. Ibid.

37. Boris Souvarine, *Staline. Aperçu historique du Bolchévisme* (Paris, 1935).

38. Andreyeva, *op. cit.*

39. For accounts of the meetings, see Robert G. Kaiser, 'How Gorbachev Stood Up to His Rivals', *Washington Post Weekly*, 20–26 June 1988, 24–25; and Giulietto Chiesa's account in Chiesa and Roy Medvedev, *Time of Change*, New York, Pantheon, 1989, 191–4.

40. 'For who among us does not know that *serious* conflicts and differences of opinion are usually decided, not by votes 'according to the rules', but by struggles and threats to 'resign'? N.[V. I.] Lenin, *Pismo k tovarishchu* (Geneva, 1904), 18.

41. 'Printsipy perestroiki: Revoliutsionnost', myshleniia i deistviia', *Pravda*, 5 April 1988, 2.

42. According to Chiesa's article, 'Storia segreta', Andreyeva in her original text had warned only of the influence in the West of 'the spiritual heirs of Trotsky, Kamenev, Preobrazhensky, Zinoviev, Bukharin, and Yagoda'. The editors retained only 'Trotsky and Yagoda', and added 'Dan, Martov, and others in the category of Russian Social Democracy' along with 'descendents of the NEPmen, the Basmachi [Moslem rebels who opposed Soviet power in Central Asia up to 1928], and kulaks who bear a grudge against socialism'. They had broadened her characterization of *glasnost* from an intra-mural party dispute into a many-sided challenge by the enemies of the party and the regime, some of whom had fought it arms in hand.

43. The idea that the bureaucracy had a class interest in the maintenance of Stalinism had already been extensively explored and endorsed in a series of articles in different publications by sociologist Anatoli Butenko. Critics quickly accused him of having been corrupted by the theories of James Burnham, which were well known to Soviet ideologues. See James Burnham, *The Managerial Revolution* (New York, John Day, 1941).

44. Mikhail Chulaki, 'Leningrad and Provincialism', *MN*, 17–24 July 1988, 7.

45. On 27 March the Party Control Commission had rehabilitated the victims of the Leningrad case.

46. Yuri Afanasyev, 'Specific Facts, Honest Assessments', *MN*, 26 June–3 July 1988, 8.

9 DROPPING THE PILOT: GORBACHEV RETIRES GROMYKO

1. Andrei Amalrik, *Will the Soviet Union Survive until 1984?* (New York, Harper and Row, 1970), 12.

2. Falin interviewed in Bonn, 1 April 1987, *Foreign Broadcast Information Service* (FBIS-USSR), 15 April 1987, R3.
3. Text of the theses, approved on 23 May, in *Pravda*, 27 May 1988, 1–3.
4. *Pravda*, 29 June 1988, 4.
5. *Pravda*, 2 July 1988, 7.
6. *Pravda*, 1 July 1988, 7. In the speech as seen on Moscow television on 30 June, Melnikov said, 'Both of us know who they are, Comrade M. S. Solomentsev...'. This was deleted in the *Pravda* account. See Michel Tatu, 'Nineteenth Party Conference', *Problems of Communism*, May–Aug. 1988, 8.
7. *Pravda*, 1 July 1988, 9.
8. Aleksandr Bovin, 'Perestroika i vneshnaia politika', *Izvestiia*, 16 June 1988, 5.
9. Report on Ligachev's speech in 'Kursom sozidaniia', *Pravda*, 5 June 1988, 2.
10. *Pravda*, 2 July 1988, 11.
11. This no doubt referred to a poll on 'social justice and privilege', *MN*, 10–17 June 1988, 10, attempting to quantify citizens' resentments about perks enjoyed by the apparat. The most glaring injustice noted, according to the poll, was privileged access to special food shops (84 per cent), followed by access to books and films (80 per cent), flats in good neighbourhoods (67 per cent), dachas (65 per cent), health care (60 per cent), preference in booking rail and bus tickets (52 per cent), and chauffeured cars (44 per cent).
12. Yevgenii Ambartsumov, 'The Poisonous Mist Disperses: Victims of the Moscow Trials Rehabilitated', *MN*, 26 June–3 July 1988, 10; Lev Razgon, 'At Long Last', *MN*, 3–10 July 1988, 10. In addition, it was announced on 10 July that Bukharin, Rykov, Tomskii, Rakovsky, and other defendants at the 1938 trial of the 'Anti-Soviet Bloc of Rights and Trotskyites' had been posthumously readmitted to the party. The list of names of the original 'Bloc of Rightists and Trotskyites' of 1932 released by the court (*Pravda*, 5 Aug. 1988, 1–2) corresponds to the list published in 'Besposhchadnyi otpor vragam leninskoi partii', *Pravda*, 11 Oct. 1932, 5, and includes, alongside Zinoviev and Kamenev, the signatories of the 'Letter of Eighteen Bolsheviks'. This confirms that the 'Ryutin–Slepkov group' and the Eighteen were the same, as I argued in my *Soviet Succession Struggles*, Ch. 6. The charges in all three Moscow Trials claimed the existence of a conspiracy originating with this affair in 1932.
13. 'Trotsky called Stalin the Revolution's gravedigger in 1926. But, when stepping up industrialization and forcing the peasants to join collective farms, Stalin followed Trotsky's leftist conception, taking it to absurd lengths.' – Ambartsumov, 'Poisonous Mist', 10.
14. Vyacheslav Dashichev, 'Vostok–zapad: poisk novykh otnoshenii', *Literaturnaia gazeta* 18 May 1988, 14. Pursuing the point, Yevgeni Ambartsumov argued that Stalin 'undermined the anti-fascist front then taking shape and ignited the democratic west's mistrust of the USSR.' *MN* (British edition), July 1988, 12. The account in Chernyaev, *Shest' let*, 171–2 and 238–9 stresses that the campaign had been targeted at Gromyko for some time.
15. *Pravda*, 26 July 1988, 4.
16. 'Socialist Democracy, Reform, and Economic Renewal at Work', *Political Affairs* (Aug. 1988), 6; 'The World We Preserve Must Be Livable', *World Marxist Review* (May 1988), 21–8.
17. *New York Times*, 1 July 1988, A4.
18. For the background to these, see Gregg Herken, *The Winning Weapon: The Atomic Bomb in the Cold War, 1945–1950* (New York, Knopf, 1981), the last chapters of which give details of the military plans.

19. Lev Bezymenskii and Valentin Falin, 'Kto razviazal kholodnuiu voinu,' *Pravda*, 29 Aug. 1988, 6.
20. 'Kto razviazal . . . ,' 6.
21. *Izvestiia*, 16 Oct. 1988, 7.
22. Roy Medvedev and Giulietto Chiesa, *Time of Change* (New York, Pantheon, 1989), 228.
23. It was suggested that Gorbachev had experienced an epiphany on his trip to Krasnoyarsk in August, ostensibly to make a speech on Far Eastern policy, when he talked with citizens and was accosted by complaints on the economy. However this event may have been staged.
24. Vadim Medvedev, 'A Modern Concept of Socialism', *Pravda International*, Vol. 2, Nos. 11–12, 1988, 7.
25. Vasili Selyunin, 'Istoki', *Novyi mir*, No. 5, 1988, 162–89.
26. Cf. n. 8, for differences with (and, I think, improvements on) Aganbegyan's approach.
27. 'Istoki', 164. Selyunin concluded with a traditional admonition to act with dispatch: 'An abyss can be crossed only in a single leap – you can't cross it in two'. Nevertheless, at this time he did not advocate a full programme of privatization, but only market pricing in a state-owned sector, a part of small business, and the cooperatives.
28. Marx, in contrast to Lenin, professed a low opinion of Aleksandr Herzen and other emigré publicists who founded a school of Russian socialism based on the village commune. 'I never held rosy views about this communistic Eldorado . . .' according to 'the imaginative lies of Citizen Herzen'. Marx to Engels, 10 Feb. 1870, *Selected Correspondence, 1846–1895* (New York, International Publishers, 1942), 283–4.
29. 'Between Lenin and Marx lies Russia. By means of Lenin, Russia enters into a dialogue with a classical universalist Marx, and through Lenin, classical Marxism enters into a skirmish with Russia'. Mikhail Gefter, 'Rossiia i Marks', *Rabochii klass i sovremennyi mir* (July–Aug. 1988), 155.
30. Ibid., 157.
31. Interview with Gefter, *Twentieth Century and Peace* (Aug. 1987), 45.
32. 'Rossiia i Marks', 170.
33. Vladimir Maksimenko, 'Sotsialisticheskaia orientatsiia: perestroika predstavlenii', *Mirovaia ekonomika i mezhdunarodnye otnosheniia* (2/1989), 102. 'The terms "national democracy", "non-capitalist development", and "socialist orientation" derive from the perspectives of the period of War Communism.' (99–100)
34. Ibid., 95.
35. Selyunin, 168–70. This argument may be found in anarchist correspondence from the period. Petr Kropotkin called the compulsory grain requisitions 'the blackest page of Bolshevism'. 'Pis'mo Kropotkina k Dmitrovskim kooperatorivam', *Probuzhdenie* (June 1929), 59.
36. Aleksandr Tsipko, 'O zonakh, zakrytykh dlia mysli', *Nauka i zhizn'* (Nov. 1988), 45–55.
37. For example, N. Spasov, 'Evoliutsiia podkhoda SShA k regionalnym problemam', *Mirovaia ekonomika i mezhdunarodnye otnosheniia* (3/1989), 28, suggesting that US opinion had taken a decisive turn toward the Soviets in the Spring and Summer of 1988.
38. For this argument, see Dov S. Zakheim, 'U.S. Aid to Perestroika Could Make Soviet Economy Lean and Mean', *Los Angeles Times*, 24 April 1989.
39. Institute on Global Conflict and Cooperation and Soviet Committee for European Security and Cooperation, Alan Sweedler and Brett Henry (eds),

Conventional Forces in Europe: Proceedings of a Conference Held in Moscow October 4–6, 1988 (1989), 15–16.

40. Interviews on Hungarian television, 4 Sep. 1988, in *FBIS-EEU*, 6 Sep. 1988, 2. Gyula Horn, under-secretary of state for the Hungarian foreign ministry, did not share this view, urging instead that such a reduction be part of a 'comprehensive' scheme, rather than a matter to be settled between the USSR and Hungary.

41. Defence Minister Yazov would report to the Military Commission of the Central Committee at the July 1991 plenum that since the time of this decree, submarine-launched ballistic missiles had been cut by 54 per cent, tanks by 44 per cent, and aircraft by 38 per cent.

10 1989: THE YEAR OF ANGER AND REMEMBERING

1. Rudolf Spielmann, *The Art of Sacrifice in Chess*, J. Du Mont trans. (New York, McKay, 1951), vii.
2. Berlinguer quoted in an interview with Antonio Rubbi, *World Marxist Review* (April 1989), 64. Rubbi's articles on New Internationalism had contributed to the debate on Eurocommunism.
3. Konstantin Zarodov, *The Political Economy of Revolution* (Moscow, Progress Publishers, 1981), 224. Togliatti had insisted that the ideas in the Yalta Memorandum had a precedent in the Popular Front of the thirties.
4. V. K. Naumov, 'IKP pered s"ezdom', *Kommunist*, 1/1989, 103.
5. Kevin Devlin, 'Soviet Pressure behind Bilak's Ouster', *RFE/RL Research*, background report, No. 30, 17 Feb. 1989.
6. These efforts ran parallel to the actions of the Italian foreign ministry, urging a new 'Marshall Plan' for East Europe, in return for disarmament and liberal changes in the bloc countries. In October it announced the extension of a line of credit of $800 million from the Italian banks to the USSR to finance trade in equipment for Soviet light industry. Foreign minister Gianni De Michelis cited a specific Italian role in the former Habsburg lands – Austria, Hungary, and Yugoslavia – an area he said 'that politically has been called the East, that geographically is in the center, but culturally is part of the west'.
7. Henry Kamm, 'Hearts and Minds Yearn for the Glories of Mitteleuropa, Wherever It is', *New York Times*, 17 July 1988, 35.
8. *Pravda*, 18 Jan. 1989, 4.
9. *Izvestiia*, 3 Feb. 1989, 6.
10. V. I. Maksimenko, 'Sotsialisticheskaia orientatsiia: perestroika predstavlenii', *Mirovaia ekonomika i mezhdunarodnye otnosheniia* (2/1989), 99.
11. *Moscow News*, 8–15 Jan. 1989, 10.
12. *Pravda*, 8 Jan. 1989, 1.
13. Dmitri Kazutin, 'Three Years Later – In Tallinn?' *Moscow News*, 18 Feb. 1988, 13.
14. *FBIS-Eastern Europe* (88–140) 21 July 1988, 25.
15. Text in Charles Gati, *The Bloc That Failed* (Bloomington and Indianapolis, University of Indiana, 1990), 209. Gati headed the American delegation to the conference. See his 'Eastern Europe on Its Own', *Foreign Affairs* (Annual for 1988–89), 103–5.
16. For an argument against the theory, see V. N. Kudryavtsev, 'Reforma politicheskoi sistemy i obshchestvennaia nauka', *Kommunist* (3/1989), 6.

17. Grosz's speech on Hungarian television, *FBIS-EEU* (88–172), 5 Sep. 1988, 19.
18. On the Varga line, see Ch. 2.
19. Interview with Bogomolov on Hungarian television, *FBIS-EEU*, 24 Feb. 1989, 30.
20. The argument may be found in Hannes Adomeit, 'Gorbachev and German Unification', *Problems of Communism* (July–Aug. 1990), 4–7; and Bruce D. Porter, 'The Coming Resurgence of Russia', *National Interest* (Spring, 1991), 18.
21. This was widely recognized in the West. Even those who began to speak of an end to the Cold War at the beginning of 1989, assumed that none of the major states involved wanted a united Germany. See, for example, Michael Mandlebaum, 'Ending Where It Began', *New York Times*, 27 Feb. 1989, A19.
22. Chebrikov interview in *Pravda*, 11 Feb. 1989, 2.
23. Interview with Vitali Korotich, by Robert Scheer, in *Interview* (Dec. 1988), 198.
24. N. Spasov, 'Evoliutsiia podkhoda SShA k regionalnym problemam', *Mirovaia ekonomika i mezhdunarodnye otnoshenia* (3/1989), 28.
25. *New York Times*, 16 May 1989, 1.
26. As Deng explained it to Zbigniew Brzezinski. See *Power and Principle* (New York, Farrar, Strauss, and Giroux, 1983), 403–25.
27. Gorbachev's speech in Krasnoyarsk, *Pravda*, 18 Sep. 1988, 3.
28. Gaston J. Sigur and Richard Armitage, 'To Play in Asia, Moscow Has to Pay', *New York Times*, 2 Oct. 1988, E25.
29. See Adi Ignatius, 'Chinese Policies at Cross Purposes', *Wall Street Journal*, 1 Feb. 1989, A8.
30. Roy Medvedev and Giulietto Chiesa, *Time of Change* (New York, Pantheon, 1989), Michael Moore trans., 147.
31. At a small meeting with Tsipko and Leonid Dobrokhotov, then with the Central Committee Ideology Department, at Big Sur, CA, May 1989.
32. Dmitri Volkogonov, 'Fenomen Stalina' (the preface to *Triumf i tragedia: Politicheskii portret J. V. Stalina*), *Literaturnaia gazeta*, 9 Dec. 1988, 13.
33. Interview with Medvedev and Volkogonov, *MN*, 19–26 Feb. 1989, 8–9.
34. Yegor Yakovlev, 'The Last Act', *MN*, 29 Jan.–5 Feb. 1989, 9.
35. V. Sirotkin, 'Uroki NEPa', *Izvestiia*, 10 March 1989, 3.
36. *MN*, 26 Feb.–5 March 1989, 15.
37. See above, Chapter 9.
38. Aleksandr Tsipko, 'Egoizm mechtatelei', *Nauka i zhizn*, 1/1989, 46. For an examination of this point, see my *Marxism and the Russian Anarchists* (San Francisco, 1977), Ch. 8.
39. Ibid., 48.
40. Aleksandr Yakovlev's speech in 'Velikaia frantsuzskaia revoliutsiia i sovre-mennost'', *Sovetskaia kul'tura*, 15 July 1989, 4. Yakovlev curiously excepted Lenin from this company of Russian revolutionaries who derived inspiration from the Jacobins of 1793. The Jacobinism of Lenin had been up to then a staple in the arguments of Selyunin and others.
41. Aleksandr Tsipko, 'O zonakh, zakrytykh dlia mysli', *Nauka i zhizn'*, (11/1988), 50.
42. Interview with Imre Pozsgay on Hungarian radio, *FBIS-EEU*, 16 Feb. 1989, 25.
43. Interview with Imre Pozsgay on Hungarian television, *FBIS-EEU*, 10 June 1989, 23.
44. See Pozsgay's remarks on 'the historic rapprochement of the two wings of the workers' movement', in a Belgrade interview, *FBIS EEU*, 23 June 1989, 47.
45. *FBIS EEU*, 10 Nov. 1988, 25.
46. Michel Tatu, 'Intervention in Eastern Europe', in Stephen S. Kaplan, *Diplo-*

macy of Power: Soviet Armed Forces as a Political Instrument (Washington, DC, Brookings, 1981), 256.

47. For a summary of the commission's early deliberations, see Thomas S. Szayna, 'Addressing "Blank Spots" in Polish–Soviet Relations', *Problems of Communism* (Nov.–Dec. 1988), 37–61.
48. I was told this by his long-time associate, Valentin Berezhkov, 9 Nov. 1989.
49. Vyacheslav Dashichev, 'Stalin in Early 1939', *MN*, 3–10 Sep. 1989, 16.
50. Aleksandr Bovin, 'Back to 1939', Ibid., 3. Novosti Press Agency, publisher of *Moscow News*, disavowed as unscientific the Dashichev article in *Pravda*, 26 Aug. 1989, but the *Moscow News* editors called up the prerogatives of *glasnost*, and ideological pluralism, and cited in support of Dashichev Western studies by Robert C. Tucker, Gerhard Weinberg, Franz Borkenau, Jiri Hochman and Jonathan Haslam. The Western citations in this exchange once again demonstrated for Russian readers who the real authorities were. In 1988, *MN* had published a letter from Marshal Zhukov to historian Vasilii Sokolov, in which Zhukov recalled that 'Stalin thought that having declared war on Great Britain and France, Germany would not be able to end the war there soon', and quoted Stalin as having said: 'At present we are economically weaker than Germany but the war it is waging may wear it out completely.' (*MN*, 15–22 May 1988, 8–9).
51. Rakowski had indicated this in a *Le Monde* interview (15 Feb. 1989, 5) in which he foresaw the possibility of the party's relinquishing its monopoly of power.
52. Quoted in John Tagliabue, 'Polish Party and Solidarity on Collision Course', *New York Times*, 31 July 1989, A3.
53. As in Vitali Zhurkin's description of the House of Europe as a new system of security based on Helsinki pact institutions, in *Pravda*, 17 May 1989, 4.
54. See *Foreign Report*, 4 May 1989, 1–3.
55. Aleksandr Bovin, 'Generals Study Detente', *MN*, 13–20 Aug. 1989, 3.
56. Interview with Reszo Nyers on Hungarian television, 30 July 1989, in *FBIS-EEU*, 31 July 1989, 16.
57. He had already been elected to this post in the old Supreme Soviet in October 1988.
58. Headed since September 1988 by Aleksandr Kapto. Its organizations and functions were outlined in Kapto's article, 'Siloi primera, siloi ubezhdeniia', *Pravda*, 20 Feb. 1989, 2, claiming a basis in Gorbachev's formulations of 6 Jan. 1989 on 'de-ideologization' of international relations. Zaikov's remarks were aimed at the lowest level of responsibility but necessarily implied criticism of Vadim Medvedev as well.
59. From the TASS report in *Pravda*, 21 July 1989, 1–4.
60. Andranik Migranyan, 'Dolgii put' k evropeiskomu domu', *Novyi mir* (July 1989), 183.
61. See the discussion in *MN*, 13–20 Aug. 1989, 3.
62. *Izvestiia*, 28 May 1989, 7.
63. G. W. F. Hegel, *Philosophy of Right*, T. M. Knox trans. (London, Oxford and New York, 1967), 267.
64. For a description of *perestroika* as a stage in 'civil society recovering', employing the categories of Fernand Braudel, see Moshe Lewin, *The Gorbachev Phenomenon* (Berkeley and Los Angeles, U.C. Press, 1988), and my review in *American Historical Review* (Feb. 1990).
65. Mikhail Gorbachev, *Perestroika*, 82.
66. A. Levikov, 'Are Leningraders Any Worse?' *MN*, 13–20 Aug. 1989, 10.

67. Ligachev's interview on the plenum in *Pravda*, 22 Sep. 1989, 6.
68. Ligachev was accused of having received bribes from Inamszhan Usmankhodzhaev, a former party first secretary in Uzbekistan, who had made them in the course of his interrogation following arrest on corruption charges in 1988. But afterward Usmankhodzhaev claimed that Gdlyan had extracted his testimony by threatening to arrest members of his family.
69. Andranik Migranyan spelled out the meaning of this position to be that the 'Finlandization' of the bloc countries would signal their freedom to develop an economic system of the Western type alongside a pro-Soviet foreign policy; the West, in return for a Soviet 'refusal to set up new leftist Totalitarian regimes' in the Third World, should agree not to try to make the bloc countries anti-Soviet. 'An Epitaph to the Brezhnev doctrine', *MN*, 27 Aug.–3 Sep. 1989, 6.
70. Statement by Gyula Thurmer on Hungarian television, 28 Aug. 1989, text in *FBIS-EEU*, 31 Aug. 1989, 33.
71. Craig Whitney, Serge Schmemann, and David Binder, 'Party Coup Turned East German Tide', *New York Times*, 19 Nov. 1989.
72. These ministries were headed by Lt.-General Czeslaw Kiszczak, General Florian Sawicki, and Marcin Swiecicki, respectively.
73. Shevardnadze interview with Fyodor Burlatsky, 'Otstavka bolshe, chem zhizn', *Literaturnaia gazeta*, 10 April 1991, 3.
74. Foreign minister Peter Varkonyi made no attempt to deny this when I put the question to him in a small meeting in January 1990.
75. Joint statement by Gorbachev and Mazowiecki, 24 Nov. 1989, *Pravda*, 25 Nov. 1989, 2.

11 FROM THE WALL TO STAVROPOL: GORBACHEV'S GERMAN POLICY

1. Modrow interview in *Die Zeit*, 20 April 1990.
2. On this bargain, see Josef Joffe, 'Once More: The German Question', *Survival*, March/April 1990, 134–5.
3. Gorbachev's press conference in Milan, *Pravda*, 3 Dec. 1989, 2.
4. Michael R. Beschloss and Strobe Talbot, *At the Highest Levels* (Boston, Little, Brown, 1993), 184. Chernyaev, 'Gorbachev and Germany', calls it his own idea.
5. Aleksandr Prokhanov, 'Tragediia tsentralizma', *Literaturnaia Rossiia*, 5 Jan. 1990, 4–5.
6. Ligachev's speech to the Central Committee plenum, *Pravda*, 7 Feb. 1990, 5–6.
7. Evgenii Ambartsumov, 'Change Creates Hope, Not Agony', *MN*, 19–26 Nov. 1989, 12.
8. Jens Reich, 'Perestroika, Socialism, Two Germanies', *MN*, 26 Nov.–3 Dec. 1989, 6. His optimism faded quickly. On his visit to the United States two months later, he treated it as a foregone conclusion that his party would not get more than ten per cent of the vote in the March elections (it got three per cent), and that in any case, socialism in the GDR had no future.
9. Gorbachev's meeting with Lithuanian party members, *TASS*, 14 Jan. 1990.
10. Ibid.
11. When asked whether the Council was an alternative to the Politburo, Shatalin answered disingenuously that 'The question is meaningless if only because half the Council members are members of the Politburo.' Interview with Shatalin, *MN*, 15–22 April 1990, 6.

12. Quoted in Giulietto Chiesa, 'The 28th Congress of the CPSU', *Problems of Communism*, July–Aug. 1990, 36.

13. Quoted in Yegor Yakovlev, 'The Presidential Council: An Insider's View', *MN*, 26 Aug.–2 Sep. 1990, 8.

14. Theodore Karasik and Brenda Horrigan, 'Gorbachev's Presidential Council', RAND paper p-7665, 14–15.

15. As with his description of the function of the Defence Council: 'The Council of Defence has always been within the Ministry of Defence – such was the tradition. But the Council of Defence has broader functions than those of the military or defence industry. This is a body of nation-wide importance, and it is more appropriate for it to be attached to the President.' *MN*, 26 Aug.–2 Sep. 1990, 9.

16. *Parliamentary Affairs* (Oct. 1990), 493.

17. *MN*, 21–28 Jan. 1990, 9.

18. Ilya Chubais, 'Democratic Platform', *MN*, 22–29 April 1990, 7.

19. Interview with Prokofiev, *MN*, 8–15 July 1990, 4.

20. Of the many cases that could be cited, M. P. Kapustin, 'Ot kakogo nasledstva my otkazyvaemsia?', *Oktiabr'*, No. 4, 1988, 186 and *passim*.

21. Aleksandr Tsipko, 'The Party Must Learn to Live in the World of Truth', *Moscow News* (British edition), 1–7 June 1990, 4.

22. Aleksandr Tsipko, *Ideia sotsializma* (Moscow, 1976), 260, quoted approvingly by Boris Kagarlitsky, *The Thinking Reed*, Brian Pearce trans. (London and New York, Verso, 1988), 287.

23. Aleksandr Tsipko, 'The Fate of the Socialist Idea', *MN*, 24 June–1 July 1990, 6.

24. Aleksandr Tsipko, 'Awakening Russia', *MN*, 8–15 July 1990, 3.

25. On *Vekhi*, see Leonard Schapiro, 'The *Vekhi* Group and the Mystique of Revolution', *Slavonic and East European Review*, Vol. 34, No. 82, 56–76.

26. *MN*, 21 Oct. 1990, 7. To round out the contrast, Georgian respondents rejected both forms of socialism (one per cent for 'democratic socialism' and 19 per cent for 'Swedish socialism') with 48 per cent favouring capitalism.

27. Ibid., 16 Dec. 1990, 9.

28. According to statistics from PlanEcon, CIA, and Goskomstat, in *New York Times*, 16 Sep. 1990.

29. *MN*, 18 Nov. 1990, 7. By Oct. 1990, the figure had fallen to 21 per cent.

30. Falin had opposed from the beginning the Gorbachev policy of acquiescing to the Two Plus Four formula. The *New York Times*, 19 May 1990, 5, reported his criticism of the economic agreement between the two Germanies making the Deutschmark the one currency, 'not an example', he said, 'of how two sovereign states should equally represent their interests.'

31. See Giulietto Chiesa, 'The 28th Congress of the CPSU', *Problems of Communism*, July–Aug. 1990, 31.

32. 'Khronika s"ezda, den' pervyi', *Pravda*, 3 July 1990, 5.

33. 'Idti dal'she putem perestroiki', Ibid., 2.

34. One hundred and one members of the Democratic Platform, in a joint declaration with the Marxist Platform, decided to stay in the party to bolster its 'democratic wing'. *TASS*, 20 July 1990, 62.

35. In Britain and France, one in four were opposed, while in Poland, only 17 per cent believed that founding a new Germany would create a lasting peace. Alan Riding, 'Survey Finds 2 in 3 Poles Opposed to German Unity', *New York Times*, 20 Feb. 1990, A8.

12 THE SECOND RUSSIAN REVOLUTION GATHERS

1. Leningrad Mayor Anatoli Sobchak supported the liberation of the security organs from the party by the bizarre reflection that 'Precisely in the thirties–forties these agencies ceased to be state bodies and turned into party bodies.' *MN*, 8–15 1990, 5.

2. As attested by several field-grade officers who were members of the *Shchit* (Shield) military trade union. See the roundtable discussion 'Military Coup in the USSR?' *MN*, 23–30 Sep. 1990, 8. A Central Committee conference of 20 Dec. 1989 ruled that every officer who joined *Shchit* was subject to dismissal from the armed forces within 24 hours.

3. Yuri Afanasyev, 'The Coming Dictatorship', *New York Review of Books*, 31 Jan. 1991, 39.

4. See Victor Yasmann, 'Elite Think Tank Prepares Post-Perestroika', *RFE-RL Reports on the USSR*, 13 May 1991.

5. In fact, the 'New Right' explicitly rejected what it called 'Stalinist internationalism'. Sergei Kurginyan, '"Novaia pravaia" – za reform, levye – eto voina', in Gleb Pavlovskii (ed.) *Pravaia al'ternativa* (Moscow, Postfactum, 1990), 10–12.

6. 'Bor'ba za perestroiku', *Pravda*, 22 Sep. 1990, 1.

7. Gorbachev's opening speech to the Central Committee plenum, text in *Pravda*, 9 Oct. 1990, 1.

8. Security Council members included Yanaev (Vice President), Pavlov (Prime Minister), Bessmertnykh (Foreign Minister), Pugo (Interior), Kryuchkov (KGB), Yazov (Defence), Boldin (Chief of Staff) with Primakov as advisor, and Bakatin, probably in charge of party supervision of the armed forces and security organs.

9. Gorbachev's speech to the Supreme Soviet, 17 Nov. 1990, text in *Pravda*, 18 Nov. 1990, 1. During the 7 Nov. parade, a lone gunman, Aleksandr Shmonov, fired two shots at Gorbachev from a distance of 50yd in front of Lenin's tomb. These went astray only because a policeman threw himself on the gunman as they were fired.

10. Ibid., 2.

11. Interview with Yazov, 'Vechernii razgovor s ministrom oborony', *Komsomol'skaia pravda*, 1 Dec. 1990, 2.

12. See Elizabeth Teague, 'Window on the *Soiuz* Group', *RFE-RL Report on the USSR*, 7 May 1991.

13. Alexander Rahr, 'The CPSU Strikes Back', *RFE-RL Report on the USSR*, 8 Feb. 1991.

14. *Otriady Militsii Osobogo Naznacheniia* (Special Purpose Police Detachments).

15. Statement by the *Moscow News* board of directors, *MN*, 20–27 Jan. 1991, 1.

16. *MN*, 27 Jan.–3 Feb. 1991, 8–9.

17. No real conclusion about this emerges from Shevardnadze's rambling, emotional account in *The Future Belongs to Freedom* (New York, Free Press, 1991), ch. 8.

18. Foreign Minister Bessmertnikh seconded this view by pointing out that, since the initiative was not addressed to Bush, he could not reject it.

19. See Mary C. Fitzgerald, 'The Soviet Army after the Gulf', *RFE-RL Reports*, 11 April 1991.

20. Poll in *MN*, 24–31 March 1991, 7.

21. *Pravda*, 1 March 1991, 1.

22. As part of a Soviet effort to weaken the French-supported group of states, and capitalizing on the perceived opening provided by the Kellogg–Briand pact of 1928. The Litvinov Protocol extended the pact's sentiments about

avoiding war to east central Europe. There followed Soviet pacts with Estonia, Latvia, Lithuania, Poland, Romania, Turkey, Persia, and Danzig. The Romanian agreement did not recognize Bessarabia, taken from Russia in 1918, as Romanian territory.
23. Interview with Alksnis, *MN*, 10–17 Feb. 1991, 7.

13   FROM THE COUP TO THE END

1. 'Slovo k narodu', *Sovetskaia Rossiia*, 23 July 1991, 1.
2. Mikhail Gorbachev, *The August Coup: The Truth and the Lessons* (New York, HarperCollins, 1991), 14.
3. *Pravda*, 5 Aug. 1991, 2.
4. Chernyaev, *Shest' let*, 477–80. Text of the Crimea article in *The August Coup*, 97–127.
5. Ibid., 119. The reference is to Gorchakov's policy of *recueillement*.
6. Ibid., 19.
7. *Pravda*, 20 Aug. 1991, 3.
8. Shevardnadze interview on CNN, 23 Aug. 1991.
9. *The August Coup*, 28. In an interview with Radio Free Europe/Radio Liberty, 18 Aug. 1996, Gorbachev told of a meeting on 30 July 1991 with Yeltsin and Nazarbaev in which he agreed to remove the heads of Defence and the KGB, replacing Prime Minister Valentin Pavlov with Nazarbaev. A KGB tape of the meeting got into Kryuchkov's hands and he used it to persuade the army chiefs to support the coup.
10. Ibid., 20.
11. Radio Free Europe/Radio Liberty Daily Report, 21 Aug. 1991.
12. Yeltsin interview on Russian television, 25 Aug., text in *New York Times*, 26 Aug. 1991, A6.
13. 'Er musste am Leben bleiben', *Der Spiegel*, 6 Oct. 1991, 203; *Izvestiia*, 10 Oct. 1991, 1.
14. Gavriil Popov warned American Ambassador Jack Matlock at a lunch on 20 June that a coup was in prospect, and that its agents were Pavlov, Kryuchkov, Yazov, and Lukyanov. See *Autopsy on an Empire* (New York, Random House, 1995), 540–41.
15. This was the 'Romanian Variant', also entailing the physical elimination of the 'Siamese twins' Gorbachev and Yeltsin, which the Committee understandably viewed with dread. See Valentin Pavlov, *Avgust iznutri: Gorbachevputsch* (Moscow, Delovoi mir, 1993), 70.
16. *Der Spiegel*, 6 Oct. 1991, 3; *Izvestiia*, 10 Oct. 1991, 7.
17. *The August Coup*, 20.
18. Interview with Anatoly Sobchak, *MN*, 1–8 Sep. 1991, 10. A slightly different account is in Sobchak, *Khozdenie vo vlast'* (Moscow, Novosti, 1991), 277.
19. P. Grachev, 'Desantniki protiv naroda ne poshli', *Krasnaia zvezda*, 31 Aug. 1991, 3.
20. Yeltsin's decree 'K grazhdanam Rossii', cites articles 64, 69, 70, and 72 of the Russian Federation Constitution of 1962 and article 62 of the Brezhnev Constitution of 1977. Text in Iurii Luzhkov, *72 chasa agonii* (Moscow, 1991).
21. Text of the Appeal as published by Postfactum News Agency, 19 Aug. 1991, on SOVSET computer network.
22. Test in Ibid.
23. RFE/RL Daily Report, 20 Aug. 1991.

24. Postfactum News Agency release, 19 Aug. 1991, as seen on SOVSET.
25. *Tri dnia: 19–21 avgusta 1991g.* (Moscow, Postfactum, 1991), 4–5.
26. RFE/RL Daily Report, 20 Aug. 1991.
27. Boris Yeltsin, *The Struggle for Russia* (New York, Random House, 1994), 58.
28. Ibid. See also Bill Keller, 'A Coup Gone Awry', *New York Times*, 25 Aug. 1991, 10.
29. Sobchak, *Khozhdenie vo vlast'*, 281.
30. Interview with General Tsaplin in *Izvestiia*, 17 Oct. 1991, 3.
31. Michael Dobbs, 'KGB Officers Tell of Key Unit Disobeying Order', *Washington Post*, 28 Aug. 1991, 3. According to General Tsaplin, Golovatov led the Group Alpha assault on the Vilnius television tower in Jan. 1991, under the command of General Achalov, with participation of the 'Z' Administration (Protection of Constitutional Order) of the KGB, which had special responsibility for monitoring riots, ethnic conflicts, foreign businesses, and the media. See interview with General Tsaplin.
32. Texts in Postfactum, 26 Aug. 1991. A coded telegram from the Secretariat of the Central Committee of the CPSU was signed at 12:05, 19 Aug. The general communication to the republics was headed 'Top Secret, file 47, 18 Aug. 1991, 55 copies, on the State of Emergency', and signed by Oleg Shenin, with the date 20 Aug. 1991.
33. Interview with an anonymous KGB officer, *MN*, 13–20 Oct. 1991, 16.
34. Quoted in Serge Schmemann, 'Gorbachev Back as Coup Fails: Yeltsin Gains New Power', *New York Times*, 22 Aug. 1991, A6.
35. Text excerpts of the special session in the *New York Times*, 24 Aug. 1991, A7.
36. See Chapter 7.
37. A claim that was stressed by Otto Latsis, 'Zapomnim vsekh, obdumaem vse', *Izvestiia*, 22 Aug. 1991, 4, and by Gorbachev, *Memoirs*, 642–4.
38. Douglas L. Clarke, 'Concern about Nuclear Weapons', RFE/RL Reports, 3 Sep. 1991.
39. Mikhail S. Gorbachev, 'My Final Hours', *Time*, 11 May 1992, 44.
40. Ibid.
41. Ibid., 43.
42. RFE/RL Daily Report, 11 Dec. 1991.

14 CONCLUSION: UTOPIA AND REPENTANCE

1. Winston S. Churchill, *The Gathering Storm* (New York, Houghton Mifflin, 1961), 9.
2. Gorbachev's speech in San Francisco, 7 May 1992.
3. Francis Fukuyama, 'Rest Easy. It's Not 1914 Anymore', *New York Times*, 9 Feb. 1992, E17, an answer to my letter, 'What New Storms for Eastern Europe?', *New York Times*, 25 Dec. 1991.
4. The seminal and, in a sense, final formulation of these ideas, as I have argued in Ch. 9, was in Mikhail Gefter, 'Rossiia i Marks', *Rabochii klass i sovremennyi mir* (July–Aug. 1988).
5. Andrei Amalrik, *Will the Soviet Union Survive until 1984?* (New York and Evanston, Harper and Row, 1970), 19.
6. Michael Meers, 'Money Flows and the Republic's Future', *Perspectives on Change*, 22 June 1992, SOVSET.

7. For example, Anton W. DePorte, *Europe Between the Superpowers: The Enduring Balance* (New Haven and London, Yale University Press).
8. Alexis de Tocqueville, *Democracy in America*, George Lawrence, trans. (New York and Garden City, Doubleday, 1969), 412–13.
9. Mikhail Gorbachev, *Perestroika: New Thinking for Our Country and the World* (New York and Cambridge, Harper and Row, 1987), 185–6.
10. Arnold Toynbee, *Civilization on Trial* (New York, Meridian, 1958), 119; and 'A Turning Point in the Cold War?', *International Affairs* (Oct. 1950), 457.
11. Mikhail Gorbachev, *The August Coup*, 81.
12. Duff Cooper, *Talleyrand* (New York, Fromm International, 1986), 237.

Index